Encyclopaedia of Mathematical Sciences
Volume 137

---

*Invariant Theory and Algebraic Transformation Groups VIII*

Venkatramani Lakshmibai
Komaranapuram N. Raghavan

# Standard Monomial Theory

## Invariant Theoretic Approach

Springer

Venkatramani Lakshmibai
Department of Mathematics
Northeastern University, Boston 02115
USA
e-mail: lakshmibai@neu.edu

Komaranapuram N. Raghavan
Institute of Mathematical Sciences
C. I. T. Campus, Taramani
Chennai, 600 113
INDIA
e-mail: knr@imsc.res.in

Founding editor of the Encyclopaedia of Mathematical Sciences: Revaz V. Gamkrelidze

ISBN 978-3-540-76756-5 e-ISBN 978-3-540-76757-2

DOI 10.1007/978-3-540-76757-2

Encyclopaedia of Mathematical Sciences ISSN 0938-0396

Library of Congress Control Number: 2007939889

Mathematics Subject Classification (2000): 13F50, 14M12, 14M15, 14M17, 14L35



*Cover design:* WMXDesign GmbH, Heidelberg, Germany

Printed on acid-free paper

9 8 7 6 5 4 3 2 1

springer.com

To the memory of
Professor C. Musili

# Preface

The goal of this book is to present the results of Classical Invariant Theory (abbreviated CIT) and those of Standard Monomial Theory (abbreviated SMT) in such a way as to bring out the connection between the two theories. Even though there are many recent books on CIT, e.g., [25, 35, 53, 97, 99], none of them discusses SMT: there is but only a passing mention of the main papers of SMT towards the end of [53]. Details about the connection are also not to be found in the comprehensive treatment of SMT [59] that is in preparation. Hence the need was felt for a book that describes in some detail this natural and beautiful connection.

After presenting SMT for Schubert varieties—especially, for those in the ordinary, orthogonal, and symplectic Grassmannians—it is shown (using SMT) that the categorical quotients appearing in CIT may be identified as "suitable" open subsets of certain Schubert varieties. Similar results are presented for certain canonical actions of the special linear and special orthogonal groups. We have also included some important applications of SMT: to the determination of singular loci of Schubert varieties, to the study of some affine varieties related to Schubert varieties—ladder determinantal varieties, quiver varieties, variety of complexes, etc.—and to toric degenerations of Schubert varieties.

Prerequisite for this book is some familiarity with commutative algebra, algebraic geometry and algebraic groups. A basic reference for commutative algebra is [27], for algebraic geometry [37], and for algebraic groups [7]. We have also included a brief review of GIT (geometric invariant theory), a reference for which is [87] (and also [96]).

We have mostly used standard notation and terminology, and have tried to keep notation to a minimum. Throughout the book, we have numbered Theorems, Lemmas, Propositions etc., in order according to their section and subsection; for example, 3.2.4 refers to fourth item in the second subsection of third section of the present chapter. The chapter number is also mentioned if the item appears in another chapter.

This book may be used for a year long course on invariant theory and Schubert varieties. The material covered in this book should provide adequate preparation for graduate students and researchers in the areas of algebraic geometry and algebraic groups to work on open problems in these areas.

*A Homage & an acknowledgement*: In the original plan for this book, Musili was one of the co-authors. Unfortunately, Musili passed away suddenly on Oct 9, 2005. We dedicate this book to the memory of Musili. We also would like to thank Ms. Bhagyavati (Musili's wife) and Ms. Lata (Musili's daughter) for providing us with the files that Musili had prepared.

We wish to thank the referees for their comments.

Boston, Trieste
October 2007

*V. Lakshmibai*
*K. N. Raghavan*

# Contents

# 1

# Introduction

## 1.1 The subject matter in a nutshell

This book aims to describe the beautiful connection between Schubert varieties and their STANDARD MONOMIAL THEORY (SMT for short) on the one hand and CLASSICAL INVARIANT THEORY (CIT for short) on the other. This connection was discovered by Lakshmibai and Seshadri [67]. Here is the opening paragraph of [67] in a slightly edited form:

> The main aim of this paper is to show how the work of De Concini and Procesi [22] on classical invariant theory can be interpreted to suggest a generalization of the Hodge-Young theory of standard monomials (cf. Hodge [41] and Hodge and Pedoe [42]). This generalization is given as a set of conjectures (which have now been proved in collaboration with C. Musili [61]). On the other hand, we also show that the results of De Concini and Procesi follow as consequences of the generalization.

SMT is the name given to this generalization of the Hodge-Young theory. And the term CIT in this book refers, for the most part, to the results of De Concini and Procesi mentioned in the quote above. These results, when the characteristic of the ground field is zero, are classical—cf. Weyl's book [115].

As we will see, determinantal varieties form the bridge connecting CIT and SMT, and Schubert varieties become relevant because they are natural compactifications of determinantal varieties.

### 1.1.1 What is CIT?

CIT concerns certain canonical actions of classical groups $G$ on affine spaces, namely, cases **A**, **B**, **C**, **D**, and **E** to be discussed in §1.2 below. It describes, in each case, a presentation for the ring of $G$-invariant polynomial functions on the affine space. This description comprises of two theorems known as the *first* and *second fundamental theorems*. The first fundamental theorem specifies a finite set of algebra generators, over the ground field, for the ring of invariants. Note that for the action of a reductive group $G$ on an affine variety, the ring of $G$-invariant polynomial functions is a finitely generated algebra (classically, i.e., in characteristic zero, this result goes back to Weyl [115]; in positive characteristic, this follows from well-known results of Nagata [93, 94] and Haboush [36]: see Chapter 9 for details). The second fundamental theorem specifies a finite set of generators for the ideal of relations among the algebra generators: that there always exists a finite set of generators for the ideal follows from Hilbert's basis theorem [79].

### 1.1.2 What is SMT?

The roots of SMT are to be found in the work of Hodge [41, 42], who described nice bases for the homogeneous co-ordinate rings of Schubert varieties of the Grassmannian in the Plücker embedding (over a field of characteristic 0). Grassmannians being precisely the homogeneous spaces that arise as quotients of special linear groups by maximal parabolic subgroups, it is natural to try to generalize Hodge's work to natural projective embeddings of other quotients $G/Q$ where $G$ is a semi-simple algebraic group and $Q$ a parabolic subgroup. In the early '70s Seshadri initiated this generalization and called it SMT.

### 1.1.3 The SMT approach to CIT

The main idea in this approach is to relate a certain subring of the ring of invariants (which will turn out to be in fact the ring of invariants) as the ring of functions on an affine variety related to a Schubert variety. This allows the use of SMT to prove the first and second fundamental theorems.

## 1.2 The subject matter in detail

In this section, we shall explain the SMT-approach [67] as well as the approach of De Concini-Procesi [22] for the actions of general linear, symplectic, and the orthogonal groups. In both approaches, one makes a guess on the ring of invariants; to be more precise, there are some obvious invariants (we call these the *basic invariants*), and one shows (in both approaches) that the ring of invariants is in fact generated (as an algebra over the base field) by these invariants.

Fix a field $K$, algebraically closed of arbitrary characteristic, and $V$ a finite dimensional vector space over $K$.

**A.** The general linear group $G := GL(V)$ of invertible linear transformations acts naturally on both $V$ and its dual $V^*$. Consider $G$ acting diagonally on $Z := V^{\oplus m} \oplus V^{*\oplus q}$; here $V^{\oplus m}$ denotes the direct sum of $m$ copies of $V$.

Let $R$ be the ring of polynomial functions on $Z$. There is a natural action of $G$ on $R$. Let $R^G$ be the subring of $R$ consisting of those polynomial functions that are invariant under $G$. The "scalar products" $(v_1, \ldots, v_m; f_1, \ldots, f_q) \mapsto \varphi_{ij} := f_j(v_i)$, for $1 \le i \le m$ and $1 \le j \le q$, are evidently invariant—these are the "basic invariants."

**B.** Let now $V$ be equipped with an alternating non-degenerate bilinear form $(\ ,\ )$—note that this forces $V$ to be even dimensional. Let $Sp(V)$ be the symplectic group, namely, the subgroup of $GL(V)$ consisting of those linear automorphisms of $V$ that preserve the form. Consider the diagonal action of the symplectic group $Sp(V)$ on $Z := V^{\oplus m}$.

Let $R$ be the ring of polynomial functions on $Z$. The "basic invariants" in this case are the bilinear products $(v_1, \ldots, v_m) \mapsto \varphi_{ij} := (v_i, v_j)$ for $1 \leq i, j \leq m$.

Now suppose that instead of an alternating form we have a symmetric form. More precisely:

**C.** Let the characteristic of $K$ be different from 2 and let $V$ be equipped with a non-degenerate symmetric bilinear form $(\ ,\ )$. Let $O(V)$ be the orthogonal group, namely, the sub group of $GL(V)$ consisting of those linear automorphisms of $V$ that preserve the form. Consider the diagonal action of the orthogonal group $O(V)$ on $Z := V^{\oplus m}$.

Let $R$ be the ring of polynomial functions on $Z$. The "basic invariants" are still the bilinear products $\varphi_{ij} := (v_i, v_j)$.

In all of the three cases **A, B, C** above, let us denote by $S$ the subalgebra of $R^G$ generated by the basic invariants. The goal is

(1) to show $S = R^G$ which will yield immediately the First Fundamental Theorem.

(2) to construct a "nice" basis for $S(= R^G)$ which will yield the "straightening relations" (see §1.5) and hence the Second Fundamental Theorem.

To explain the proof of De Concini-Procesi, it suffices to quote in a slightly edited form from [22]:

> The line of the proof is the following: we have an algebraic group $G$ acting on an affine variety $E$ with coordinate ring $R$ and we have a subring $S$ of $R^G$, namely the one generated by the basic invariants, which we want to show equals $R^G$. First we show that on an open set $U \subseteq W$ (where an element $s$ of $S$ is invertible) the group action is a product action; thus the localized invariant ring $R^G[1/s]$ turns out to be $S[1/s]$. Then we have to find a way to cancel $s$: i.e., if $sa \in S$ and $a \in R^G$ we must show that $a \in S$. This is accomplished by finding an explicit basis of the ring $S$ and deducing the cancellation result from this. This part is the main contribution of the paper.

### 1.2.1 Proof of (1), (2) by SMT approach (cf. [67])

Let us denote the basic invariants by $f_1, \cdots, f_N$. We have a natural map

$$\psi : X \to \mathbb{A}^N,\ x \mapsto (f_i(x))$$

In all of the three cases **A, B, C** above, it is not difficult to see that $\psi(X)(= Spec\ S)$ gets identified with a determinantal variety $\mathbb{D}$: In each case, let $G$ denote the group in question, namely, $G = GL(V),\ Sp(V),\ O(V)$ in cases **A**, **B**, **C** respectively. Let $n$ be the dimension of $V$.

• In **A**, $\mathbb{D}$ is the subvariety $D_{n+1}(M_{m,q})$ of $M_{m,q}(K)$ (the space of $m \times q$ matrices with entries in $K$) consisting of matrices of rank at most $n$, i.e., the matrices all of whose $(n+1)$-minors vanish.

• In **B**, $\mathbb{D}$ is the subvariety $D_{n+1}(Sk\, M_m)$ of $Sk\, M_m(K)$ (the space of skew symmetric $m \times m$ matrices with entries in $K$) consisting of matrices of rank at most $n$.

• In **C**, $\mathbb{D}$ is the subvariety $D_{n+1}(Sym\, M_m)$ of $Sym\, M_m(K)$ (the space of symmetric $m \times m$ matrices with entries in $K$) consisting of matrices of rank at most $n$.

Now $f_1, \cdots, f_N$ being $G$-invariants, the morphism $\psi$ goes down to a morphism

$$\psi_{\mathbb{D}} : Spec\, R^G \to \mathbb{D}(= Spec\, S)$$

The main idea behind the proof of the First Fundamental Theorem in [67] is to show that $\psi_{\mathbb{D}}$ satisfies the hypotheses of ZMT (Zariski's Main Theorem):

(i) $\psi_{\mathbb{D}}$ is surjective with finite fibers

(ii) $\psi_{\mathbb{D}}$ is birational

(iii) $\mathbb{D}$ is normal.

It would then follow by ZMT that $\psi_{\mathbb{D}}$ is in fact an isomorphism, and we would obtain that the inclusion $S \hookrightarrow R^G$ is in fact an equality. The verifications in [67] of (i) & (ii) turn out to be rather straight forward in view of certain geometric invariant theoretic considerations (see Chapter 10 for details). Thus to conclude that $S = R^G$, proving normality of $Spec\, S(= \mathbb{D})$ is the only non-trivial part. Here is where the Schubert variety connection is used in the approach in [67]. To make this more precise, in case **A** (respectively, **B**, **C**) above, it turns out that $M_{m,q}(K)$ (respectively $Sk\, M_m(K),\ Sym\, M_m(K)$) gets identified with the "opposite cell" $\mathcal{O}^-$ in the Grassmannian $G_{q,m+q}$ (respectively the orthogonal Grassmannian $SO(2m)/P_m$, the symplectic Grassmannian $Sp_{2m}/P_m$ ); and $\mathbb{D}$ gets identified with $\mathcal{O}^- \cap X$, for a suitable Schubert variety $X$ in $G_{q,m+q}$ (respectively in $SO(2m)/P_m$, $Sp_{2m}/P_m$). Hence the normality of $Spec\, S(= \mathbb{D})$ follows, once we know the normality of Schubert varieties. Normality of Schubert varieties is a consequence of SMT.

Using the SMT-basis for the homogeneous co-ordinate ring of $X$, we obtain a basis for $\mathbb{D}$ (by the process of dehomogenization). This in turn yields the Second Fundamental Theorem.

Thus using Schubert varieties and their SMT, we obtain in one stroke, the proofs of the first and second fundamental theorems.

At the time of the appearance of [22], SMT was developed only for "minuscule" $G/P$'s. To be more precise, in his thesis written under Seshadri's guidance, Musili [88] had extended Hodge's results to arbitrary characteristics. Soon after, Seshadri [111] had generalized Hodge's results to quotients by *minuscule*[1] maximal

[1] A maximal parabolic subgroup $P$ is *minuscule* if the Weyl group translates of a highest weight vector span the space of global sections on $G/P$ of the ample generator of the Picard group of $G/P$. The geometry of a minuscule $G/P$, i.e., when $P$ is minuscule, is very similar to that of a Grassmannian.

parabolics. As is easily seen, the $G/P$'s appearing in cases **A, B** are minuscule; and the bases for $R^G$ as given by [22] on the one hand and SMT on the other are the same.

But the $G/P$ arising in case **C** is not minuscule. Given this, the following problem cried out for an answer: shouldn't there be an approach to CIT in case **C** similar to that of cases **A**, **B** above? Analysing carefully the work of De Concini and Procesi in case **C**, Lakshmibai and Seshadri [67] arrived at a conjectural SMT for Schubert varieties in quotients of classical semi-simple groups by maximal parabolic subgroups (technically speaking a little more generally but we will ignore that here for the sake of simplicity). These conjectures were later proved by them in collaboration with C. Musili in [61, 62]. Taking for granted such an SMT, case **C** too can be handled in a fashion entirely analogous to cases **A** and **B**.

### 1.2.2 $SL_n(K)$, $SO_n(K)$ actions

Replacing the general linear group in case **A** discussed in §1.2 by the special linear group, consider

**D.** The special linear group $SL(V)$ acting diagonally on $V^{\oplus m} \oplus V^{*\oplus q}$.

The basic invariants in this case are the scalar products $\varphi_{ij} := f_j(v_i)$ (as for the general linear group—cf. case **A** above) and the determinants $u(I) := \det[v_{i_1}, \ldots, v_{i_n}]$, $\xi(J) := \det[f_{j_1}, \ldots, f_{j_n}]$, where $1 \leq i_1 < \ldots < i_n \leq m$, $1 \leq j_1 < \ldots < j_n \leq q$—these appear only in the case when $m \geq n$.

Replacing in case **C** the orthogonal group by the special orthogonal group, consider

**E.** The diagonal action of the special orthogonal group $SO(V)$ on $V^{\oplus m}$.

The basic invariants in this case are the bilinear products as above and the determinants $u(I) := \det[v_{i_1}, \ldots, v_{i_n}]$, where $1 \leq i_1 < \ldots < i_n \leq m$—these appear only if $m \geq n$.

De Concini and Procesi [22] treat CIT in cases **A**, **B**, and **C** fully. In cases **D** and **E**, they prove the first fundamental theorem, but there are no details about the second fundamental theorem.

As for cases **D** and **E**, the invariant rings are not—not in any obvious way at least—rings of functions on open parts of Schubert varieties. So the approach to CIT described in §1.2.1 does not work in exactly the same way. Nevertheless Schubert varieties remain relevant. And an SMT theoretic approach to CIT in cases D and E has recently been worked out in [63, 72] respectively.

In Case **D**, the normality of the ring $S$ generated by the basic invariants is deduced by degenerating its spectrum to a toric variety. As a first step towards the degeneration, a basis for $S$ is constructed. Straightening relations are then written down and the degeneration is carried out using the straightening relations. The Cohen-Macaulayness of $S$ also follows immediately as a corollary of the degeneration. That the poset structure of Schubert varieties in the Grassmannian form a distributive lattice is used in the proof crucially.

In case **E**, the ring $S$ generated by basic invariants arises as the ring of functions on a branched covering of degree 2 over the symmetric determinantal variety $D_{n+1}(Sym\ M_m)$. The normality of this ring is proved by showing that it is non-singular in codimension one and also Cohen-Macaulay (in view of Serre's criterion—see [79] for example). The first property follows by noting that the branch locus is contained in the singular locus of $D_{n+1}(Sym\ M_m)$ and that $D_{n+1}(Sym\ M_m)$ being an open part of a Schubert variety is normal (and in particular non-singular in codimension one). The Cohen-Macaulay property however takes some work to prove. As a first step, one obtains a basis for the ring $S$. Then a deformation argument due to De Concini and Lakshmibai [21] is applied. They used it originally to show that Schubert varieties in quotients of classical groups by maximal parabolics are arithmetically Cohen-Macaulay for the embedding given by the ample generator of the Picard group.

This finishes the discussion of the SMT approach to CIT. There is a lot more that has happened in SMT beyond what is described above. A brief history of SMT is given in §1.4 below. But the scope of this book is confined only to that part of SMT that has been discussed above. In this book we treat in detail all cases A–E via SMT.

## 1.3 Why this book?

The main subject matter of this book, namely the connection between Schubert varieties on the one hand and CIT on the other described in §1.2 above, has not come until now within the scope of any book. The books on CIT—and there are quite a few recent ones among them, e.g., [25, 35, 53, 97, 99]—come no where near discussing this connection. In fact, except in [53] where there is a quick mention of the main papers of SMT towards the end, the books are totally silent about the connection.

On the other hand, looking from the SMT side, the book [59] under preparation aims at being a comprehensive account of SMT and is authored by the main players in the creation of that theory. But here too there is very little discussion on the connection between Schubert varieties and CIT. The connection is of course mentioned in the book's introduction which recounts in detail the history of the development of SMT. But the emphasis of the book being elsewhere, this connection is not treated in the book itself.

The raison d'etre for the present book is thus clear.

As seen in §1.2.1, the fact that determinantal varieties arise as open sets of Schubert varieties plays a crucial role in the connection between CIT and SMT. Although the relationship between determinantal varieties and Schubert varieties is quite classical, except for [40, 51, 89], there is not much in the literature about it. Further, to the best of our knowledge, [67] is the first work in the literature which discusses the relationship between determinantal varieties in the space of symmetric and skew-symmetric matrices on the one hand and Schubert varieties on the other. These aspects are treated in detail in the present book.

## 1.4 A brief history of SMT

We have discussed above the origins of SMT and how the work of De Concini and Procesi on CIT influenced the development of SMT. What follows is a brief account of the history of SMT. The book [59] under preparation aims to give a comprehensive account of SMT. Its introduction gives a detailed account of the history of SMT. But, since the present book will in all likelihood appear before [59], the following account may still be of some value.

We may divide SMT into four stages. The first stage consists of the work done up to the appearance of the paper of De Concini and Procesi on CIT. As explained above, this stage comprises of the work of Hodge [41,42], Musili [88], and Seshadri [111].

The second stage consists of the work done in roughly over a decade after the appearance of De Concini-Procesi's paper [22] on CIT. This stage begins with the paper of Lakshmibai and Seshadri [67] which describes SMT conjecturally for Schubert varieties in quotients of classical groups by maximal parabolic subgroups. These conjectures are proved in [61,62]. Finally, in [69], SMT is established for quotients of classical groups by parabolic subgroups (maximal or otherwise).

Classical groups having been satisfactorily addressed, the third stage attempts to handle exceptional groups: [55] handles $G_2$, [64] handles $E_6$, and [70] handles the case of the affine Kac Moody group $\hat{sl}_2$. Finally, as a culmination of all the work, the conjectures for a general SMT (for Schubert varieties in $G/Q$ for an arbitrary parabolic subgroup $Q$ of a symmetrizable Kac-Moody Group $G$) are formulated by Lakshmibai-Seshadri in [71].

Solving the conjectures of [71] required new ideas. This is what Littelmann has done in his work. His papers [74,75,77] form the fourth and final stage of SMT. An important idea of Littelmann is to view "standard tableaux" (the indexing set for the SMT basis) as certain paths—the so-called *Lakshmibai-Seshadri paths*. Littelmann uses the theory of quantum groups. The technique of [103] is also crucially used. Thus SMT is now complete thanks to Littelmann's work!

Our goal being the description of the connection between SMT and CIT, it suffices for us to consider SMT for quotients of classical groups by maximal parabolic subgroups. So in this book we develop SMT in the spirit of [61,62,67]. In particular, we do not discuss Littelmann's work or quantum group theory. For details of Littelmann's work, one could refer to the original papers [74,75,77], or the book [59].

## 1.5 Some features of the SMT approach

Let us now discuss some features and advantages of the Schubert variety theoretic approach to CIT. This approach is more conceptual than that of [22], the reason being that Schubert varieties provide a powerful inductive tool facilitating the proofs. It also opens doors for generalization to other groups/situations.

An important feature in the Schubert-variety-theoretic approach is the qualitative description of quadratic straightening relations—a straightening relation is the expression for a non-standard monomial as a linear sum of standard monomials. In this approach, the generation of the space of all monomials by standard monomials hinges on the qualitative description of certain quadratic straightening relations; further, the proof of the linear independence of standard monomials is also carried out by using these quadratic relations (and induction on the dimension of a Schubert variety). In the context of CIT, in this approach, such relations are first established on Schubert varieties, and then are specialized to rings of invariants.

Even though, there are some combinatorial description of relations of any degree on a determinantal variety in [26, 101], one cannot deduce the required qualitative description of the quadratic relations from the relations found in loc.cit; the reason for this is that in [26, 101], a typical relation (on a determinantal variety) expresses a given monomial as a linear sum of monomials (not necessarily standard) which are greater than the given monomial for a suitable order (which is different from the order used in SMT) on the set of all monomials. Thus the Schubert-variety-theoretic approach seems to be indispensable from this perspective. Also, the Schubert-variety-theoretic approach seems to be the only approach which can yield the desired qualitative description of the quadratic relations - a crucial step in SMT. Similar remarks apply to the symplectic and the orthogonal group actions also.

Yet another important advantage in this approach is that we obtain the proof of Cohen-Macaulayness for these rings of invariants, something one does not get in the approach of [22]. The rings of invariants in cases **A**, **B**, **C** described in §1.2 above being identified with open subsets of suitable Schubert varieties, the Cohen-Macaulayness for these rings of invariants follows at once from the Cohen-Macaulayness properties for Schubert varieties. The proof, via SMT, of the Cohen-Macaulayness of Schubert varieties in the Grassmannian, namely those that are related to the categorical quotients in case **A**, is proved in [88] (see also Chapter 4). The proof, via SMT, of the general case of Schubert varieties in $G/P$, where $G$ is semi-simple and $P$ a maximal parabolic subgroup of classical type (in particular for those related to the categorical quotients in cases **A**, **B**, **C** is proved by De Concini-Lakshmibai [21]. While the proof in [88] uses commutative algebra arguments, that in [21] uses "deformation technique". Chirivi [17] has extended the latter technique to the case when $G$ is any semi-simple algebraic group and $P$ is any parabolic subgroup.

The deformation technique consists in constructing a flat family over $\mathbb{A}^1$, with the given variety as the generic fiber (corresponding to $t \in K$ invertible). If the special fiber (corresponding to $t = 0$) is Cohen-Macaulay, then one may conclude the Cohen-Macaulayness of the given variety. Hodge algebras (cf. [20]) are typical examples where the deformation technique affords itself very well. Deformation technique is also used in [12, 33, 46]. The philosophy behind these works is that if there is a "standard monomial basis" for the co-ordinate ring of the given variety, then the deformation technique will work well in general (using the "straightening relations").

In recent times, among the several techniques of proving the Cohen-Macaulayness of algebraic varieties, particularly those that are related to Schubert varieties, two techniques have proved to be quite effective, namely, Frobenius splitting technique and deformation technique. Frobenius splitting technique is used in [104], for example, for proving the (arithmetic) Cohen-Macaulayness of Schubert varieties; further, in [83], the normality of Schubert varieties is proved using the Frobenius splitting technique. This technique is also used in [80, 81, 84] for proving the Cohen-Macaulayness of certain varieties.

## 1.6 The organization of the book

Turning now to the organization of the book, we have tried to make it self-contained keeping in mind the needs of prospective graduate students and young researchers. After reviewing some basics in Algebraic Geometry and Algebraic Groups in Chapters 2 and 3, we first present SMT for the Grassmannian and its Schubert varieties (Musili's thesis (cf. [88])) in Chapter 4. We then discuss the relationship between determinantal varieties and Schubert varieties in the Grassmannian in Chapter 5. Similar relationships between determinantal varieties in the space of symmetric (respectively skew-symmetric) matrices and Schubert varieties in the symplectic (respectively orthogonal) Grassmannian are established in Chapter 6 (respectively Chapter 7). The main results of SMT are stated in Chapter 8 and are proved in the Appendix. Chapter 9 is a review of GIT. In Chapter 10, using the results of Chapter 5, we describe a basis for the ring of invariants for the $GL_n(K)$-action (cf. Example **A** above) thus giving a Standard Monomial Theoretic proof for DeConcini-Procesi's results in this case. We also discuss the $SL_n(K)$-action on the space considered in (A) above, and describe a basis for the corresponding ring of invariants in Chapter 11. Chapters 6 and 7 describe similar results for the symplectic and orthogonal group actions (cf. Examples **B, C** above) at the same time giving a Standard Monomial Theoretic proof for DeConcini-Procesi's results in these cases. In Chapter 12, we discuss the $SO_n(K)$-action on the space considered in (C) above, and describe a basis for the corresponding ring of invariants; we further deduce some related results on the moduli space $M$ of equivalence classes of semistable rank 2 vector bundles on a smooth projective curve of genus $> 2$. We also describe a characteristic-free basis for the ring of invariants for the (diagonal) adjoint action of $SL_2(K)$ on $\underbrace{sl_2(K) \oplus \cdots \oplus sl_2(K)}_{m \text{ copies}}$.
In Chapter 13, we discuss some important applications of SMT; we first present a discussion of singular loci of Schubert varieties. Next, we discuss the relationship of ladder determinantal varieties, Quiver varieties and variety of complexes to Schubert varieties, and deduce results for these varieties; in this chapter, as another application of SMT, we present the results of toric degenerations of Schubert varieties in the Grassmannian.

# 2

# Generalities on algebraic varieties

In this chapter, we recollect some basic facts on commutative rings and algebraic varieties. For details, we refer the reader to [27, 37, 79].

## 2.1 Some basic definitions

**Definition 2.1.0.1** *The* Krull dimension*: (or simply the* dimension*) of a commutative ring R (with 1) is the supremum of the lengths of chains of prime ideals of R and it is denoted by dim R. It need not be finite even for Noetherian rings. It is however finite for Noetherian* local *rings.*

If $R$ is a Noetherian local ring with its maximal ideal $\mathbf{m}$ and residue field $\mathbf{k} = R/\mathbf{m}$, then the least cardinality of a set of generators of $\mathbf{m}$ equals $\dim_{\mathbf{k}}(\mathbf{m}/\mathbf{m}^2)$; it is called the *embedding dimension* of $R$, denoted $e(R)$, and is an upper bound for the dimension of $R$.

In case $R$ is an integral domain with its field of fractions (or quotients) $K = Q(R)$ and $R$ is also a finitely generated algebra over a field $F$, then $K$ is a finitely generated field extension of $F$ and $\dim R$ is precisely the *transcendence degree* of $K$ over $F$. (This is not obvious but follows as an immediate consequence of the *Noether's Normalisation Lemma* - see [86] for a statement and proof of Noether's Normalisation Lemma.)

The localisation of a ring $R$ at a prime ideal $\mathbf{p}$ is denoted by $R_{\mathbf{p}}$. The dimension of $R_{\mathbf{p}}$ is called the *height* of $\mathbf{p}$ and is denoted by $\mathrm{ht}_R(\mathbf{p})$. For a ring $R$, dimension of $R$ is the supremum of the heights of its maximal ideals and in particular, for a local ring it is the height of its maximal ideal.

**Definition 2.1.0.2** *A* regular local ring *is a Noetherian local ring R for which the Krull dimension is the same as its embedding dimension, i.e., dim(R) = e(R). We say that a Noetherian ring is* regular *if the localisations at all of its maximal ideals are regular local rings.*

**Definition 2.1.0.3** Normal domains: *If $R$ is an integral domain and $K = Q(R)$ (its field of fractions), then we say that $R$ is* normal *if it is* integrally closed *in $K$, i.e., an element of $K$ which is a root of a* monic *polynomial with coefficients in $R$ must be already in $R$.*

Normality is a *local property* in the sense that $R$ is normal if and only if $R_{\mathbf{p}}$ (resp.$R_{\mathbf{m}}$) is normal for all prime ideals $\mathbf{p}$ (resp. maximal ideals $\mathbf{m}$) of $R$. The ring of quotients $S^{-1}R$ of a regular (resp. normal) ring $R$ at any multiplicatively closed subset $S$ is regular (resp. normal).

## 2.2 Algebraic varieties

### 2.2.1 Affine varieties

Let $K$ be an algebraically closed field of arbitrary characteristic. For a non–negative integer $n \in \mathbb{Z}^+$, the linear space $K^n$ is called the *Affine* $n$–space over $K$ and is denoted by $\mathbb{A}^n$ (by convention, $\mathbb{A}^0 = (0)$, a point). Let $A = K[X_1, \cdots, X_n]$ be the ring of polynomial functions on $\mathbb{A}^n$. The field $K(X_1, \cdots, X_n)$ which is the field of fractions of $A$ is called the field of *rational functions* on $\mathbb{A}^n$ and is also denoted by $K(\mathbb{A}^n)$.

For a point $P = (a_1, \cdots, a_n) \in \mathbb{A}^n$, the $a_i$'s are called the coordinates of $P$. An element $f \in A$ defines a $K$-valued function on $\mathbb{A}$ by evaluation, namely, $P \mapsto f(P) = f(a_1, \cdots, a_n)$. The subring $\mathcal{O}_P = \mathcal{O}_P(\mathbb{A}^n) := \{f/g \in K(\mathbb{A}^n) \mid g(P) \neq 0\}$ which is a *local ring* is called the local ring of $\mathbb{A}^n$ at $P$.

Recall (see [86] for example) that the ideal $\mathbf{m} =< X_1 - a_1, \cdots, X_n - a_n >$ is maximal in $A$ for all $(a_1, \cdots, a_n) \in K^n$ (for every field $K$, algebraically closed or not). The converse is true only for algebraically closed fields $K$ (which is known as the **Hilbert's Nullstellensatz** - see [86] for further details). Thus points of $\mathbb{A}^n$ for $K$ algebraically closed can be identified with the set $\mathrm{Spm}(A)$ of all maximal ideals of the ring $A$. Given $f \in A$, of total degree $d \geq 1$, the set of zeros of $f$ in $K^n$, namely,

$$V(f) := \{(a_1, \cdots, a_n) \in \mathbb{A}^n \mid f(a_1, \cdots, a_n) = 0\}$$

is called the *(affine) hypersurface* whose equation is $f$ and $d$ is also called the degree of $V(f)$. If $d = 1$, $V(f)$ is called a *hyperplane*. It can be seen that a point $P = (a_1, \cdots, a_n) \in V(f) \iff f \in \mathbf{m}_P$ where $\mathbf{m}_P =< X_1 - a_1, \cdots, X_n - a_n >$, i.e., $V(f)$ can be identified with the set of all maximal ideals of the quotient ring $A/ < f >$.

Given $T \subseteq A$, the set $V(T) = \cap_{f \in T} V(f)$ of all **common zeros** of elements of $T$ is called an *affine algebraic* (or simply an *algebraic*) subset of $\mathbb{A}^n$. The family of all algebraic subsets satisfy the axioms of closed sets and the corresponding topology is called the *Zariski topology* on $\mathbb{A}^n$.

#### Coordinate Rings

Let $J(T)$ be the ideal generated by $T$ in $A$ and $\sqrt{J(T)}$, the *radical* of $J(T)$: recall the definition of the radical of an ideal $J$, denoted $\sqrt{J}$,

$$\sqrt{J} := \{f \in K[X_1, \cdots, X_n] | f^r \in J \text{ for some } r \geq 1\}$$

We say that $J$ is a *radical ideal* if $J = \sqrt{J}$.

Since $A$ is Noetherian, any ideal in $A$ is finitely generated. Let then $J(T) =< f_1, \cdots, f_m >$ and $\sqrt{J(T)} =< g_1, \cdots, g_\ell >$. Now it follows that

$$V(T) = \bigcap_{f\in T} V(f) = \bigcap_{j=1}^{m} V(f_j) = \bigcap_{f\in J(T)} V(f) = \bigcap_{f\in\sqrt{J(T)}} V(f) = \bigcap_{j=1}^{\ell} V(g_j),$$

i.e., every algebraic subset is a finite intersection of hypersurfaces. In fact, we have

$$V(T) = \bigcap_{f\in J(T)} V(f) = V(f_1, \cdots, f_m) = V(g_1, \cdots, g_\ell) = \bigcap_{f\in\sqrt{J(T)}} V(f).$$

Given a subset $Y \subseteq \mathbb{A}^n$, let $I(Y) = \{f \in A^n \mid f(y) = 0, \ \forall\, y \in Y\}$, which is the largest ideal in $A$ vanishing on $Y$, called *the ideal of* $Y$. We have the following basic facts:

$$V(T) \supseteq V(S) \iff \sqrt{J(T)} \subseteq \sqrt{J(S)}$$
$$V(I(Y)) = \overline{Y}$$
$$I(V(T)) = \sqrt{J(T)}$$

Thus $\sqrt{J(T)}$ is the ideal of $V(T)$. If $V = V(T)$ and $I = \sqrt{J(T)}$, the ring $A/I$ is called the *coordinate ring* of $V$ and is denoted by $K[V]$. We also write $V =$ Spec $K[V]$ and the points of $V$ can be identified with the set of all maximal ideals of $K[V]$, i.e., with the set of all maximal ideals in $A$ containing the ideal $I$ of $V$.

### The Local Rings $\mathcal{O}_P(V)$

For $P = (a_1, \cdots, a_n) \in V$, the ideal $\mathbf{m}_P =< X_1 - a_1, \cdots, X_n - a_n >$ (mod $I$) is called the maximal ideal of $V$ at $P$ and the local ring $\mathcal{O}_P(V) := \mathcal{O}_P(\mathbb{A})/I =$ the localization of $K[V]$ at the maximal ideal $\mathbf{m}_P$ is called the *local ring* of $V$ at $P$, or also the *stalk* at $P$.

An algebraic subset $V$ whose ideal is a prime ideal $\mathbf{p}$ is called an *affine variety* (defined by $\mathbf{p}$) so that its coordinate ring $K[V] = A/\mathbf{p}$ is an integral domain. The field $Q(K[V])$ of fractions of $K[V]$ is called the field of *rational functions* on $V$ and it is denoted by $K(V)$.

**Definition 2.2.1.1** *An affine variety $V$ is said to be* non-singular *or* smooth at a point *$P \in V$ if its local ring $\mathcal{O}_P(V)$ at $P$ is regular. We say that $V$ is non-singular or smooth if it is so at all of its points.*

**Definition 2.2.1.2** *An affine variety $V$ is said to be* normal at a point *$P \in V$ if its local ring $\mathcal{O}_P(V)$ at $P$ is a normal domain. We say that $V$ is* normal *if it is so at all of its points.*

**Definition 2.2.1.3** *An affine variety $V$ is said to be* factorial at a point $P \in V$ *if its local ring $\mathcal{O}_P(V)$ at $P$ is a unique factorization domain (i.e., every element in $\mathcal{O}_P(V)$ has a unique factorization as a product of prime elements in $\mathcal{O}_P(V)$). We say that $V$ is* factorial *if it is so at all of its points.*

**Definition 2.2.1.4** *The* topological dimension *of an algebraic subset $V$ is defined as the supremum of the lengths of chains of algebraic varieties contained in $V$ and it is denoted by dim $V$.*

The topological dimension of $V$ is the same as the Krull dimension of its coordinate ring $K[V]$. For example, one can see that the dimension of $\mathbb{A}^n$ is $n$ and dimension of $V(f)$ is $n-1$ (if degree of $f$ is positive).

## Projective Varieties

**Definition 2.2.1.5** ***Projective Space:*** *Given a non-zero vector space $L$ over $K$, the set of all lines (i.e., one dimensional subspaces) in $L$ is called the* projective space *associated to $L$ and is denoted by $\mathbb{P}(L)$. It is also called the projective $n$–space if $L$ is of dimension $n+1$ over $K$.*

Choosing a basis of $L$ and writing $L = K^{n+1}$ $(n \geq 0)$, (with respect to that basis), we write $\mathbb{P}^n = \mathbb{P}^n_K$ for $\mathbb{P}(K^{n+1})$ or simply $\mathbb{P}$ if there is no ambiguity about $K$ and $n$.

**Convention:** $\mathbb{P}^n$ = a point if $n = 0$.

A point $P$ in $\mathbb{P}^n$ has $n+1$ **homogeneous** coordinates $(a_0, a_1, \cdots, a_n)$, $a_j \in K$, in the sense that (1) not all $a_j$'s are zero and (2) both $(a_0, a_1, \cdots, a_n)$ and $(\lambda a_0, \lambda a_1, \cdots, \lambda a_n)$ represent the same point for all $\lambda \in K^\star$.

Let $S = K[X_0, X_1, \cdots, X_n]$ be the polynomial ring in $n+1$ variables over $K$. Unlike the affine case, an element $f \in S$ does not define a function on $\mathbb{P}^n$. Nevertheless, we can talk about the vanishing or non-vanishing of a *homogeneous polynomial* $f \in S$ at a point $P = (a_0, a_1, \cdots, a_n) \in \mathbb{P}^n$ because we have $f(\lambda(a_0, a_1, \cdots, a_n)) = \lambda^d f((a_0, a_1, \cdots, a_n))$ if $d$ is the total degree of $f$. This allows us to define the *(projective) hypersurface* $V_+(f)$ whose *equation* is a homogeneous polynomial $f$ of total degree $d$, i.e., $V_+(f) = \{P = (a_0, a_1, \cdots, a_n) \in \mathbb{P}^n \mid f(P) = 0\}$. As before, the degree of $f$ is called the degree of $V_+(f)$. It is called a *hyperplane* if $d = 1$, etc.

Given $T \subseteq S$ but consisting of only homogeneous polynomials of possibly different degrees, the set $V_+(T) = \cap_{f \in T} V_+(f)$ of all *common zeros* of elements of $T$ is called a *projective algebraic* (or simply *an algebraic*) subset of $\mathbb{P}^n$. The family of all algebraic subsets satisfy the axioms of closed sets and the corresponding topology is again called the *Zariski topology* on $\mathbb{P}^n$.

Now keep the natural gradation on $S = \sum_{m \geq 0} S_m$ (where $S_0 = K$) and proceed exactly as before replacing "ideals" by "homogeneous ideals", "generators" of ideals by "homogeneous generators", etc., and define *projective varieties* as $V_+(\mathbf{p})$ for homogeneous prime ideals $\mathbf{p}$ of $S$. The graded ring $S/\mathbf{p}$ is called the *homogeneous coordinate ring* of $V_+(\mathbf{p})$. We write $V_+(\mathbf{p})$ (or just $V_+$) = Proj($S/\mathbf{p}$). Consider the affine variety $V = V(\mathbf{p}) = \operatorname{Spec} S/\mathbf{p}$ defined by $\mathbf{p}$ in $K^{n+1}$. We note that the origin $(0, \cdots, 0)$ of $K^{n+1}$ is a point of the affine variety $V$ because $\mathbf{p}$ is a homogeneous ideal of $S$. The affine variety $V$ is called the *cone over* $V_+$, and origin is the *vertex* of the cone.

Nonsingularity, normality, factoriality etc. are defined for a projective variety in the same way as for the affine varieties. Some of the basic facts are the following:

1. A variety which is both affine and projective is a single point.
2. Every projective variety in $\mathbb{P}^n$ has a finite open covering by suitable affine subvarieties of $K^n$.
3. The affine variety $V$ is a *cone* over the projective variety $V_+$ as the *base* and origin as the *vertex*.
4. $\dim(V) = 1 + \dim(V_+)$.
5. The linear space $K^{n+1}$ is the cone over $\mathbb{P}^n$.
6. The projective variety $V_+$ is non–singular (resp. normal, factorial) at all of its points $\Leftrightarrow$ $V$ is non–singular (resp. normal, factorial) at all points outside the vertex.
7. $V$ is non–singular at its vertex $\Leftrightarrow V = K^m$ is linear $\Leftrightarrow V_+ = \mathbb{P}^{m-1}$ for some $m \geq 1$.
8. $V$ is normal (resp. factorial) at its vertex $\Leftrightarrow$ $V$ is so at all of its points $\Leftrightarrow$ $S/\mathbf{p}$ is so.

**Definition 2.2.1.6** ***Projective Normality:*** *A projective variety $V_+$ is said to be* projectively *(or* arithmetically*)* normal *(resp.* factorial*) if its cone $V$ is normal (resp. factorial) at its vertex, i.e., the stalk at the vertex is a normal domain (resp. a unique factorization domain).*

The following example shows that projective normality is a property of the particular projective embedding of the variety (unlike the affine varieties). The projective line $\mathbb{P}^1$ is obviously projectively normal since its cone is the affine plane $K^2$ (which is non-singular). However, it can be also embedded in $\mathbb{P}^3$ as the **quartic curve**, namely,

$$V_+ = \{(a^4, a^3b, ab^3, b^4) \in \mathbb{P}^3 \mid (a, b) \in \mathbb{P}^1\},$$

i.e., $V_+ = V_+(XT - YZ, TY^2 - XZ^2)$, but the coordinate ring of its cone $V$ which is $K[X, Y, Z, T]/(XT - YZ, TY^2 - XZ^2)$ is not normal.

Using the language of Weil divisors, Cartier divisors and line bundles on $V_+$ (as in the affine case), we define the divisor class group $\mathrm{Cl}(V_+)$ and the Picard group $\mathrm{Pic}(V_+)$ of $V_+$.

In $\mathbb{P}^n = V_+(\mathbf{0})$, any hypersurface (of degree $d$) is a Cartier divisor. Two hypersurfaces define the same divisor class if and only if they have the same degree. The line bundle corresponding to the divisor class of a hyperplane is called the *hyperplane bundle* or the *tautological line bundle* on $\mathbb{P}^n$ and it is denoted by $\mathbf{O}_{\mathbb{P}}(1)$. The line bundle corresponding to the divisor class of a hypersurface of degree $d$ is then the $d^{\text{th}}$ tensor power of $\mathbf{O}_{\mathbb{P}}(1)$ and it is denoted by $\mathbf{O}_{\mathbb{P}}(d)$ $(= \mathbf{O}_{\mathbb{P}}(1)^{\otimes d})$. The space $H^0(\mathbb{P}^n, \mathbf{O}_{\mathbb{P}}(d))$ of *global sections* of $\mathbf{O}_{\mathbb{P}}(d)$ turns out to be $S_d$, the homogeneous component of $S$ of degree $d$ (see [37] for details).

For a projective subvariety $X = V_+(\mathbf{p})$ of $\mathbb{P}^n$, the restriction of $\mathbf{O}_{\mathbb{P}}(1)$ to $X$ gives a line bundle, called the *(induced) hyperplane bundle* on $X$ (for this embedding of $X$ in $\mathbb{P}^n$) and is denoted by $\mathbf{O}_X(1)$. It follows that the space of sections $H^0(X, \mathbf{O}_X(d))$ contains the homogeneous component $(S/\mathbf{p})_d$ but need not be equal (it is well-known that equality holds for $d$ sufficiently large). However, if $X$ is projectively normal, then equality holds for all $d$.

**More general varieties:** Let $X$ be a locally closed (i.e., an open subset of a closed subset) in the ambient space $Z$ ($Z$ being $\mathbb{A}^n$ (respectively $\mathbb{P}^n$) for some $n \in \mathbb{N}$). Then $X$ has a natural variety structure and one refers to $X$ as a *locally closed subvariety* of $Z$ or also as a *quasi-affine* (respectively *quasi-projective*) variety. All of the above concepts extend naturally to these more general varieties. In the sequel, by an "algebraic variety", we shall mean an affine or quasi affine or projective or quasi projective variety.

# 3

# Generalities on algebraic groups

This chapter is a review of generalities on algebraic groups - root systems, Weyl groups, parabolic subgroups etc. For details on root systems and Weyl groups, we refer the reader to [44, 45]. For details on algebraic groups, we refer the reader to [7,43,47].

## 3.1 Abstract root systems

Fix a vector space $V$ over $\mathbb{R}$ with a non-degenerate positive definite inner product $(\cdot,\cdot)$. Define the *reflection* relative to a nonzero vector $\alpha \in V$ to be the linear transformation on $V$ given by

$$s_\alpha(v) = v - \frac{2(v,\alpha)}{(\alpha,\alpha)}\alpha.$$

Note that $s_\alpha$ maps $\alpha$ to $-\alpha$ and fixes the hyperplane perpendicular to $\alpha$. Denote

$$\langle v,\alpha\rangle = \frac{2(v,\alpha)}{(\alpha,\alpha)} \quad \text{and} \quad \alpha^* = \frac{2\alpha}{(\alpha,\alpha)}$$

Note that we have $(v,\alpha^*) = \langle v,\alpha\rangle$.

**Remark 3.1.0.1.** Note that $\alpha^*$ is also written as $\alpha^\vee$ by some authors in the literature. In the sequel we shall use both versions $\langle v,\alpha\rangle$ and $(v,\alpha^*)$ interchangeably.

An abstract *root system* $R$ in $V$ is defined by the following axioms:

1. $R$ is finite, spans $V$ and does not contain 0.
2. If $\alpha \in R$, the only multiples of $\alpha$ in $R$ are $\pm\alpha$.
3. For each $\alpha \in R$, $R$ is stable under $s_\alpha$.
4. **Integrality:** For $\alpha, \beta \in R$, $s_\alpha(\beta) - \beta = -\langle\beta,\alpha\rangle\alpha$ is an integral multiple of $\alpha$, i.e., $\langle\beta,\alpha\rangle$ is an integer.

The positive integer $\ell = \dim(V)$ is called the *rank* of $R$. A subset $S = \{\alpha_1, \ldots, \alpha_\ell\}$ of $R$ is called a *basis of the root system* if $S$ is a vector space basis of $V$ and if every $\alpha \in R$ can be expressed as $\alpha = \sum c_i \alpha_i$ with all the coefficients $c_i$ being integers with the same sign. Using this property, $R$ can be partitioned into *positive roots* $R^+$ and *negative roots* $R^-$ with respect to $S$. The elements of $S$ are called *simple roots*. The integers $\langle \alpha_i, \alpha_j \rangle$, $1 \le i, j \le \ell$, are called the *Cartan integers* and they determine the root system completely, up to isomorphism.

The information given by the Cartan integers is summarized in a *Dynkin diagram*. The Dynkin diagram is a graph with vertices indexed by the simple roots and the number of edges between $\alpha_i$ and $\alpha_j$ for $i \neq j$ is $\langle \alpha_i, \alpha_j \rangle \langle \alpha_j, \alpha_i \rangle$ with an arrow pointing to the smaller of the two roots if they have different lengths. All irreducible root systems are completely classified by their Dynkin diagrams.

**Example.** Let $V$ be the hyperplane in $\mathbb{R}^3$ consisting of $\{(x_1, x_2, x_3) \in \mathbb{R}^3 \mid \sum x_i = 0\}$ equipped with the usual inner product. Let $\{e_1, e_2, e_3\}$ denote the standard basis for $\mathbb{R}^3$. Let $R = \{\pm(e_1 - e_2), \pm(e_2 - e_3), \pm(e_1 - e_3)\}$. Then $R$ is a root system (of rank 2), known as of "Type $A_2$" in $V$. The set $S := \{\alpha_1, \alpha_2\} = \{e_1 - e_2, e_2 - e_3\}$ is a basis of the root system $R$. The Cartan integer $\langle \alpha_i, \alpha_j \rangle$ is 2 or $-1$ according as $i = j$ or $i \neq j$.

The *Weyl group* $W$ of the root system $R$ is the subgroup of $GL(V)$ generated by the $s_\alpha$ for $\alpha \in R$. It follows from the axioms of root systems that $W$ is also finite; further, $W$ is a Coxeter group, and is (minimally) generated by the *simple reflections* $s_{\alpha_1}, \ldots, s_{\alpha_\ell}$, denoted $s_1, \ldots, s_\ell$ corresponding to the simple roots. For $w \in W$, the *length* of $w$, denoted $\ell(w)$, is defined to be the minimum of the length of an expression for $w$ as a product of simple reflections; if $\ell(w) = p$, then an expression $w = s_{a_1} s_{a_2} \ldots s_{a_p}$ is called a *reduced expression* for $w$ and the (ordered) sequence $s_{a_1}, s_{a_2}, \ldots, s_{a_p}$ or simply $a_1, a_2, \ldots, a_p$ is called a *reduced word* for $w$. Note that for a given $w$, there can be many reduced expressions. For instance, in the above example, we have $W =$ the symmetric group $S_3$ with simple reflections $s_1 :=$ the transposition $(1, 2)$ and $s_2 :=$ the transposition $(2, 3)$. The two reduced expressions $s_1 s_2 s_1$, $s_2 s_1 s_2$ give the same permutation $w = (321)$. Here and throughout the text we will denote a permutation $w$ as $(w_1 w_2 \ldots w_n)$ (here, 1 is mapped to $w_1$, 2 to $w_2$, etc).

The following facts about Weyl groups are often used.

**FACTS:**

- There is a unique element in $W$ of *longest length*, denoted by $w_0$.
- The set of all reflections in $W$, is obtained by conjugating all simple reflections.
- For $w \in W$, $\ell(w)$ equals $\#\{\beta \in R^+ \mid w^{-1}(\beta) < 0\}$.

For details, see [44, 45].

Any vector $\lambda \in V$ such that $\langle \lambda, \alpha \rangle \in \mathbb{Z}$ for every $\alpha \in R$ is called an abstract *weight*. By the integrality axiom, roots are weights. The set of all weights form a lattice $\Lambda$, called the *weight lattice* which contains the lattice generated by $R$ as an abelian subgroup of finite index. The lattice generated by $R$ is usually called the *root lattice*. If $S = \{\alpha_1, \ldots, \alpha_\ell\}$ is a basis of $R$, then $\Lambda$ has a corresponding basis consisting of the *fundamental weights*: $\varpi_1, \ldots, \varpi_\ell$, defined by the condition that $\langle \varpi_i, \alpha_j \rangle = \delta_{ij}$

(Kronecker delta) for all $i, j \in \{1, \cdots, \ell\}$. A weight $\lambda = \sum c_i \varpi_i$ is said to be *dominant* if all $c_i$ are non-negative integers, and is said to be *regular* if $c_i \neq 0$, for all $i$. In the sequel, $\Lambda^+$ will denote the set of all dominant weights in $\Lambda$.

The *Weyl chambers* in $V$ are the connected components in the complement of the union of hyperplanes orthogonal to the roots. These chambers are in one-to-one correspondence with the bases of the root system. The chamber containing the dominant weights with respect to $S$ is also called the *dominant chamber*.

## 3.2 Root systems of algebraic groups

### 3.2.1 Linear algebraic groups

An affine algebraic subvariety $G$ of $K^n$ which is also a group such that its group operations are "morphisms" (i.e., polynomial maps) in the category of affine varieties is called an *affine* or *linear algebraic group* or an *affine group variety* over $K$ ([7,43]). The group $\mathrm{GL}_n(K)(\subset M_n(K) = K^{n^2})$ of all non–singular $n \times n$ matrices over $K$, called the *general linear* group, is a *universal* example of a linear algebraic group over $K$; this is closed in $K^{n^2+1}$ with defining ideal being the principal ideal generated by $(Z \cdot \det(X_{ij}) - 1)$ in $K[\{X_{ij}\}_{1 \leq i,j \leq n}, Z]$. This is indeed universal, thanks to a theorem of Chevalley, which says that any affine algebraic group over $K$ is a closed subgroup of $\mathrm{GL}_n(K)$ for some $n$ (see [7] for a proof of Chevalley's result). Another equally universal example is the *special linear* group $\mathrm{SL}_n(K) = \{A \in K^{n^2} \mid \det(A) = 1\}$ whose defining ideal is the principal ideal generated by $(\det(X_{ij}) - 1)$. There are other *classical* examples like the *symplectic* and *orthogonal* groups, and also other examples, the so-called *exceptional* groups.

Recall (cf. [7,43]:
A *torus* is an algebraic group isomorphic to a product of copies of $K^*$. Maximal tori exist in a given algebraic group (by reasons of dimensions); further, all maximal tori are conjugates and hence have the same dimension. Let $l$ be the dimension of a maximal torus; $l$ is defined as the *rank* of $G$. A *Borel subgroup* in an algebraic group $G$ is a maximal connected solvable subgroup; further, all Borel subgroups in $G$ are conjugates (and hence have the same dimension). The *radical* of $G$, denoted $R(G)$, is the connected component through the identity element of the intersection of all the Borel subgroups in $G$. An algebraic group $G$ is *semisimple*, if $R(G)$ is trivial. The group $SL_n(K)$ is an example of a semisimple algebraic group.

For the rest of this chapter, we shall suppose $G$ to be a connected, semisimple algebraic group (defined over the algebraically closed base field $K$). Let us fix a maximal torus $T$ in $G$.

Let $X(T) := \mathrm{Hom}_{\text{alg. gp.}}(T, K^*)$ be the character group of $T$ (written additively). Let $V$ be any finite dimensional $T$-module. Then the *complete reducibility* of $T$ implies (in fact, equivalent to the fact) that $V$ is the direct sum of *weight spaces*, i.e., $V = \oplus_{\chi \in X(T)} V_\chi$, where $V_\chi = \{v \in V : tv = \chi(t)v, \ \forall t \in T\}$. We say $\chi \in X(T)$ is a *weight in* $V$ if $V_\chi \neq 0$. Let $\Phi(T, V)$ be the set of non-zero weights in $V$.

Denote the Lie algebra of $G$ by $\mathfrak{g}$ or Lie $G$. For any $g \in G$, consider $\mathrm{Inn}(g) : G \to G$ given by $x \mapsto gxg^{-1}$. This induces the *adjoint action* on the Lie algebra $\mathrm{Ad}(g)$ : Lie $G \to$ Lie $G$. For $V =$ Lie $G$ as a $T$-module via the adjoint action, the corresponding $\Phi(T, \text{Lie } G)$ is called the *root system of G relative to T*; it is an abstract root system. One considers $\Phi(T, \text{Lie } G)$ to be a root system contained in the vector space $X(T) \otimes_{\mathbb{Z}} \mathbb{R}$ of rank $l$ $(= \dim T)$. We shall denote this root system by $R$. The Weyl group $W$ associated to $\Phi(T, \text{Lie } G)$ is isomorphic to $N(T)/T$ where $N(T)$ is the normalizer of $T$ in $G$.

Let $B$ be a Borel subgroup of $G$ such that $B \supset T$. The subset $\Phi(T, \text{Lie } B)$ of $\Phi(T, \text{Lie } G)$ is called the set of *positive roots of R relative to B*, denoted $R^+$. The indecomposable elements in $R^+$, i.e., the set of all $\alpha$ such that $\alpha$ cannot be expressed as a positive sum of other elements in $R^+$, is a set of simple roots for the root system $R$ which we shall denote by $S$. Thus, choosing $B$ is equivalent to choosing a set of simple roots $S$ in $R$. The fundamental weights $\{\varpi_i : 1 \leq i \leq l\}$ are defined as in Section 3.1. We have the inclusions

$$\text{root lattice} \subseteq X(T) \subseteq \text{weight lattice}.$$

The elements in $X(T)$ which lie in the dominant chamber with respect to $S$ are called *dominant characters*. Also, if $G$ is simply connected, then $X(T)$ equals the weight lattice $\Lambda$.

**Example**. Let $G = SL_n(K)$ be the special linear group of rank $n - 1$, $T$ the subgroup of diagonal matrices and $B$ the subgroup of upper triangular matrices. Then $T$ (resp. $B$) is a maximal torus (resp. a Borel subgroup). Further, $N(T)$ is the subgroup consisting of matrices with exactly one nonzero entry in each row and column, and $W = N(T)/T$ is the symmetric group $S_n$. We have

$$\text{Lie } G = \{n \times n \text{ matrices with trace } 0\}.$$

We further have,

$$\text{Lie}G = \mathfrak{h} \oplus \mathfrak{n}_+ \oplus \mathfrak{n}_-,$$

where

$$\mathfrak{h} = \{\text{ diagonal matrices in Lie}G\},$$

$$\mathfrak{n}_+ = \oplus_{1 \leq i < j \leq n}(E_{ij}), \ \mathfrak{n}_- = \oplus_{1 \leq j < i \leq n}(E_{ij})$$

(here $E_{ij}$ is the elementary $n \times n$ matrix with 1 at the $(i, j)$-th place and zeros elsewhere). Hence the roots may be identified with $\epsilon_i - \epsilon_j$, $1 \leq i, j \leq n$, $i \neq j$, where $\epsilon_i - \epsilon_j$ is the character of $T$ which sends $\mathrm{diag}(t_1, \cdots .t_n)$ in $T$ to $t_i t_j^{-1}$. Further, $R^+$ may be identified with $\{\epsilon_i - \epsilon_j,\ 1 \leq i < j \leq n\}$, and the fundamental weight $\omega_i$, $1 \leq i \leq n - 1$ may be identified with the character of $T$ sending $\mathrm{diag}(t_1, \cdots .t_n)$ in $T$ to $t_1 \cdots t_i$. In this example, $G$ is simply connected (and hence $X(T)$ coincides with the weight lattice).

**Root Subgroups**

The Lie algebra of $G$, denoted by $\mathfrak{g}$, contains a one-dimensional subalgebra $\mathfrak{g}_\alpha$ for each root $\alpha \in R$. Each $\mathfrak{g}_\alpha$ is a $T$-weight space of weight $\alpha$; $\mathfrak{g}_\alpha$ is called a *root space*. Furthermore, if $\mathfrak{h}$ denotes the Lie algebra of $T$, then $\mathfrak{g}$ decomposes as a direct sum of $T$-weight spaces

$$\mathfrak{g} = \mathfrak{h} \oplus \bigoplus_{\alpha \in R} \mathfrak{g}_\alpha.$$

For each $\beta \in R$, there exists a unique connected $T$-stable subgroup $U_\beta$ of $G$ (i.e., $U_\beta$ is normalized by $T$) with $\mathfrak{g}_\beta$ as its Lie algebra. The subgroup $U_\beta$ is called a *root subgroup*.

Let $\mathbb{G}_a$ denote the additive group $K$. For $\beta \in R$, there exists an isomorphism $\theta_\beta : \mathbb{G}_a \cong U_\beta$ such that for all $t \in T$, $x \in \mathbb{G}_a$, $t\theta_\beta(x)t^{-1} = \theta_\beta(\beta(t)x)$. Let $B_u$ denote the unipotent part of $B$, i.e., the subgroup of $B$ consisting of all the unipotent elements in $B$. Then we have an identification of varieties

$$B_u \cong \prod_{\beta \in R^+} U_\beta,$$

the product being taken in any fixed order. Further, $B_u$ is normalized by $T$, and $B$ is the semi-direct product of $B_u$ and $T$.

**Example**. In the example of Section 3.2, $(E_{i,j})$, the one dimensional span of $E_{i,j}$ for $1 \le i, j \le n$, $i \neq j$, are all the root spaces. The root subgroup $U_{i,j}$, $1 \le i, j \le n$, $i \neq j$ is the subgroup of $G$ consisting of matrices with 1's along the diagonal and the only possible non-zero entry occurs at the $(i, j)$-th place. $B_u$ is the subgroup of $G$ consisting of unipotent upper triangular matrices.

### 3.2.2 Parabolic subgroups

A *parabolic subgroup* of $G$ is any closed subgroup $Q \subset G$ such that $G/Q$ is a projective variety (or equivalently, a complete variety, since $G/Q$ is quasi projective (see [7, 43])).

**Fact:** A subgroup $Q$ of $G$ is parabolic if and only if it contains a Borel subgroup.

Let $Q$ be a parabolic subgroup containing $B$. Let $R(Q)$ be the radical of $Q$, namely, the connected component through the identity element of the intersection of all the Borel subgroups of $Q$. Let $R_u(Q)$ be the unipotent radical of $Q$ (the subgroup of unipotent elements of $R(Q)$), and let $R_Q^+$ be the subset of $R^+$ defined by $R^+ \setminus R_Q^+ = \{\beta \in R^+ \mid U_\beta \subset R_u(Q)\}$. Let $R_Q^- = -R_Q^+ (= \{-\beta, \beta \in R_Q^+\})$, $R_Q = R_Q^+ \cup R_Q^-$, and $S_Q = S \cap R_Q$. Then $R_Q$ is a subroot system of $R$ called the *root system associated to* $Q$, with $S_Q$ as a set of simple roots and $R_Q^+$ (resp. $R_Q^-$) as the set of positive (resp. negative) roots of $R_Q$ relative to $S_Q$.

Conversely, given a subset $J$ of $S$, there exists a parabolic subgroup $Q$ of $G$ such that $Q$ contains $B$. Namely, $Q$ is generated by $B$ and $\{U_{-\alpha} \mid \alpha \in R^+ \text{ and } \alpha = \sum_{\beta \in J} a_\beta \beta\}$.

Thus the set of parabolic subgroups containing $B$ is in bijection with the power set of $S$. Note that for $Q = B$ (resp. $G$), $S_Q$ is the empty set (resp. the whole set $S$).

The subgroup of $Q$ generated by $T$ and $\{U_\alpha \mid \alpha \in R_Q\}$ is called the *Levi subgroup* associated to $S_Q$, and is denoted $L_Q$. We have that $Q$ is the semidirect product of $R_u(Q)$ and $L_Q$ (see [7,43] for details).

The set of *maximal parabolic subgroups* containing $B$ is in one-to-one correspondence with $S$. Namely given $\alpha \in S$, the parabolic subgroup $Q$ where $S_Q = S \setminus \{\alpha\}$ is a maximal parabolic subgroup, and conversely. In the sequel, we shall also denote the maximal parabolic subgroup $Q$, where $S_Q = S \setminus \{\alpha_i\}$ by $P_i$.

### The Weyl Group of a Parabolic Subgroup

Given a parabolic subgroup $Q$, let $W_Q$ be the subgroup of $W$ generated by $\{s_\alpha \mid \alpha \in S_Q\}$. $W_Q$ is called the *Weyl group* of $Q$. Note that $W_Q \cong N_Q(T)/T$, where $N_Q(T)$ is the normalizer of $T$ in $Q$.

**Definition 3.2.2.1** ***The set $W_Q^{\min}$ of minimal representatives of*** $W/W_Q$*: In each coset $wW_Q \in W/W_Q$, there exists a unique element of minimal length. Let $W_Q^{min}$ be the* set of minimal length representatives of $W/W_Q$. *We have*

$$W_Q^{min} = \{w \in W \mid l(ww') = l(w) + l(w'),\ \textit{for all } w' \in W_Q\}.$$

In other words, each element $w \in W$ can be written uniquely as $w = u \cdot v$ where $u \in W_Q^{\min}$, $v \in W_Q$ and $l(w) = l(u) + l(v)$. The set $W_Q^{\min}$ may be also be characterized as

$$W_Q^{\min} = \{w \in W \mid w(\alpha) > 0,\ \text{for all } \alpha \in S_Q\}.$$

(Here by a root $\beta$ being $> 0$ we mean $\beta \in R^+$.) In the literature, $W_Q^{\min}$ is also denoted by $W^Q$; in the sequel, we shall use both of the notations.

**Definition 3.2.2.2** ***The set $W_Q^{\max}$ of maximal representatives of*** $W/W_Q$*: In each coset $wW_Q$ there exists a unique element of maximal length. Let $W_Q^{max}$ be the* set of maximal length representatives *of $W/W_Q$:*

$$W_Q^{max} = \{w \in W \mid w(\alpha) < 0\, \textit{for all } \alpha \in S_Q\}.$$

*Further, if we denote by $w_Q$ the unique element of maximal length in $W_Q$, then we have*

$$W_Q^{max} = \{ww_Q \mid w \in W_Q^{min}\}.$$

## 3.3 Schubert varieties

The projective variety $G/Q$ is called a *generalised flag* variety. This terminology stems from the fact that for the case of $G = GL_n(K)$, and $Q = B$, the (Borel) sub group of upper triangular matrices, it is indeed a flag variety; namely, let

$$\mathcal{F} = \{\texttt{all flags } F : V_0 \subset V_1 \subset \cdots \subset V_n = K^n,\ \texttt{dim}\, V_i = i\}$$

We have a transitive action of $GL_n(K)$ on $\mathcal{F}$, and $B$ is the isotropy at the standard flag $\mathcal{F}_0$ (with $V_i$ = the span of $\{e_1, \cdots, e_i\}$, where $\{e_1, \ldots, e_n\}$ denotes the standard basis of $K^n$). Thus we obtain a bijection between $GL_n(K)/B$ and $\mathcal{F}$. Hence $\mathcal{F}$ acquires a projective variety structure, and $\mathcal{F}$ with this variety structure is called *the flag variety*.

Returning to the case of a general $G/Q$, we have that for the induced action of $T$ on $Q$, there are only finitely many fixed points $\{e_w := wW_Q \mid w \in W/W_Q\}$. For $w \in W/W_Q$, the $B$-orbit $C_Q(w) := Be_w$ in $G/Q$ is a locally closed subset of $G/Q$, called the *Bruhat* or *Schubert cell*. The Zariski closure of $C_Q(w)$ with the canonical reduced structure is the *Schubert variety* associated to $w$, and is denoted $X_Q(w)$. Thus Schubert varieties in $G/Q$ are indexed by $W_Q^{\min}$ (or also by $W_Q^{\max}$). Note that if $Q = B$, then $W_Q = \{id\}$, and the Schubert varieties in $G/B$ are indexed by the elements of $W$.

**Remark 3.3.0.1.** We use the notation $X(w)$ for the Schubert variety in $G/Q$ indexed by $w \in W$ when $Q = B$ or when the parabolic subgroup is understood from context.

**Dimension of $X_B(w)$:** Let $w \in W$. The isotropy subgroup in $G$ at the ($T$-fixed) point $e_w$ (in $G/B$) is $wBw^{-1}$; hence, the isotropy subgroup in $B_u$ at $e_w$ is generated by the root subgroups $\{U_\beta, \beta \in R^+ \mid U_\beta \subset wBw^{-1}\}$, i.e., $\{U_\beta, \beta \in R^+ \mid w^{-1}(\beta) > 0\}$. Hence we get an identification

$$C_B(w) \cong \prod_{\{\beta \in R^+ \mid w^{-1}(\beta) < 0\}} U_\beta$$

Now $\ell(w)$ equals $\#\{\beta \in R^+ \mid w^{-1}(\beta) < 0\}$ (cf. §3.1, **Facts**). Thus $C_B(w)$ is isomorphic to the affine space $K^{\ell(w)}$. Hence we obtain

$$dim\, X_B(w)\, (= dim\, C_B(w)) = \ell(w)$$

Note that for $w = id$, $X(id) = C(id) = \{\text{a point}\}$, the unique point cell.

**Dimension of $X_Q(w)$:** Let $w \in W/W_Q$. Denoting the unique representative for $w$ in $W_Q^{\min}$ (resp. $W_Q^{\max}$) by $w_Q^{\min}$ (resp. $w_Q^{\max}$), we have that under the canonical projection $\pi_Q : G/B \to G/Q$, $X_B(w_Q^{\min})$ maps birationally onto $X_Q(w)$, and $X_B(w_Q^{\max}) = \pi_Q^{-1}(X_Q(w))$. Hence we obtain

$$dim\, X_Q(w) = dim\, X_B(w_Q^{\min}) = \ell(w_Q^{\min})$$

**Big cell and opposite big cell:** We have, $G/B = X(w_0)$, $w_0$ being the element in $W$ of largest length. The cell $C_B(w_0)$ is the unique cell of maximal dimension $(= \ell(w_0) = \#R^+)$; it is affine, open and dense in $G/B$, called the *big cell* of $G/B$. It is denoted as $\mathcal{O}$. Let $B^-$ be the Borel subgroup *opposite* to $B$ (note that $B^- = w_0Bw_0^{-1} = w_0Bw_0$, since $w_0^{-1} = w_0$). The $B^-$-orbit $B^-e_{id}$ is again affine, open and dense in $G/B$, and is called the *opposite big cell* of $G/B$. It is an affine neighborhood of $e_{id}$, and is denoted as $\mathcal{O}^-$. More generally, for $\tau \in W$, $\tau B^-\tau^{-1} \cdot e_\tau$

is affine, open and dense in $G/B$. It is an affine neighborhood of $e_\tau$, and is denoted as $\mathcal{O}_\tau^-$.

**Opposite cell in $X_B(w)$:** Let $w \in W$. Then $Y_B(w) := X_B(w) \cap \mathcal{O}^-$ is called the *opposite cell* in $X_B(w)$ (though for a general $w$, it need not be a cell). It is a closed sub variety of the affine space $\mathcal{O}^-$ ($\cong \mathbb{A}^N$, $N$ being the dimension of $G/B$ equal to $\# R^+$), and hence is affine. In particular, it is an open neighborhood of $e_{id}$ in $X_B(w)$. More generally, for $\tau, w \in W, \tau \leq w$, $X_B(w) \cap \mathcal{O}_\tau^-$ is an open neighborhood of $e_\tau$ in $X_B(w)$.

**Remark 3.3.0.2.** For a general $G/Q$, $Q$ being any parabolic subgroup, the above concepts - big cell, opposite big cell, opposite cell in a Schubert variety etc., are defined in the obvious way.

**Bruhat decomposition:** We have

$$G/Q = \dot{\bigcup_{w \in W^Q}} Be_w Q(mod\ Q),$$

$$X_Q(y) = \dot{\bigcup_{\{w \in W^Q,\ e_w \in X_Q(y)\}}} Be_w Q(mod\ Q),\ y \in W^Q$$

**The Bruhat-Chevalley order:** There is a partial order on $W^Q$, known as the *Bruhat-Chevalley order* (or just *Bruhat order*), induced by the partial order on the set of Schubert varieties given by inclusion, namely, for $w_1, w_2 \in W^Q$, $w_1 \geq w_2 \iff X_Q(w_1) \supseteq X_Q(w_2)$. Taking $Q = B$, we obtain a partial order on $W$ (the Weyl group of $G$). This partial order plays an important role in the theory of Schubert varieties. Chevalley has shown (cf. Chevalley [15]) that this order can be stated equivalently in terms of the Weyl group, namely $w_1 \geq w_2 \in W$ if every (equivalently, any) reduced expression for $w_1$ contains a sub-expression which is a reduced expression for $w_2$. For example, $s_1s_2s_3s_2 \geq s_1s_3s_2$.

Another description for Bruhat-Chevalley order is given by the following: For $w \in W$, let $X(w^{(d)})$ denote the projection of $X(w)$ under $G/B \to G/P_d$, $1 \leq d \leq l$; then for $u, v \in W$, $u \leq v$ if and only if $X(u^{(d)}) \subseteq X(v^{(d)})$ for all $1 \leq d \leq l$.

### 3.3.1 Weyl and Demazure modules

Let $\mathfrak{g}_{\mathbb{C}} = \text{Lie}\ G_{\mathbb{C}}$, and $U(\mathfrak{g}_{\mathbb{C}})$ be the universal enveloping algebra of $\mathfrak{g}_{\mathbb{C}}$. Let $U^+(\mathfrak{g}_{\mathbb{C}})$ be the subalgebra of $U(\mathfrak{g}_{\mathbb{C}})$ generated by $\{X_\alpha, \alpha \in S\}$, and $U_{\mathbb{Z}}^+(\mathfrak{g}_{\mathbb{C}})$ be the Kostant $\mathbb{Z}$-form of $U^+(\mathfrak{g}_{\mathbb{C}})$) (recall (cf. [52]) that $U_{\mathbb{Z}}^+(\mathfrak{g}_{\mathbb{C}})$ is the $\mathbb{Z}$-subalgebra of $U^+(\mathfrak{g}_{\mathbb{C}})$ generated by $\{\frac{X_\alpha^n}{n!}, \alpha \in R^+, n \in \mathbb{N}\}$).

Let $\lambda$ be a dominant weight, and $V(\lambda)$ be the irreducible $G$-module over $\mathbb{C}$. Fix a highest-weight vector $u_\lambda$ (i.e., annulled by $X_\beta$ for all $\beta \in R^+$) in $V(\lambda)$ of weight $\lambda$ (for the $T$-action); we have that the weight $\lambda$ in $V(\lambda)$ has multiplicity one. For $w \in W$, fix a representative $n_w$ for $w$ in $N_T(G)$ (the normalizer of $T$ in $G$), and set $u_{w,\lambda} = n_w \cdot u_\lambda$, known as *an extremal weight vector*; it is a weight vector in $V(\lambda)$

(for the $T$-action) of weight $w(\lambda)$, and is unique up to scalars. Having fixed $\lambda$, we shall denote $u_\lambda$ (resp. $u_{w,\lambda}$) by just $u$ (resp. $u_w$). Set $V_{w,\mathbb{Z}}(\lambda) = U^+_{\mathbb{Z}}(\mathfrak{g}_{\mathbb{C}})u_w$. For any field $K$, let $V_{w,\lambda} = V_{w,\mathbb{Z}}(\lambda) \otimes K,\ w \in W$. Then $V_K(\lambda) := V_{w_0,\lambda}\ (= V_{w_0,\mathbb{Z}}(\lambda) \otimes K)$ is the *Weyl module* with *highest weight* $\lambda$, and for $w \in W,\ V_{w,\lambda}$ is the *Demazure module* corresponding to $w$ (and $\lambda$). Note, the Bruhat-Chevalley order also appears in the context of Demazure modules as follows: Given a dominant weight $\lambda$, let $P_\lambda$ be the stabilizer of $K \cdot u$, where $u$ as above, is a highest-weight vector $u$ in $V_K(\lambda)$; note that the Weyl group $W_{P_\lambda}$ of $P_\lambda$ is the subgroup of $W$ generated by $\{s_\alpha, \alpha \in S,\ s_\alpha(\lambda) = \lambda\}$. The extremal weight vectors in $V_K(\lambda)$ are indexed by $W^{min}_{P_\lambda}$; further, for $w_1, w_2 \in W^{min}_{P_\lambda}$, we have, $w_1 \leq w_2 \Leftrightarrow V_{w_1,\lambda} \subseteq V_{w_2,\lambda}$, equivalently, $u_{w_1}$ is in $V_{w_2,\lambda}$.

Fix a dominant weight $\lambda$. Then $V_K(\lambda)^*$ is again a Weyl module; it is the Weyl module with highest weight $i(\lambda)$, $i$ being the *Weyl involution* $i := -w_0$. Note that $i(\lambda) = -w_0(\lambda)$ is the negative of the lowest weight $w_0(\lambda)$ in $V_K(\lambda)$. More generally, the weights in $V_K(\lambda)^*$ are the negatives of the weights in $V_K(\lambda)$. For $w \in W$, as above, let $u_w$ be an extremal weight vector (unique up to scalars) in $V_K(\lambda)$. Let $p_w$ be an extremal weight vector (again unique up to scalars) in $V_K(\lambda)^*$ of weight $-w(\lambda)$. There exists a canonical $G$-invariant pairing $(,)$ on $V_K(\lambda) \times V_K(\lambda)^*$; further, we have,

$$(u_w, p_{w'}) = \delta_{w,w'}$$

### 3.3.2 Line bundles on $G/Q$

For the study of $G/B$, there is no loss in generality in assuming that $G$ is simply connected; in particular, as remarked in §3.2, we have that the character group $X(T)$ coincides with the weight lattice $\Lambda$. Henceforth, we shall suppose that $G$ is simply connected. Let $X = G/B$, and consider the canonical projection $\pi : G \to X$. Then $\pi : G \to X$ is a principal $B$-bundle with $B$ as both the structure group and fiber. Let $\mathbb{G}_m$ denote the multiplicative group $K^*$ and let $\lambda \in X(T)$. Then $\lambda$ defines a character $\lambda_B : B \to \mathbb{G}_m$ obtained by composing the natural map $B \to T$ with $\lambda : T \to \mathbb{G}_m$ (recall that $B$ is the semi-direct product of $T$ and $B_u$ (= the unipotent part of $B$)). We shall denote $\lambda_B$ also by just $\lambda$. Now $\lambda : B \to \mathbb{G}_m$ gives rise to an action of $B$ on $K(= \mathbb{G}_a)$, namely $b \cdot k = \lambda(b)k, b \in B, k \in K$. Set $E = G \times K/\sim$, where $\sim$ is the equivalence relation defined by $(gb, b \cdot k) \sim (g, k),\ g \in G, b \in B, k \in K$. Then $E$ is the total space of a line bundle, say $L(\lambda)$, over $X$. Thus we obtain a map

$$L : X(T) \to \mathrm{Pic}(G/B), \quad \lambda \mapsto L(\lambda),$$

where $\mathrm{Pic}(G/B)$ is the Picard group of $G/B$ (namely, the group of isomorphism classes of line bundles on $G/B$). We recall (cf. Chevalley [16]) that the above map is in fact an isomorphism of groups since $G$ is simply connected.

On the other hand, consider the prime divisors $X(w_0 s_i),\ 1 \leq i \leq l$ on $G/B$. Let $L_i = \mathcal{O}_{G/B}(X(w_0 s_i))$ be the line bundle defined by $X(w_0 s_i),\ 1 \leq i \leq l$. The Picard group $\mathrm{Pic}(G/B)$ is a free abelian group generated by the $L_i$'s, and under the isomorphism $L : X(T) \cong \mathrm{Pic}(G/B)$, we have $L(\omega_i) = L_i,\ 1 \leq i \leq l$ (see

Chevalley [16]). Thus for $\lambda = \sum_{i=1}^{l} \langle \lambda, \alpha_i \rangle \omega_i$, we have $L(\lambda) = \otimes_{i=1}^{l} L_i^{\otimes \langle \lambda, \alpha_i \rangle}$. We shall write $L(\lambda) \geq 0$ if $\lambda \geq 0$, i.e., if $\lambda$ is dominant.

As above, let $E$ denote the total space of the line bundle $L(\lambda)$, over $X (= G/B)$. Let $\sigma : E \to X$ be the canonical map $\sigma(g, c) = gB$. Let

$$M_\lambda = \{ f \in k[G] \mid f(gb) = \lambda(b) f(g),\ g \in G, b \in B \}.$$

Then $M_\lambda$ can be identified with the *space of sections* $H^0(G/B, L(\lambda)) := \{ s : X \to E \mid \sigma \circ s = \mathrm{id}_X \}$ as follows. Let $f \in M_\lambda$. To $f$, we associate a section $s : X \to E$ by setting $s(gB) = (g, f(g))$. To see that $s$ is well defined, consider $g' = gb$, $b \in B$. Then $(g', f(g')) = (gb, f(gb)) = (gb, \lambda(b) f(g)) = (gb, b \cdot f(g)) \sim (g, f(g))$. From this, it follows that $s$ is well defined. Conversely, given $s \in H^0(G/B, L(\lambda))$, consider $gB \in G/B$. Let $s(gB) = (g', f(g'))$, where $g' = gb$ for some $b \in B$ (note that $g'B = gB$, since $\sigma \circ s = \mathrm{id}_X$). Now the point $(g', f(g'))$ may also be represented by $(g, \lambda(b)^{-1} f(gb))$ (since $(g', f(g')) = (gb, f(gb)) \sim (g, \lambda(b)^{-1} f(gb))$. Thus given $g \in G$, there exists a unique representative of the form $(g, f(g))$ for $s(gB)$. This defines a function $f : G \to K$. Further, this $f$ has the property that for $b \in B$, $f(g) = \lambda(b)^{-1} f(gb)$, i.e., $f(gb) = \lambda(b) f(g)$, $b \in B$, $g \in G$. Thus we obtain an identification

$$M_\lambda = H^0(G/B, L(\lambda)).$$

It can be easily checked that the above identification preserves the respective $G$-module structures.

The structure of the line bundle $L(\lambda)$ can often be described in terms of the weight $\lambda$. For example, Chevalley has proven the following facts (see [16] and also [47]):

1. $H^0(G/B, L(\lambda)) \neq 0$ if and only if $\lambda$ is dominant.
2. $L(\lambda)$ is ample $\Leftrightarrow$ $L(\lambda)$ is very ample $\Leftrightarrow$ $\lambda$ is dominant and regular.

(A line bundle $L$ on an algebraic variety $X$ is *very ample* if there exists an immersion $i : X \hookrightarrow \mathbb{P}^N$ such that $i^*(\mathcal{O}_{\mathbb{P}^N}(1)) = L$. A line bundle $L$ on $X$ is *ample* if $L^m$ is very ample for some positive integer $m \geq 1$. See [37] for details.)

## Summary

Let $\lambda \in \Lambda^+$, i.e., $\lambda$ is a dominant weight. We have (see Jantzen [47] for example),

$$H^0(G/B, L(\lambda)) \cong V_K(\lambda)^*$$

where $V_K(\lambda)^*$, as above, is the dual Weyl module corresponding to $\lambda$. Now using the fact that the multiplicity of the weight $i(\lambda)$ in $V_K(\lambda)^*$ equals one, we obtain that $H^0(G/B, L(\lambda))$ is $G$-indecomposable. The indecomposability of $H^0(G/B, L(\lambda))$ together with the complete reducibility of $G$ in characteristic 0 implies that if $\mathrm{char} K = 0$, then $H^0(G/B, L(\lambda))$ is irreducible. Thus we obtain geometric realizations of irreducible $G$-modules in characteristic 0.

We further have the following facts (see [47] for details):
1. There exists a bijection between $\Lambda^+$ and the set of isomorphism classes of finite dimensional irreducible $G$-modules (in any characteristic).
2. $H^0(G/B, L(\lambda)) = \{f \in K[G] \mid f(gb) = \lambda(b)f(g), g \in G, b \in B\}$.
3. $H^0(G/B, L(\lambda)) \neq 0$ if and only if $\lambda \in \Lambda^+$.
4. $L(\lambda)$ is ample $\Longleftrightarrow$ $L(\lambda)$ is very ample $\Longleftrightarrow$ $\lambda$ is dominant and regular.
5. Let $\lambda$ be dominant. There exists a unique $B$-stable line in $H^0(G/B, L(\lambda))$, and the $G$-submodule generated by this line is the irreducible $G$-module with highest weight $i(\lambda)$, $i(= -w_0)$ being the Weyl involution.
6. Let $\lambda$ be dominant. Then $H^0(G/B, L(\lambda))$ is an indecomposable $G$-module.
7. Let $\lambda$ be dominant. Then the weights in $H^0(G/B, L(\lambda))$ are the negatives of the weights in the Weyl module $V_K(\lambda)$; in particular, the extremal weights in (the Weyl module) $H^0(G/B, L(\lambda))$ are given by $\{-w(\lambda), w \in W\}$. In the sequel, the extremal weight vector in $H^0(G/B, L(\lambda))$ of weight $-w(\lambda)$ (which is unique up to scalars) will be denoted by $p_w$.
8. Let $\lambda$ be dominant. We have a natural identification

$$H^0(X(w), L_\lambda) \cong V^*_{w,\lambda}, \quad w \in W$$

9. Let $\lambda = \varpi$ be a fundamental weight with $P$ as the maximal parabolic subgroup. Then $L(\varpi)$ is the ample generator of $\operatorname{Pic} G/P (\cong \mathbb{Z})$.
10. In characteristic 0, $H^0(G/B, L_\lambda)$ is $G$-irreducible with highest weight $i(\lambda)$ and conversely. Thus in characteristic 0, $\{H^0(G/B, L_\lambda),\ \lambda \in \Lambda^+\}$ gives (up to isomorphism) all finite-dimensional irreducible $G$-modules.

### 3.3.3 Equations defining a Schubert variety

Let $G$ be semi-simple, and $P_i$ where $i \leq l$ ($l$ being the rank $G$) be a maximal parabolic subgroup with $\varpi_i$ as the associated fundamental weight. Let $L_i$ be the ample generator of $\operatorname{Pic} G/P_i (\cong \mathbb{Z})$. Now $L_i$ induces a projective embedding $G/P_i \hookrightarrow \mathbb{P}(V_K(\varpi_i))$; in fact, for $\tau \in W^{P_i}$, the point $\tau P_i$ in $G/P_i$ gets identified with the point in $\mathbb{P}(V_K(\varpi_i))$ representing the one-dimensional span of an extremal weight vector in $V_K(\varpi_i)$ of weight $\tau(\varpi_i)$.

Let $w \in W$, and $w_i$ be the projection of $w$ under $W \to W/W_{P_i}$. We have the following:

**Fact 1:** ([10, Theorem 3.5.2], [47, II.14.15]) The restriction map $H^0(G/P_i, L_i^m) \to H^0(X(w_i), L_i^m)$, $m \geq 0$ is surjective.

**Fact 2:** ([105], [10, Theorem 3.5.2]) Fix a basis $D_i$ for the kernel of the surjective restriction map from $H^0(G/P_i, L_i)$ to $H^0(X(w_i), L_i)$. Then the ideal sheaf of $X(w_i)$ in $G/P_i$ is generated by $D_i$.

Denoting

$$R = \bigoplus_{m \in \mathbb{Z}_+} H^0(G/P_i, L_i^m)$$

$$R(w_i) = \bigoplus_{m \in \mathbb{Z}_+} H^0(X(w_i), L_i^m)$$

Facts 1 & 2 imply that the restriction map $R \to R(w_i)$ is surjective, and that the kernel is generated (as an ideal) by $D_i$.

We further have,

**Fact 3:** ([50]) The ideal sheaf of $X(w)$ in $G/B$ is generated by $\{D_i,\ 1 \le i \le l\}$. Denoting

$$A = \bigoplus_{\underline{a}} H^0(G/B, \bigotimes_{i=1}^{l} L_i^{a_i})$$

$$A(w) = \bigoplus_{\underline{a}} H^0(X(w), \bigotimes_{i=1}^{l} L_i^{a_i}),$$

where $\underline{a} = (a_1, \ldots, a_l) \in \mathbb{Z}_+^l$, Fact 3 implies that the restriction map $A \to A(w)$ is surjective, and its kernel is generated as an ideal by $\{D_i,\ 1 \le i \le l\}$, i.e., by elements of total degree 1.

**Remark 3.3.3.1.** In the literature, one describes this by saying "the ideal sheaf of a Schubert variety in $G/B$ is generated in degree one".

# 4

# Grassmannian Variety

In this chapter, we study the Schubert varieties in the Grassmannian. After introducing the Grassmannian and its Schubert varieties in §1 and §2, we present the Standard monomial theory for Schubert varieties ([41, 42, 88]) in the Grassmannian in §3 and §4. We then present a proof of the "vanishing theorems" for Schubert varieties in §5, and deduce normality and Cohen-Macaulayness for Schubert varieties in §6.

## 4.1 The Plücker embedding

Let us fix the integers $1 \leq d < n$ and let $V = K^n$. The *Grassmannian* $G_{d,n}$ is the set of $d$-dimensional subspaces $U \subset V$. Let $U$ be an element of $G_{d,n}$, and $a_1, \ldots, a_d$ a basis of $U$, where

$$a_j = \begin{pmatrix} a_{1j} \\ a_{2j} \\ \cdots \\ a_{nj} \end{pmatrix}, \text{ with } a_{ij} \in k, \text{ for } 1 \leq i \leq n,\ 1 \leq j \leq d.$$

(here, each vector $a_j$ is written as a column vector with respect to the standard basis of $K^n$). Thus the basis $\{a_1, \ldots, a_d\}$ gives rise to an $n \times d$ matrix $A = (a_{ij})$ of rank $d$, whose columns are the vectors $a_1, \ldots, a_d$. Let $a'_1, \ldots, a'_d$ be another basis giving rise to an $n \times d$ matrix $A'$. Then there exists a change of basis matrix $C \in GL_d(K)$, such that $A' = AC$. Conversely, given two $n \times d$ matrices $A$ and $A'$ of rank $d$, such that $A' = AC$, for some matrix $C \in GL_d(K)$, the columns of both $A$ and $A'$ are $d$ linearly independent vectors in $V$, generating one and the same $d$-dimensional subspace of $V$. Identifying an $n \times d$ matrix with a point in the affine space $\mathbb{A}^{nd}$, we see that $G_{d,n}$ can be viewed as $(\mathbb{A}^{nd} \setminus Z)/\sim$, where $Z$ is the set of $n \times d$ matrices of rank less than $d$, and the equivalence relation $\sim$ is defined by

$$A \sim A' \quad \text{if there exists a } C \in GL_d(K) \text{ such that } A' = AC.$$

(Note in particular that $\mathbb{P}^{n-1} = G_{1,n}$.)

Thus denoting $M^0_{n\times d} = \mathbb{A}^{nd} \setminus Z$, $G_{d,n}$ is simply the orbit space $M^0_{n\times d}/GL_d(K)$, for the action of $GL_d(K)$ on $M^0_{n\times d}$ by multiplication on the right.

### 4.1.1 The partially ordered set $I_{d,n}$

Define the set

$$I_{d,n} := \{\underline{i} = (i_1, \ldots, i_d) | 1 \leq i_1 < \cdots < i_d \leq n\},$$

Define a partial order $\geq$ on $I_{d,n}$ by defining $\underline{i} \geq \underline{j} \Leftrightarrow i_t \geq j_t, \forall t$.

### 4.1.2 Plücker embedding and Plücker coordinates

Let $N = \#I_{d,n}$; then $N = \binom{n}{d}$, and the coordinates of the affine space $\wedge^d V = K^N$ will be indexed by the set $I_{d,n}$. For a non-zero $v \in K^N$, we shall denote by $[v]$ the point determined by $v$ in $\mathbb{P}^{N-1}$. Let

$$X = \underbrace{V \oplus \cdots \oplus V}_{d \text{times}} = K^{nd}.$$

The exterior product map

$$\wedge^d : X \to \wedge^d V$$

sends the element $(a_1, \ldots, a_d)$ to the point $a_1 \wedge \cdots \wedge a_d$ in $\wedge^d V$. Identifying $\wedge^d V$ with $K^N$, the $\underline{i}$-th coordinate of $a_1 \wedge \cdots \wedge a_d$ is given by the determinant of the $d$-minor of the $n \times d$ matrix $A = (a_{ij})$ (where the columns of the matrix $A = (a_{ij})$ are the vectors $a_1, \ldots, a_d$), with row indices $i_1, \ldots, i_d$. Note that $a_1 \wedge \cdots \wedge a_d = 0$ if and only if the matrix $A$ belongs to the set $Z$ of matrices of rank less than $d$. Moreover, it is easily seen that given $(a_1, \ldots, a_d)$, $(a'_1, \ldots, a'_d)$ in $X$, such that the associated $n \times d$ matrices $A$, $A'$, are equivalent (i.e., there exists a matrix $C \in GL_d(K)$ such that $A' = AC$), then

$$a'_1 \wedge \cdots \wedge a'_d = \det(C) a_1 \wedge \cdots \wedge a_d.$$

This shows that the map $\wedge^d$ induces a well defined map

$$p : G_{d,n} \to \mathbb{P}(\wedge^d V) = \mathbb{P}^{N-1}$$

called the *Plücker map*. For $\underline{i} \in I_{d,n}$, we shall denote the $\underline{i}$-th component of $p$ by $p_{\underline{i}}$, or by $p_{i_1,\ldots,i_d}$, where $\underline{i} = (i_1, \ldots, i_d)$; the $p_{\underline{i}}$'s, with $\underline{i} \in I_{d,n}$, are called the *Plücker coordinates*. For $U$ in $G_{d,n}$, as above, let $A_{n,d}$ be the matrix representing $U$; then $p_{i_1,\ldots,i_d}(U) = \det(A_{i_1,\ldots,i_d})$, where $A_{i_1,\ldots,i_d}$ denotes the $d \times d$ sub matrix of $A$ consisting of the rows with indices $i_1, \ldots, i_d$.

Given a point $U$ in $G_{d,n}$ represented by the $n \times d$ matrix $A$, and numbers $1 \leq i_1, \ldots, i_d \leq n$ (not necessarily distinct, nor in increasing order), $A_{i_1,\ldots,i_d}$ will denote the matrix whose rows are the rows of $A$ with indices $i_1, \ldots, i_d$, taken in the order in which they appear in $A$. Also, we define $p_{i_1,\ldots,i_d}(U) = \det(A_{i_1,\ldots,i_d})$. It is easy to see that $p_{i_1,\dots,i_d} = 0$ if $i_1,\dots,i_d$ are not distinct, and $p_{i_1,\dots,i_d} = (-1)^{l(\sigma)} p_{i_{\sigma(1)},\dots,i_{\sigma(d)}}$ if $i_1,\dots,i_d$ are distinct and $\sigma$ is the permutation of the set $\{1,\dots,d\}$, such that $\sigma(1) < \cdots < \sigma(d)$.

For each $\underline{i} \in I_{d,n}$ consider the point $e_{\underline{i}}$ of $G_{d,n}$ represented by the $n \times d$ matrix whose entries are all 0, except the ones in the $i_j$-th row and $j$-th column, which are equal to 1, $1 \le j \le d$. Clearly, for $\underline{i}, \underline{j} \in I_{d,n}$,

$$p_{\underline{i}}(e_{\underline{j}}) = \begin{cases} 1, & \text{if } \underline{i} = \underline{j}; \\ 0, & \text{otherwise.} \end{cases}$$

**Theorem 4.1.2.1** *The Plücker map is injective.*

*Proof.* Let $U, U'$ be two points in $G_{d,n}$ such that $p(U) = p(U')$. Then there exists $\underline{l} = (l_1,\dots,l_d) \in I_{d,n}$ such that $p_{\underline{l}}(U) = p_{\underline{l}}(U') \neq 0$, and we may assume that $p_{\underline{l}}(U) = p_{\underline{l}}(U') = 1$. If $U$ and $U'$ are represented by the $n \times d$ matrices $A$, respectively $A'$, we have $\det(A_{l_1,\dots,l_d}) = \det(A'_{l_1,\dots,l_d}) = 1$. Replacing $A$ and $A'$ with $AC$, respectively $A'C'$, where $C = A^{-1}_{l_1,\dots,l_d}$, $C' = A'^{-1}_{l_1,\dots,l_d}$, we can assume that $A_{l_1,\dots,l_d} = A'_{l_1,\dots,l_d} = \mathrm{Id}_d$. For $1 \le i \le n$ and $1 \le j \le d$, we have

$$\begin{aligned} a_{ij} &= \det A_{l_1,\dots,l_{j-1},i,l_{j+1},\dots,l_d} = p_{l_1,\dots,l_{j-1},i,l_{j+1},\dots,l_d}(U) \\ &= p_{l_1,\dots,l_{j-1},i,l_{j+1},\dots,l_d}(U') = \det A'_{l_1,\dots,l_{j-1},i,l_{j+1},\dots,l_d} \\ &= a'_{ij}. \end{aligned}$$

Thus $A = A'$, hence $U = U'$, which shows that $p$ is injective.

In view of Theorem 4.1.2.1, we identify $G_{d,n}$ with $\mathrm{Im}\, p$ and consider $G_{d,n} \subset \mathbb{P}^{N-1}$ set theoretically. Next we will show that $G_{d,n}$ is closed in $\mathbb{P}^{N-1}$.

### 4.1.3 Plücker quadratic relations

**Theorem 4.1.3.1** *The Grassmannian $G_{d,n} \subset \mathbb{P}^{N-1}$ consists of the zeroes in $\mathbb{P}^{N-1}$ of the following quadratic polynomials:*

$$\sum_{\lambda=1}^{d+1} (-1)^{\lambda} p_{i_1,\dots,i_{d-1},j_\lambda} p_{j_1,\dots,\widehat{j_\lambda},\dots,j_{d+1}} \qquad (*)$$

*where $i_1,\dots,i_{d-1}$ and $j_1,\dots,j_{d+1}$ are any numbers between 1 and $n$.*
*(A relation obtained by equating an expression given by (*) to 0 is usually referred to as a* Plücker relation*).*

*Proof.* Fix $1 \le i_1,\dots,i_{d-1}, j_1,\dots,j_{d+1} \le n$. Let $U \in G_{d,n}$ be represented by the $n \times d$ matrix $A$. Then (*) evaluated at $U$ equals

$$\sum_{\lambda=1}^{d+1} (-1)^{\lambda} \det(A_{i_1,\dots,i_{d-1},j_\lambda}) \det(A_{j_1,\dots,\widehat{j_\lambda},\dots,j_{d+1}}).$$

Expanding the first determinants along their last rows, the above expression equals

$$\sum_{\lambda=1}^{d+1}(-1)^{\lambda}\sum_{\mu=1}^{d}(-1)^{d+\mu}a_{j_\lambda,\mu}\det(A^{\widehat{\mu}}_{i_1,\ldots,i_{d-1}})\det(A_{j_1,\ldots,\widehat{j_\lambda},\ldots,j_{d+1}}),$$

where $A^{\widehat{\mu}}_{i_1,\ldots,i_{d-1}}$ is obtained from $A_{i_1,\ldots,i_{d-1}}$ by deleting the $\mu$-th column. The last expression can be written in the form

$$\sum_{\mu=1}^{d}(-1)^{d+\mu}\det(A^{\widehat{\mu}}_{i_1,\ldots,i_{d-1}})\sum_{\lambda=1}^{d+1}(-1)^{\lambda}a_{j_\lambda,\mu}\det(A_{j_1,\ldots,\widehat{j_\lambda},\ldots,j_{d+1}}),$$

or

$$\sum_{\mu=1}^{d}(-1)^{d+\mu}\det(A^{\widehat{\mu}}_{i_1,\ldots,i_{d-1}})(-1)\det({}^{\mu}A_{j_1,\ldots,\widehat{j_\lambda},\ldots,j_{d+1}})$$

where ${}^{\mu}A_{j_1,\ldots,\widehat{j_\lambda},\ldots,j_{d+1}}$ is obtained from the matrix $A_{j_1,\ldots,\widehat{j_\lambda},\ldots,j_{d+1}}$ by adding the $\mu$-th column of $A$ as its first column. Note that this matrix has two identical columns, and therefore its determinant is 0. This implies that the last expression is 0 and hence we obtain that the point $U$ in $G_{d,n}$ is a zero of the polynomials given by (*).

Conversely, let $q = (q_{\underline{i}})_{\underline{i}\in I_{d,n}}$ be a zero of the polynomials given by (*). Let $q_{l_1,\ldots,l_d} = 1$, for some $(l_1,\ldots,l_d)\in I_{d,n}$. For $1\le i\le n$, $1\le j\le d$ let

$$a_{ij} = q_{l_1,\ldots,l_{j-1},i,l_{j+1},\ldots,l_d}.$$

Let $A = (a_{ij})$. Note that $A_{l_1,\ldots,l_d} = \mathrm{Id}_d$. Clearly, $\mathrm{rank}A = d$ (since $\det A_{l_1,\ldots,l_d} = 1$). For $\underline{j}\in I_{d,n}$, let us denote $p_{\underline{j}}(A) = \det A_{j_1,\ldots,j_d}$. We shall now show that for $\underline{j}\in I_{d,n}$, we have $p_{\underline{j}}(A) = q_{\underline{j}}$, from which it will follow that the point $U$ in $G_{d,n}$ represented by $A$ satisfies $p_{\underline{j}}(U) = q_{\underline{j}}$, $\underline{j}\in I_{d,n}$, i.e., $q\in G_{d,n}$.

We prove this by decreasing induction on $\#\{\underline{l}\cap\underline{j}\}$, the starting point of induction being $\#\{\underline{l}\cap\underline{j}\} = d-1$ (i.e. $\underline{l}$ and $\underline{j}$ differ in just one place), and the result in this case follows from our definition of $A$. In the Plücker relation $q_{\underline{l}}q_{\underline{j}} + \sum \pm q_{\underline{l}'}q_{\underline{j}'} = 0$, $\underline{l}'$ differs from $\underline{l}$ in just one place (namely the $d$-th place), while $\#\{\underline{j}'\cap\underline{l}\} > \#\{\underline{j}\cap\underline{l}\}$ (since some entry in $\{\underline{j}\setminus\underline{l}\}$ has been replaced by $l_d$). Thus we obtain $q_{\underline{l}'} = p_{\underline{l}'}(A)$, $q_{\underline{j}'} = p_{\underline{j}'}(A)$. Also, from the first part, the $d$-minors of $A$ satisfy the Plücker relations. Hence we obtain $p_{\underline{l}}(A)q_{\underline{j}}(A) = q_{\underline{l}}q_{\underline{j}}$, which implies $p_{\underline{j}}(A) = q_{\underline{j}}$ (since $p_{\underline{l}}(A) = q_{\underline{l}} = 1$).

### 4.1.4 More general quadratic relations

The Plücker coordinates satisfy a more general type of quadratic relations described as follows: Let $\underline{i},\underline{j}\in I_{d,n}$. Suppose that $\underline{i},\underline{j}$ are not comparable, say, $\underline{i}\not\geq\underline{j}$. This implies that there exists a $t$, $t\le d$ such that $i_r\ge j_r$, $1\le r\le t-1$, and $i_t < j_t$. Let $[\underline{i},\underline{j}]$ denote the set of permutations $\sigma$ of the set $\{i_1,\cdots,i_t,j_t,\cdots,j_d\}$ such that $i_1^\sigma<\cdots<i_t^\sigma$; $j_t^\sigma<\cdots<j_d^\sigma$. Then the Plücker coordinates $p_{\underline{i}}$, $p_{\underline{j}}$ satisfy the

following relation:

$$Q_{\underline{i},\underline{j}} : \sum_{\sigma} \operatorname{sign}(\sigma)\, p_{\underline{i}^\sigma} p_{\underline{j}^\sigma} = 0, \tag{**}$$

where $\underline{i}^\sigma = (i_1^\sigma, \cdots, i_t^\sigma, i_{t+1}, \cdots, i_d) \uparrow$, $\underline{j}^\sigma = (j_1, \cdots, j_{t-1}, j_t^\sigma, \cdots, j_d^\sigma) \uparrow$, and $\sigma$ runs over $[\underline{i}, \underline{j}]$ (here, for a $d$-tuple $(l_1, \cdots, l_d)$, $l_i \in \mathbb{Z}$, $(l_1, \cdots, l_d) \uparrow$ denotes the $d$-tuple obtained from $(l_1, \cdots, l_d)$ by arranging the entries in the ascending order). The proof of this is very similar to the above proof.

For $\sigma \neq id$, $\underline{i}^\sigma$ is either $= 0$ (namely, if there is a repetition in $\{i_1^\sigma, \cdots, i_t^\sigma, i_{t+1}, \cdots, i_d\}$) or $\underline{i}^\sigma > \underline{i}$ (since in $\{i_1, \cdots, i_t, j_t, \cdots, j_d\}$ any $j >$ any $i$, and $i_r \geq j_r$, $1 \leq r \leq t-1$); similarly, for $\sigma \neq id$, $\underline{j}^\sigma$ is either $= 0$ (namely, if there is a repetition in $\{j_1, \cdots, j_{t-1}, j_t^\sigma, \cdots, j_d^\sigma\}$) or $\underline{j}^{\sigma} < \underline{j}$. Hence (**) may be reformulated as

$$p_{\underline{i}} p_{\underline{j}} = \sum \pm p_\alpha p_\beta, \tag{***}$$

where $\alpha, \beta$ run over a certain subset of $I_{d,n}$ such that $\alpha \geq \underline{i}$ and $\beta < \underline{j}$.

### 4.1.5 The cone $\widehat{G_{d,n}}$ over $G_{d,n}$

Let $A = K[G_{d,n}]$, the homogeneous coordinate ring of $G_{d,n}$ for the Plücker embedding. Let $\widehat{G_{d,n}} = \operatorname{Spec} A$, the cone over $G_{d,n}$. Let $X = V \oplus \cdots \oplus V$ ($d$ copies). With respect to the standard basis $\{e_1, \ldots, e_n\}$ of $V (= K^n)$, $X$ may be identified with $M_{n \times d}$, the space of $n \times d$ matrices with entries in $K$. With this identification, we have $K[X] = K[x_{ij},\ 1 \leq i \leq n, 1 \leq j \leq d]$. Consider the canonical morphism $\pi : X \to \wedge^d V$, $\pi(u_1, \ldots, u_d) = u_1 \wedge \ldots \wedge u_d$. The morphism $\pi$ corresponds to the algebra homomorphism $\pi^* : K[x_{\underline{i}},\ \underline{i} \in I_{d,n}] \to K[x_{ij}]$, $\pi^*(x_{\underline{i}}) = p_{\underline{i}}$. The image of $\pi$ is precisely $\widehat{G_{d,n}}$, the cone over $G_{d,n}$ (note that $\pi(X)$ is precisely the set of decomposable vectors in $\wedge^d V$). Thus $K[G_{d,n}]$ gets identified with the subring of $K[x_{ij}]$ generated by $\{p_{\underline{i}} \mid \underline{i} \in I_{d,n}\}$.

### 4.1.6 Identification of $G/P_d$ with $G_{d,n}$

Let $G = SL_n(K)$. Let $P_d$ be the maximal parabolic subgroup with $S \setminus \{\alpha_d\}$ as the associated set of simple roots. Then

$$P_d = \left\{ A \in G \;\middle|\; A = \begin{pmatrix} * & * \\ 0_{(n-d)\times d} & * \end{pmatrix} \right\},$$

$$W_{P_d} = S_d \times S_{n-d}.$$

For the natural action of $G$ on $\mathbb{P}(\wedge^d V)$, we have, the isotropy at $[e_1 \wedge \cdots e_d]$ is $P_d$ while the orbit through $[e_1 \wedge \cdots e_d]$ is $G_{d,n}$. Thus we obtain a surjective morphism $\pi : G \to G_{d,n}$, $g \mapsto g \cdot a$, where $a = [e_1 \wedge \cdots e_d]$. Further, the differential $d\pi_e : \operatorname{Lie} G \to T(G_{d,n})_a$ (= the tangent space to $G_{d,n}$ at $a$) is easily seen to be surjective. Hence we obtain an identification $f_d : G/P_d \cong G_{d,n}$ (cf. Borel [7],

Proposition 6.7). With this identification, we have $H^0(G/P_d, L_{\omega_d}) = (\wedge^d V)^*$ (here $\omega_d$ is the fundamental weight associated to $P_d$ and $L_{\omega_d}$ is the associated line bundle on $G/P_d$; note that $L_{\omega_d}$ is the ample generator of Pic $G/P_d$ ($\cong \mathbb{Z}$)).

## 4.2 Schubert varieties of $G_{d,n}$

For $1 \leq t \leq n$, let $V_t$ be the subspace of $V$ spanned by $\{e_1, \ldots, e_t\}$. For each $\underline{i} \in I_{d,n}$, the *Schubert variety associated to* $\underline{i}$ is defined to be

$$X_{\underline{i}} = \{U \in G_{d,n} \mid \dim(U \cap V_{i_t}) \geq t,\ 1 \leq t \leq d\}.$$

For the action of $G$ on $\mathbb{P}(\wedge^d V)$ the $T$-fixed points are precisely the points corresponding to the $T$-eigenvectors in $\wedge^d V$. Now

$$\wedge^d V = \bigoplus_{\underline{i} \in I_{d,n}} K e_{\underline{i}}, \quad \text{as } T\text{-modules},$$

where for $\underline{i} = (i_1, \cdots, i_d)$, $e_{\underline{i}} = e_{i_1} \wedge \cdots \wedge e_{i_d}$. Thus the $T$-fixed points in $\mathbb{P}(\wedge^d V)$ are precisely $\left[e_{\underline{i}}\right]$, $\underline{i} \in I_{d,n}$, and these points, obviously, belong to $G_{d,n}$. Further, the Schubert variety $X_{\underline{i}}$ associated to $\underline{i}$ is simply the Zariski closure of the $B$-orbit $B\left[e_{\underline{i}}\right]$ through the $T$-fixed point $\left[e_{\underline{i}}\right]$ (with the canonical reduced structure), $B$ being the subgroup of upper triangular matrices in $G$.

### 4.2.1 Bruhat decomposition

Let $\underline{i} \in I_{d,n}$. Let $C_{\underline{i}} = B[e_{\underline{i}}]$ be the Schubert cell associated to $\underline{i}$. The $C_{\underline{i}}$'s provide a cell decomposition of $G_{d,n}$. Let $X = V \oplus \cdots \oplus V$ be as above. Let $\pi : X \rightarrow \wedge^d V$, $(u_1, \ldots, u_d) \mapsto u_1 \wedge \ldots \wedge u_d$, and $p : \wedge^d V \setminus \{0\} \rightarrow \mathbb{P}(\wedge^d V)$, $u_1 \wedge \ldots \wedge u_d \mapsto [u_1 \wedge \ldots \wedge u_d]$. Let $v_{\underline{i}}$ denote the point $(e_{i_1}, \ldots, e_{i_d}) \in X$. As above, identifying $X$ with $M_{n \times d}$, $v_{\underline{i}}$ gets identified with the $n \times d$ matrix whose entries are all zero except the ones in the $i_j$-th row and $j$-th column, $1 \leq j \leq d$, which are equal to 1. We have $B \cdot v_{\underline{i}} = \{A \in M_{n \times d} \mid x_{ij} = 0, i > i_j, \text{ and } \prod_t x_{i_t t} \neq 0\}$. Denoting $\overline{B \cdot v_{\underline{i}}}$ by $D_{\underline{i}}$, we have $D_{\underline{i}} = \{A \in M_{n \times d} \mid x_{ij} = 0, i > i_j\}$. Further, $\pi(B \cdot v_{\underline{i}}) = p^{-1}(C_{\underline{i}})$, $\pi(D_{\underline{i}}) = \widehat{X_{\underline{i}}}$. From this, we obtain

**Theorem 4.2.1.1** *$X_{\underline{i}} \subseteq X_{\underline{j}}$ if and only if $\underline{i} \leq \underline{j}$.*

In particular, we have, $X_{\underline{j}} = \dot{\cup}_{\underline{i} \leq \underline{j}} B e_{\underline{i}}$ (Bruhat Decomposition).

Thus, under the set-theoretic bijection between the set of Schubert varieties and the set $I_{d,n}$, the partial order on the set of Schubert varieties given by inclusion induces the partial order $\geq$ on $I_{d,n}$. Note that $I_{d,n}$ can be identified with $\mathcal{S}_n / \mathcal{S}_d \times \mathcal{S}_{n-d}$ (here, $\mathcal{S}_n$ is the permutation group on $n$ symbols). In the sequel, we shall identify $I_{d,n}$ with the set of "minimal representatives" of $\mathcal{S}_n / \mathcal{S}_d \times \mathcal{S}_{n-d}$ in $\mathcal{S}_n$; to be very precise, a $d$-tuple $\underline{i} \in I_{d,n}$ will be identified with the element $(i_1, \ldots, i_d, j_1, \ldots, j_{n-d}) \in \mathcal{S}_n$, where $\{j_1, \ldots, j_{n-d}\}$ is the complement of $\{i_1, \ldots, i_d\}$ in $\{1, \ldots, n\}$ arranged in increasing order.

### 4.2.2 Dimension of $X_{\underline{i}}$

Let $E_{\underline{i}} = B_u \cdot v_{\underline{i}}$, where $B_u$ is the subgroup of unipotent upper triangular matrices in $G(= SL_n(K))$. Let $U_d$ be the subgroup of unipotent upper triangular matrices in $GL_d(K)$. Then $E_{\underline{i}}$ is $U_d$-stable and the canonical surjective map $\pi_0 : E_{\underline{i}} \to C_{\underline{i}}$, $Av_{\underline{i}} \mapsto A \cdot [e_{\underline{i}}]$, $A \in B_u$, is in fact the orbit map (note that $C_{\underline{i}} = B \cdot [e_{\underline{i}}] = B_u \cdot [e_{\underline{i}}]$, as $[e_{\underline{i}}]$ is a $T$-fixed point). For $M \in E_{\underline{i}}$, it is easily checked that the isotropy subgroup in $U_d$ at $M$ is {id}. Thus all the $U_d$-orbits in $E_{\underline{i}}$ are isomorphic to $U_d$. Thus $\pi_0 : E_{\underline{i}} \to C_{\underline{i}}$ is a principal fibration with fiber $U_d$. Hence we obtain

$$\dim C_{\underline{i}} = \dim E_{\underline{i}} - \frac{1}{2}d(d-1) = \sum_{t=1}^{d} i_t - d - \frac{1}{2}d(d-1) = \sum_{t=1}^{d} i_t - \sum_{t=1}^{d} t = \sum_{t=1}^{d} (i_t - t).$$

Thus we obtain

**Proposition 4.2.2.1** $\dim X_{\underline{i}} = \sum_{1 \le t \le d} (i_t - t)$.

**Remark 4.2.2.2.** We have $G_{d,n} = X_{\underline{i}}$, where $\underline{i} = (n+1-d, n+2-d, \dots, n)$. In particular, $\dim G_{d,n} = d(n-d)$.

**Remark 4.2.2.3.** Each $\underline{i} \in I_{d,n}$ determines a $\lambda = (i_d - d, \dots, i_2 - 2, i_1 - 1)$ and $\dim X_{\underline{i}}$ is simply $|\lambda| (= \sum \lambda_i,\ \lambda_i = i_{d+1-i} - (d+1-i))$.

### 4.2.3 Further results on Schubert varieties

**Lemma 4.2.3.1** *Let $\underline{i}, \underline{j} \in I_{d,n}$. Then*

$$p_{\underline{j}}|_{X_{\underline{i}}} \ne 0 \iff \underline{i} \ge \underline{j}.$$

*Proof.* From §4.1.2, we have $p_{\underline{j}}|_{X_{\underline{i}}} \ne 0 \iff e_{\underline{j}} \in X_{\underline{i}}$. In view of Bruhat decomposition, we have $e_{\underline{j}} \in X_{\underline{i}} \iff \underline{j} \le \underline{i}$. The result follows from this.

**Lemma 4.2.3.2** *Let $X_1$, $X_2$ be two Schubert varieties in $G_{d,n}$. Then $X_1 \cap X_2$ is irreducible, i.e. $X_1 \cap X_2$ is a Schubert variety (set-theoretically).*

*Proof.* Let $X_1 = X(\tau_1)$, $X_2 = X(\tau_2)$, where $\tau_1 = (a_1, \dots, a_d)$, $\tau_2 = (b_1, \dots, b_d)$. Let $X_1 \cap X_2 = \cup X(w_i)$, where $w_i < \tau_1$, $w_i < \tau_2$. Let $c_j = \min\{a_j, b_j\}$, $1 \le j \le d$, and $\tau = (c_1, \dots, c_d)$. Then, clearly $\tau \in I_{d,n}$, and $\tau < \tau_i$, $i = 1, 2$. We have $w_i \le \tau$, and hence $X_1 \cap X_2 = X(\tau)$.

**Lemma 4.2.3.3** *Let $X_1$, $X_2$ be two distinct Schubert divisors in a Schubert variety $X(\tau)$. Then $X_1 \cap X_2$ is irreducible of codimension 2 in $X$.*

*Proof.* Let $X_1 = X(\tau_1)$, $X_2 = X(\tau_2)$. Let $\tau = (c_1, \dots, c_d)$, $\tau_1 = (a_1, \dots, a_d)$, $\tau_2 = (b_1, \dots, b_d)$. Since $X(\tau_i)$ has codimension 1 in $X(\tau)$, the entries in $\tau_i$ are the same as the entries in $\tau$, but for one place, where the entry is one less than the corresponding entry in $\tau$, $i = 1, 2$. Let then $j$ such that $a_t = c_t, t \ne j, a_j = c_j - 1$, and $k$ such that $b_t = c_t, t \ne k, b_k = c_k - 1$. Let $w = (f_1, \dots, f_d)$, where $f_t = c_t$, $t \ne j, k$, $f_j = c_j - 1$, $f_k = c_k - 1$. Then $X_1 \cap X_2 = X(w)$, and $\dim X(w) = \dim X(\tau) - 2$ (clearly, in view of Proposition 4.2.2.1).

## 4.3 Standard monomial theory for Schubert varieties in $G_{d,n}$

### 4.3.1 Standard monomials

Let $R_0$ be the homogeneous coordinate ring of $G_{d,n}$ for the Plücker embedding, and for $\tau \in I_{d,n}$, let $R(\tau)$ be the homogeneous coordinate ring of the Schubert variety $X(\tau)$ (we shall denote $X_{P_d}(\tau)$ by just $X(\tau)$). In this section, we present a Standard monomial theory for $X(\tau)$ (cf. [41,42,88]), and as a consequence obtain a basis for $R(\tau)_m$, $m \in \mathbb{Z}^+$, prove the vanishing theorems, arithmetic normality and arithmetic Cohen-Macaulayness for $X(\tau)$. Standard monomial theory consists in constructing explicit bases for $R(\tau)$, $\tau \in I_{d,n}$.

**Definition 4.3.1.1** *A monomial $f = p_{\tau_1} \cdots p_{\tau_m}$ is said to be* standard *if*

$$\tau_1 \geq \cdots \geq \tau_m. \tag{*}$$

*Such a monomial is said to be* standard on $X(\tau)$, *if in addition to condition (*), we have $\tau \geq \tau_1$.*

**Remark 4.3.1.2.** Note that in the presence of condition (*), the standardness of $f$ on $X(\tau)$ is equivalent to the condition that $f\big|_{X(\tau)} \neq 0$. Thus given a standard monomial $f = p_{\tau_1} \cdots p_{\tau_m}$, $f\big|_{X(\tau)}$ is either 0 or remains standard on $X(\tau)$.

### 4.3.2 Linear independence of standard monomials

**Theorem 4.3.2.1** *The standard monomials on $X(\tau)$ of degree $m$ are linearly independent in $R(\tau)$.*

*Proof.* We proceed by induction on $\dim X(\tau)$.

If $\dim X(\tau) = 0$, $p_{\text{id}}$ is the only standard monomial on $X(\tau)$, and the result is obvious. Let $\dim X(\tau) > 0$. Let

$$\sum_{i=1}^{r} c_i F_i, \ c_i \in k, \tag{$*$}$$

be a linear relation of standard monomials $F_i$ of degree $m$. Let $F_i = p_{w_{i1}} \cdots p_{w_{im}}$. We divide the proof into the following two cases:

**Case 1:** $w_{i1} = \tau, \forall i$.

First observe that $X(\tau)$ being an integral scheme, $p_\tau$ is a non-zero divisor in $R(\tau)$; we cancel $p_\tau$ and obtain a linear relation among monomials of degree $m - 1$. The required result follows by induction on $m$ (note that when $m = 1$, the result follows by linear independence of $\{p_{\underline{i}}, \underline{i} \in I_{d,n}\}$).

**Case 2:** $w_{i1} < \tau$ for some $i$. For simplicity, assume that $w_{11} < \tau$, and $w_{11}$ is a minimal element of $\{w_{j1} \mid w_{j1} < \tau\}$. Let us denote $w_{11}$ by $\phi$. Then for $i \geq 2$, $F_i|_{X(\phi)}$ is either 0, or is standard on $X(\phi)$. Hence restricting ($*$) to $X(\phi)$, we obtain a nontrivial standard sum on $X(\phi)$ being zero, which is not possible (by induction hypothesis). Hence we conclude that $w_{i1} = \tau$ for all $i$, $1 \leq i \leq m$. Canceling $p_\tau$, we obtain a linear relation among standard monomials on $X(\tau)$ of degree $m - 1$. Using induction on $m$, the required result follows.

### 4.3.3 Generation by standard monomials

**Theorem 4.3.3.1** *Let $F = p_{w_1} \dots p_{w_m}$ be any monomial in the Plücker coordinates of degree $m$. Then $F$ is a linear combination of standard monomials of degree $m$.*

*Proof.* For $F = p_{w_1} \dots p_{w_m}$, define

$$N_F = l(w_1)N^{m-1} + l(w_2)N^{m-2} + \dots l(w_m),$$

where $N \gg 0$, say $N > d(n-d)(= \dim G_{d,n})$ and $l(w) = \dim X(w)$. If $F$ is standard, there is nothing to prove. Let $t$ be the first violation of standardness, i.e., $p_{w_1} \dots p_{w_{t-1}}$ is standard, but $p_{w_1} \dots p_{w_t}$ is not. Hence $w_{t-1} \not\geq w_t$, and using the quadratic relation (cf. the relation (***) in §4.1.4)

$$p_{w_{t-1}} p_{w_t} = \sum_{\alpha,\beta} \pm p_\alpha p_\beta, \tag{*}$$

$F$ can be expressed as $F = \sum F_i$, with $N_{F_i} > N_F$ (since $\alpha > w_{t-1}$ for all $\alpha$ on the right hand side of $(*)$). Now the required result is obtained by decreasing induction on $N_F$ (the starting point of induction, i.e. the case when $N_F$ is the largest, corresponds to standard monomial $F = p_\theta^m$, where $\theta = (n+1-d, n+2-d, \cdots, n)$, in which case $F$ is clearly standard; note that $X(\theta) = G_{d,n}$).

Combining Theorems 4.3.2.1 and 4.3.3.1, we obtain

**Theorem 4.3.3.2** *Standard monomials on $X(\tau)$ of degree $m$ give a basis for $R(\tau)_m$.*

As a consequence of Theorem 4.3.3.2, we have a qualitative description of a typical quadratic relation on a Schubert variety $X(w)$ as given by the following

**Proposition 4.3.3.3** *Let $w, \tau, \phi \in I_{d,n}$, $w > \tau, \phi$. Further let $\tau, \phi$ be non-comparable (so that $p_\tau p_\phi$ is a non-standard degree 2 monomial on $X(w)$). Let*

$$p_\tau p_\phi = \sum_{\alpha,\beta} c_{\alpha,\beta} p_\alpha p_\beta, \; c_{\alpha,\beta} \in k^* \tag{*}$$

*be the expression for $p_\tau p_\phi$ as a sum of standard monomials on $X(w)$. Then for every $\alpha, \beta$ on the right hand side we have, $\alpha >$ both $\tau$ and $\phi$, and $\beta <$ both $\tau$ and $\phi$.*

Such a relation as in (*) is called a *straightening relation.*

*Proof.* Among all the $\alpha$'s choose a minimal one, call it $\alpha_0$. Restricting (*) to $X(\alpha_0)$, we have, the restriction of the right hand side is a non-zero sum of standard monomials on $X(\alpha_0)$ (note that the restriction of $p_\alpha p_\beta$ to $X(\alpha_0)$ is non-zero if and only if $\alpha = \alpha_0$ - by the minimality of $\alpha_0$, and there is at least one term, namely $p_{\alpha_0} p_\beta$ whose restriction to $X(\alpha_0)$ is non-zero). Hence in view of linear independence of standard monomials on $X(\alpha_0)$, we obtain that the restriction of the left hand side to $X(\alpha_0)$ is non-zero. From this we conclude $\alpha_0 \geq$ both $\tau$ and $\phi$; in fact, $\alpha_0 >$ both $\tau$ and $\phi$ (for, if $\alpha_0$ equals one of them, say, $\alpha_0 = \tau$, then this would imply (by weight considerations) that $\beta = \phi$ which in turn would imply $\tau$ and $\phi$ are comparable (since

$\alpha$ and $\beta$ are), a contradiction. Any $\alpha$ is $\geq$ some minimal $\alpha_0$. Hence $\alpha$ > both $\tau$ and $\phi$.

Now replacing $\tau$ and $\phi$ by $w_0\tau$ and $w_0\phi$, we conclude from above $w_0\beta >$ both $w_0\tau$ and $w_0\phi$, i.e., $\beta <$ both $\tau$ and $\phi$.

### 4.3.4 Equations defining Schubert varieties in the Grassmannian

Let $\tau \in I_{d,n}$. Let $\pi_\tau$ be the map $R_0 \to R(\tau)$ (the restriction map). Let $\ker \pi_\tau = J_\tau$. Let $Z_\tau = \{$all standard monomials $F \mid F$ starts with $p_\phi$ for some $\phi \not\leq \tau\}$. We shall now give a set of generators for $J_\tau$ in terms of Plücker coordinates.

**Lemma 4.3.4.1** *Let $I_\tau = (p_w, \tau \not\geq w)$ (ideal in R). Then $Z_\tau$ is a basis for $I_\tau$.*

*Proof.* Let $F \in I_\tau$. Then writing $F$ as a linear combination of standard monomials

$$F = \sum a_i F_i + \sum b_j G_j,$$

where in the first sum each $F_i$ starts with $p_{\tau_{i1}}$, with $\tau \not\geq \tau_{i1}$, and in the second sum each $G_j$ starts with $\phi_{j1}$, with $\tau \geq \phi_{j1}$. This implies (cf. Lemma 4.2.3.1) that $\sum a_i F_i \in I_\tau$, and hence we obtain

$$\sum b_j G_j \in I_\tau.$$

This now implies that considered as an element of $R(\tau)$, $\sum b_j G_j$ is equal to 0 (note that $I_\tau \subset J_\tau$). Now the linear independence of standard monomials on $X(\tau)$ implies that $b_j = 0$ for all $j$. The required result now follows.

**Proposition 4.3.4.2** *Let $\tau \in I_{d,n}$. Then $R(\tau) = R_0/I_\tau$.*

*Proof.* We have, $R(\tau) = R_0/J_\tau$ (where $J_\tau$ is as above). We shall now show that the inclusion $I_\tau \subset J_\tau$ is in fact an equality. Let $F \in R_0$. Writing $F$ as a linear combination of standard monomials

$$F = \sum a_i F_i + \sum b_j G_j,$$

where in the first sum each $F_i$ starts with $p_{\tau_{i1}}$, with $\tau \not\geq \tau_{i1}$, and in the second sum each $G_j$ starts with $\phi_{j1}$, with $\tau \geq \phi_{j1}$, we have, $\sum a_i F_i \in I_\tau$, and hence we obtain
$F \in J_\tau$
$\Longleftrightarrow \sum b_j G_j \in J_\tau$ (since $\sum a_i F_i \in I_\tau$, and $I_\tau \subset J_\tau$)
$\Longleftrightarrow \pi_\tau(F)(= \sum b_j G_j)$ is zero
$\Longleftrightarrow \sum b_j G_j$ (= a sum of standard monomials on $X(\tau)$) is zero on $X(\tau)$
$\Longleftrightarrow b_j = 0$ for all $j$ (in view of the linear independence of standard monomials on $X(\tau)$)
$\Longleftrightarrow F = \sum a_i F_i$
$\Longleftrightarrow F \in I_\tau$.
Hence we obtain $J_\tau = I_\tau$.

**Equations defining Schubert varieties:**

Let $\tau \in I_{d,n}$. We have (cf. Lemma 4.2.3.1), the kernel of $(R_0)_1 \to (R(\tau))_1$ has a basis given by $\{p_w, \tau \not\geq w\}$. Combining this fact with Proposition 4.3.4.2 and Lemma 4.3.4.1 we obtain that the ideal $J_\tau$ (= the kernel of the restriction map $R_0 \to R(\tau)$) is generated by $\{p_w \mid \tau \not\geq w, w \in I_{d,n}\}$. Hence $J_\tau$ is generated by the kernel of $(R_0)_1 \to (R(\tau))_1$. Thus we obtain that $X(\tau)$ is scheme-theoretically (even at the cone level) the intersection of $G_{d,n}$ with all hyperplanes in $\mathbb{P}(\wedge^d k^n)$ containing $X(\tau)$. Further, as a closed subvariety of $G_{d,n}$, $X(\tau)$ is defined (scheme-theoretically) by the vanishing of $p_w, \tau \not\geq w,\ w \in I_{d,n}$.

## 4.4 Standard monomial theory for a union of Schubert varieties

In this section, we prove results similar to Theorems 4.3.2.1, 4.3.3.1, 4.3.3.2 for a union of Schubert varieties.

Let $X_i$ be Schubert varieties in $G_{d,n}$. Let $X = \cup X_i$.

**Definition 4.4.0.1** *A monomial $F$ in the Plücker coordinates is* standard *on the union $X = \cup X_i$ if it is standard on some $X_i$.*

### 4.4.1 Linear independence of standard monomials on $X = \cup X_i$

**Theorem 4.4.1.1** *Monomials standard on $X = \cup X_i$ are linearly independent.*

*Proof.* If possible, let

$$\sum_{i=1}^{r} a_i F_i = 0 \qquad (*)$$

be a nontrivial relation among standard monomials on $X$. Let $F_i = p_{\tau_{i1}} \cdots p_{\tau_{ir_i}}$. For simplicity, denote $\tau_{11}$ by just $\tau$. Then restricting $(*)$ to $X(\tau)$, we obtain a nontrivial relation among standard monomials on $X(\tau)$, which is a contradiction (note that for any $i$, $F_i|_{X(\tau)}$ is either 0 or remains standard on $X(\tau)$; further, $F_1|_{X(\tau)}$ is non-zero).

### 4.4.2 Standard Monomial Basis

**Theorem 4.4.2.1** *Let $X = \cup_{i=1}^{r} X(\tau_i)$, and $S$ the homogeneous coordinate ring of $X$. Then the standard monomials on $X$ give a basis for $S$.*

*Proof.* For $\tau \in I_{d,n}$, let $I_\tau$ be as in Lemma 4.3.4.1. Let us denote $I_t = I_{\tau_t}$, $1 \leq t \leq r$. We have $R(\tau_t) = R_0/I_t$ (cf. Proposition 4.3.4.2). Let $S = R_0/I$. Then $I = \cap I_t$ (note that being the intersection of reduced ideals, $I$ is also reduced, and hence the set theoretic equality $X = \cup X_i$ is also scheme theoretic). Let $F \in I$. Further, let us write $F$ as a linear combination of standard monomials

$$F = \sum c_k F_k.$$

Then in view of Lemma 4.3.4.1, we obtain that each $F_k$ starts with $p_{\tau_{k1}}$, where $\tau_{k1} \not\leq \tau_i$, $1 \leq i \leq r$. A typical element in $R_0/I$ may be written as $\pi(f)$, for some $f \in R_0$, where $\pi$ is the canonical projection $R_0 \to R_0/I$. Let us write $f$ as a sum of standard monomials

$$f = \sum a_j G_j + \sum b_l H_l,$$

where each $G_j$ (resp. $H_l$) starts with $p_\tau$ such that $\tau \not\leq \tau_i$, $1 \leq i \leq r$ (resp. $\tau \leq \tau_i$ for some $i$). We have $\pi(f) = \sum b_l H_l$ (since $\sum a_j G_j \in I$). Thus we obtain that $S$ (as a vector space) is generated by monomials standard on $X$. This together with the linear independence of standard monomials on $X$ implies the required result.

### 4.4.3 Consequences

**Theorem 4.4.3.1** *Let $X_1$, $X_2$ be two Schubert varieties in $G_{d,n}$. Then*

*(1) $X_1 \cup X_2$ is reduced.*

*(2) $X_1 \cap X_2$ is reduced.*

*Proof.* (1). For a closed subscheme $Y$ in $G_{d,n}$, let $I(Y)$ denote the ideal defining $Y$ in $G_{d,n}$. We have $I(X_1 \cup X_2) = I(X_1) \cap I(X_2)$, and hence $I(X_1 \cup X_2)$ is reduced (since $I(X_1)$, $I(X_2)$ are reduced).

(2) Let $X_1 = X(w_1)$, $X_2 = X(w_2)$. Let $A$ be the homogeneous coordinate ring of $X_1 \cap X_2$. Let $A = R_0/I$. Then $I = I_1 + I_2$. Let $X_1 \cap X_2 = \cup_{i=1}^r X(\tau_i)$ (set theoretically). Let $F \in I$. Then in the expression for $F$ as a linear combination of standard monomials

$$F = \sum a_j F_j,$$

each $F_j$ starts with $p_{\tau_{j1}}$, where $\tau_{j1} \not\leq$ either $w_1$ or $w_2$. If $B = R_0/\sqrt{I}$, then by Theorem 4.4.2.1, under $\pi : R_0 \to B$, $\ker\pi$ consists of all $f$ such that $f = \sum c_k f_k$, $f_k$ being standard monomials such that each $f_k$ starts with $p_{\phi_{k1}}$, where $\phi_{k1} \not\leq \tau_i$, for all $i$, $1 \leq i \leq r$. Hence $\phi_{k1} \not\leq$ either $w_1$ or $w_2$. Hence $\sqrt{I} = I$, and the required result follows from this.

**Theorem 4.4.3.2** *(Pieri's formula)*

$$X(\tau) \cap \{p_\tau = 0\} = \cup_{i=1}^r X(\tau_i), \quad \textit{scheme theoretically},$$

*where the varieties $X(\tau_i)$, $1 \leq i \leq r$ are all the Schubert divisors in $X(\tau)$.*

*Proof.* Let $X = \cup_{i=1}^r X(\tau_i)$, and let $B$ be the homogeneous coordinate ring of $X$. Let $B = R(\tau)/I$. Clearly, $(p_\tau) \subseteq I$, $(p_\tau)$ being the principal ideal in $R(\tau)$ generated by $p_\tau$. Let $f \in I$. Writing $f$ as

$$f = \sum b_i G_i + \sum c_j H_j,$$

where each $G_i$ is a standard monomial in $R(\tau)$ starting with $p_\tau$ and each $H_j$ is a standard monomial in $R(\tau)$ starting with $p_{\theta_{j1}}$, where $\theta_{j1} < \tau$, we have, $\sum b_i G_i \in I$. This now implies $\sum c_j H_j$ is zero on $\cup_{i=1}^r X(\tau_i)$. But now $\sum c_j H_j$ being a sum of standard monomials on $\cup_{i=1}^r X(\tau_i)$, we have by Theorem 4.4.1.1, $c_j = 0$, for all $j$. Thus we obtain $f = \sum b_i G_i$, and hence $f \in (p_\tau)$. This implies $I = (p_\tau)$. Hence we obtain $B = R(\tau)/(p_\tau)$, and the result follows from this.

## 4.5 Vanishing theorems

**Cohomology of sheaves** Let $X$ be an algebraic scheme, $\mathcal{A}b(X)$ the category of sheaves of abelian groups on $X$. Let $\mathcal{A}b$ denote the category of abelian groups. Let $\Gamma(X,\cdot)$ be the global section functor. It is left exact, i.e., if $0 \to \mathcal{F} \to \mathcal{G} \to \mathcal{H} \to 0$ is a short exact sequence of sheaves (i.e., $0 \to \mathcal{F}_x \to \mathcal{G}_x \to \mathcal{H}_x \to 0$ is exact for every $x \in X$), then $0 \to \Gamma(X,\mathcal{F}) \to \Gamma(X,\mathcal{G}) \to \Gamma(X,\mathcal{H})$ is exact. The cohomology functors $H^i(X,\cdot)$ are defined to be the right derived functors of $\Gamma(X,\cdot)$. For any sheaf $\mathcal{F}$, the groups $H^i(X,\mathcal{F})$ are defined as the *cohomology groups* of $\mathcal{F}$. Thus $H^i(X,\mathcal{F})$ are just the cohomology groups of the complex obtained by applying $\Gamma(X,\cdot)$ to an injective resolution of $\mathcal{F}$ (see [37] for details).

In the discussion below, we will be using the following well-known fact from cohomology theory (see [112] for example):

**Proposition 4.5.0.1** *Let $0 \to \mathcal{F} \to \mathcal{G} \to \mathcal{H} \to 0$ be a short exact sequence of sheaves on an algebraic scheme $X$. This gives rise to a cohomology long exact sequence*

$$\cdots \to H^{i-1}(X,\mathcal{H}) \to H^i(X,\mathcal{F}) \to H^i(X,\mathcal{G}) \to H^i(X,\mathcal{H}) \to \cdots$$

Let now $X$ be a union of Schubert varieties. Let $S(X,m)$ be the set of standard monomials on $X$ of degree $m$, and $s(X,m)$ the cardinality of $S(X,m)$. If $X = X(\tau)$ for some $\tau$, $S(X,m)$ and $s(X,m)$ will also be denoted by just $S(\tau,m)$, respectively $s(\tau,m)$.

**Lemma 4.5.0.2** (1) *Let $Y = Y_1 \cup Y_2$. Then*

$$s(Y,m) = s(Y_1,m) + s(Y_2,m) - s(Y_1 \cap Y_2, m).$$

(2) $s(\tau,m) = s(\tau,m-1) + s(H_\tau,m)$, *where $H_\tau$ is the union of all the Schubert divisors in $X_\tau$.*

(1) and (2) are easy consequences of the results of the previous section.

Let $L = L(\omega_d)$. We shall denote the restriction of $L$ to a subvariety $X$ of $G_{d,n}$ also by just $L$.

**Proposition 4.5.0.3** *Let $r$ be an integer $\leq d(n-d)$ $(= \dim G_{d,n})$. Suppose that all Schubert varieties $X$ in $G_{d,n}$ of dimension at most $r$ have the following two properties:*

(1) $H^i(X,L^m) = 0$, *for* $i \geq 1$, $m \geq 0$.

(2) *The set $S(X,m)$ is a basis for $H^0(X,L^m)$, $m \geq 0$.*

*Then all unions and intersections of Schubert varieties of dimension at most $r$ satisfy* (1) *and* (2).

*Proof.* We will prove the result by induction on $r$. Let $S_r$ denote the set of Schubert varieties $X$ in $G_{d,n}$ of dimension at most $r$. Let $Y = \cup_{j=1}^{t} X_j$, $X_j \in S_r$. Let $Y_1 = \cup_{j=1}^{t-1} X_j$, and $Y_2 = X_t$. Consider the exact sequence

$$0 \to \mathcal{O}_Y \to \mathcal{O}_{Y_1} \oplus \mathcal{O}_{Y_2} \to \mathcal{O}_{Y_1 \cap Y_2} \to 0,$$

where $\mathcal{O}_Y \to \mathcal{O}_{Y_1} \oplus \mathcal{O}_{Y_2}$ is the map $f \mapsto (f|_{Y_1}, f|_{Y_2})$ and $\mathcal{O}_{Y_1} \oplus \mathcal{O}_{Y_2} \to \mathcal{O}_{Y_1 \cap Y_2}$ is the map $(f, g) \mapsto (f - g)|_{Y_1 \cap Y_2}$. Tensoring with $L^m$, we obtain the long exact sequence

$$\to H^{i-1}(Y_1 \cap Y_2, L^m) \to H^i(Y, L^m) \to H^i(Y_1, L^m) \oplus H^i(Y_2, L^m) \to H^i(Y_1 \cap Y_2, L^m) \to$$

Now $Y_1 \cap Y_2$ is reduced (cf. Theorem 4.4.3.1) and $Y_1 \cap Y_2 \in S_{r-1}$. Hence, by the induction hypothesis (1) and (2) hold for $Y_1 \cap Y_2$. In particular, if $m \geq 0$, then (2) implies that the map $H^0(Y_1, L^m) \oplus H^0(Y_2, L^m) \to H^0(Y_1 \cap Y_2, L^m)$ is surjective. Hence we obtain that the sequence

$$0 \to H^0(Y, L^m) \to H^0(Y_1, L^m) \oplus H^0(Y_2, L^m) \to H^0(Y_1 \cap Y_2, L^m) \to 0$$

is exact. This implies $H^0(Y_1 \cap Y_2, L^m) \to H^1(Y, L^m)$ is the zero map; we have, $H^1(Y, L^m) \to H^1(Y_1, L^m) \oplus H^1(Y_2, L^m)$ is also the zero map (since by induction $H^1(Y_1, L^m) = 0 = H^1(Y_2, L^m)$). Hence we obtain $H^1(Y, L^m) = 0,\ m \geq 0$, and for $i \geq 2$, the assertion that $H^i(Y, L^m) = 0,\ m \geq 0$ follows from the long exact cohomology sequence above (and induction hypothesis). This proves the assertion (1) for $Y$.

To prove assertion (2) for $Y$, we observe

$$\begin{aligned} h^0(Y, L^m) &= h^0(Y_1, L^m) + h^0(Y_2, L^m) - h^0(Y_1 \cap Y_2, L^m) \\ &= s(Y_1, m) + s(Y_2, m) - s(Y_1 \cap Y_2, m). \end{aligned}$$

Hence Lemma 4.5.0.2 and induction hypothesis imply that

$$h^0(Y, L^m) = s(Y, L^m).$$

This together with linear independence of standard monomials on $Y$ proves assertion (2) for $Y$.

**Proposition 4.5.0.4** *Let $X$ be a Schubert variety in $G_{d,n}$. Let $Y = \cup_{j=1}^t X_j$ be a union of Schubert divisors in $X$. Let $L$ be as in Proposition 4.5.0.3. Suppose $H^i(X_j, L^m) = 0$, for $m < 0$, $0 \leq i \leq \dim Y - 1 (= \dim X_j - 1)$. Then $H^i(Y, L^m) = 0$, for $m < 0$, $0 \leq i \leq \dim Y - 1$.*

*Proof.* We will prove the result by induction on $t$ and $\dim Y$ (=$\dim X - 1$). As in the proof of Proposition 4.5.0.3, we set $Y_1 = \cup_{j=1}^{t-1} X_j$, $Y_2 = X_t$, and consider the long exact cohomology sequence

$$\to H^{i-1}(Y_1 \cap Y_2, L^m) \to H^i(Y, L^m) \to H^i(Y_1, L^m) \oplus H^i(Y_2, L^m) \to H^i(Y_1 \cap Y_2, L^m) \to$$

We have, $Y_1 \cap Y_2 = \cup_{j=1}^{t-1} X_j \cap X_t$. By Lemma 4.2.3.3, $X_j \cap X_t,\ 1 \leq j \leq t-1$ is irreducible of codimension 1 in $X_j$ (and $X_t$). Hence by induction on $\dim Y$, we have, $H^i(Y_1 \cap Y_2, L^m) = 0$, for $m < 0, 0 \leq i \leq \dim Y_1 - 2 (= \dim Y - 2)$; further, by induction on $t$, $H^i(Y_1, L^m) = 0,\ m < 0,\ 0 \leq i < \dim Y_1$, and by hypothesis, $H^i(Y_2, L^m) = 0,\ m < 0,\ 0 \leq i < \dim Y_1$. The required result now follows.

**Theorem 4.5.0.5** (Vanishing Theorems) *Let $L = L_{\omega_d}$. Note that $L$ is the restriction to $G_{d,n}$ of the tautological bundle on $\mathbb{P}(\Lambda^d V)$. Let $X$ be a Schubert variety in $G_{d,n}$. Denoting the restriction of $L$ to $X$ also by just $L$, we have*

*(a)* $H^i(X, L^m) = 0$ *for* $i \geq 1$, $m \geq 0$.
*(b)* $H^i(X, L^m) = 0$ *for* $0 \leq i < \dim X$, $m < 0$.
*(c) The set* $S(X, m)$ *is a basis for* $H^0(X, L^m)$.

*Proof.* We prove the result by induction on $m$ and $\dim X$.

If $\dim X = 0$, $X$ is just a point, and the result is obvious. Assume now that $\dim X \geq 1$. Let $X = X(\tau)$. Let $X_1, \ldots, X_s$ be all the Schubert divisors in $X$, and $Y = \bigcup_{i=1}^{s} X_i$. Then by Pieri's formula (cf. Theorem 4.4.3.2), we have,

$$Y = X(\tau) \cap \{p_\tau = 0\} \quad \text{(scheme-theoretically)}.$$

Hence the sequence

$$0 \to \mathcal{O}_X(-1) \to \mathcal{O}_X \to \mathcal{O}_Y \to 0$$

is exact (note that $\mathcal{O}_X(-1) = L^{-1}$, and that the map $L^{-1} \to \mathcal{O}_X$ is given by $p_\tau$ which is a global section of $L$). Tensoring it with $L^m$, and taking cohomology, we obtain the long exact cohomology sequence

$$\cdots \to H^{i-1}(Y, L^m) \to H^i(X, L^{m-1}) \to H^i(X, L^m) \to H^i(Y, L^m) \to \cdots.$$

Let $m \geq 0$, $i \geq 2$. Then the induction hypothesis on $\dim X$ implies (in view of Proposition 4.5.0.3) that $H^i(Y, L^m) = 0$, $i \geq 1$. Hence we obtain that the sequence $0 \to H^i(X, L^{m-1}) \to H^i(X, L^m)$, $i \geq 2$, is exact. If $i = 1$, again the induction hypothesis implies the surjectivity of $H^0(X, L^m) \to H^0(Y, L^m)$. This in turn implies that the map $H^0(Y, L^m) \to H^1(X, L^{m-1})$ is the zero map, and hence we obtain that the sequence $0 \to H^1(X, L^{m-1}) \to H^1(X, L^m)$ is exact. Thus we obtain that $0 \to H^i(X, L^{m-1}) \to H^i(X, L^m)$, $m \geq 0$, $i \geq 1$ is exact. But $H^i(X, L^m) = 0$, $m \gg 0$, $i \geq 1$ (cf. [109]). Hence we obtain

$$H^i(X, L^m) = 0 \text{ for } i \geq 1, \ m \geq 0, \tag{1}$$

and

$$h^0(X, L^m) = h^0(X, L^{m-1}) + h^0(Y, L^m) \tag{2}$$

where $h^0(X, L^m) = \dim H^0(X, L^m)$. In particular, assertion (a) follows from (1). For $m \geq 1$, the induction hypothesis on $m$ implies that $h^0(X, L^{m-1}) = s(X, m-1)$ (note that when $m = 1$, $H^0(X, L^{m-1}) = H^0(X, \mathcal{O}_X) = k$, and hence $h^0(X, L^{m-1}) = s(X, m-1)$). On the other hand, the induction hypothesis on $\dim X$ implies (in view of Proposition 4.5.0.3) that $h^0(Y, L^m) = s(Y, m)$. Hence we obtain

$$h^0(X, L^m) = s(X, m-1) + s(Y, m). \tag{3}$$

Now (3) together with part 2 of Lemma 4.5.0.2 implies that $h^0(X, L^m) = s(X, m)$. Hence (c) follows in view of the linear independence of standard monomials on $X(\tau)$ (cf. Theorem 4.3.2.1).

To prove (b), consider the long exact cohomology sequence

$$\cdots \to H^{i-1}(Y, L^m) \to H^i(X, L^{m-1}) \to H^i(X, L^m) \to \cdots$$

We have, $H^i(Y, L^m) = 0,\ m \in \mathbb{Z},\ 1 \le i \le \dim Y - 1$ (by induction hypothesis and Propositions 4.5.0.3 and 4.5.0.4). Hence we obtain the exact sequence

$$0 \to H^i(X, L^{m-1}) \to H^i(X, L^m),\ 2 \le i < \dim X.$$

But $H^i(X, L^m) = 0, m \gg 0$ (cf. [109]). Hence we obtain, $H^i(X, L^m) = 0, 2 \le i < \dim X, m \in \mathbb{Z}$ (in particular for $m < 0$).

It remains to prove (b) for $i = 0, 1$. Let then $m < 0$. Induction hypothesis together with Proposition 4.5.0.4 implies that $H^0(Y, L^m) = 0$. Hence we obtain the exact sequence

$$0 \to H^1(X, L^{m-1}) \to H^1(X, L^m),\ m < 0.$$

Thus we obtain inclusions

$$H^1(X, L^{m-1}) \subset H^1(X, L^m) \subset \cdots \subset H^1(X, L^{-1}).$$

But now the isomorphism $H^0(X, \mathcal{O}_X) \cong H^0(Y, \mathcal{O}_Y)$ together with the fact that $H^1(X, \mathcal{O}_X) = 0$ implies that $H^1(X, L^{-1}) = 0$. This now implies (in view of the above inclusions) that $H^1(X, L^m) = 0, m < 0$. Now $H^0(Y, L^m) = 0$ implies

$$H^0(X, L^{m-1}) \cong H^0(X, L^m),\ m < 0. \qquad (4)$$

Again the isomorphism $H^0(X, \mathcal{O}_X) \cong H^0(Y, \mathcal{O}_Y)$ implies that $H^0(X, L^{-1}) = 0$, and it follows that $H^0(X, L^m) = 0,\ m < 0$ (note that in view of (4), we have, $H^0(X, L^m) \cong H^0(X, L^{-1}) = 0$).

**Corollary 4.5.0.6** *We have*

1. $R = \bigoplus_{m \in \mathbb{Z}^+} H^0(X, L^m)$, *where* $X = G_{d,n}$.
2. $R(\tau) = \bigoplus_{m \in \mathbb{Z}^+} H^0(X(\tau), L^m)$, *for all* $\tau \in I_{d,n}$.

*Proof.* The assertions follow immediately from Theorems 4.3.3.2 and 4.5.0.5.

## 4.6 Arithmetic Cohen-Macaulayness, normality and factoriality

In this section, we present a proof of the arithmetic Cohen-Macaulayness and arithmetic normality of Schubert varieties in the Grassmannian. We also give a characterization of factorial Schubert varieties. We follow the treatment as in [88,89].

Let $X \hookrightarrow \mathbb{P}^N$ be a closed sub scheme which we shall suppose to be equidimensional (i.e., all irreducible components of $X$ are of the same dimension) of dimension $d$. Let $I$ be the homogeneous ideal defining $X$ and $A = K[x_0, \ldots, x_N]/I$ the homogeneous coordinate ring of $X$. Let $L$ denote the tautological line bundle on $\mathbb{P}^N$, as well as its restriction to $X$. Let (0) denote the vertex of the cone $\widehat{X}$, and $A_{(0)}$ denote the local ring $\mathcal{O}_{\widehat{X},(0)}$ at the vertex of $\widehat{X}$.

**Lemma 4.6.0.1** *Let $X$ be as above. Then $\widehat{X}$ is Cohen-Macaulay if and only if $\widehat{X}$ is Cohen-Macaulay at its vertex.*

*Proof.* Proof of the implication $\Rightarrow$ is clear. Let then the vertex (0) be a Cohen-Macaulay point of $\widehat{X}$. Then $\widehat{X}$ is Cohen-Macaulay in a neighbourhood $U$ of (0). It is clear that each fiber of the natural projection $\pi : \widehat{X} \setminus \{(0)\} \to X$ meets $U$. But these fibers are simply the $\mathbb{G}_m$-orbits in $\widehat{X} \setminus \{(0)\}$, and hence $\widehat{X}$ is Cohen-Macaulay at each point of $\pi^{-1}(x)$ for all $x \in X$.

**Theorem 4.6.0.2** *The Schubert variety $X(\tau)$ in $G_{d,n}$ is arithmetically Cohen-Macaulay, i.e., the cone $\widehat{X(\tau)}$ is Cohen-Macaulay.*

*Proof.* Let us denote $X = X(\tau), d = dimX$. Then in view of the above Lemma, it suffices to show that $\widehat{X}$ is Cohen-Macaulay at its vertex, i.e., to show that $A_{(0)}$ is Cohen-Macaulay. In view of Serre-Grothendieck Criterion, we are reduced to show that $H^i_{(0)}(\widehat{X}, \mathcal{O}_{\widehat{X}}) = 0$, for all $i \leq d$ (here, $H^i_{(0)}(\widehat{X}, \mathcal{O}_{\widehat{X}})$ denotes the local cohomology; also, recall (see [79] for example) Serre-Grothendieck Criterion that $A_{(0)}$ is Cohen-Macaulay if and only if $H^i_{(0)}(\widehat{X}, \mathcal{O}_{\widehat{X}}) = 0$, for all $i \leq d$). To prove this, let us denote $U = \widehat{X} \setminus \{(0)\}$. Then we have a long exact sequence

$$\cdots \to H^i_{(0)}(\widehat{X}, \mathcal{O}_{\widehat{X}}) \to H^i(\widehat{X}, \mathcal{O}_{\widehat{X}}) \to H^i(U, \mathcal{O}_U) \to \cdots$$

We have that $H^i(\widehat{X}, \mathcal{O}_{\widehat{X}}) = 0, i \geq 1$ (since $\widehat{X}$ is affine), and that the restriction map $H^0(\widehat{X}, \mathcal{O}_{\widehat{X}}) \to H^0(U, \mathcal{O}_U)$ is injective (since $U$ is a dense open subset of $X$). Hence we obtain the following:

1. $H^0_{(0)}(\widehat{X}, \mathcal{O}_{\widehat{X}}) = 0$
2. $0 \to H^0(\widehat{X}, \mathcal{O}_{\widehat{X}}) \to H^0(U, \mathcal{O}_U) \to H^1_{(0)}(\widehat{X}, \mathcal{O}_{\widehat{X}}) \to 0$ is exact
3. $H^{i-1}(U, \mathcal{O}_U) \cong H^i_{(0)}(\widehat{X}, \mathcal{O}_{\widehat{X}}),\ i \geq 2$

On the other hand, the natural map $U \to X$ realizes $U$ as the scheme over $X$ defined by the sheaf of $\mathcal{O}_X$-algebras $\oplus_{r\in\mathbb{Z}} L^r$. Hence we have isomorphisms

$$(*) \qquad H^i(U, \mathcal{O}_U) = \oplus_{r\in\mathbb{Z}} H^i(X, L^r),\ i \geq 0$$

We have

$$(4) \qquad H^i_{(0)}(\widehat{X}, \mathcal{O}_{\widehat{X}}) = 0, i = 0, 1$$

(for $i = 0$, this is just (1) above; for $i = 1$, this follows from (2) and Corollary 4.5.0.6 (note that by Corollary 4.5.0.6, we have $H^0(\widehat{X}, \mathcal{O}_{\widehat{X}})(= R(\tau) = A) = \oplus_{r\in\mathbb{Z}^+} H^0(X, L^r)(= \oplus_{r\in\mathbb{Z}} H^0(X, L^r)$, since $H^0(X, L^r) = 0$ for $r < 0$ (cf. Theorem 4.5.0.5 (b))).

Let then $2 \leq i \leq d$. In view of (*) and Theorem 4.5.0.5 (a),(b), we have

$$H^{i-1}(U, \mathcal{O}_U) = 0,\ 2 \leq i \leq d$$

This together with (3) implies

$$(5) \qquad H^i_{(0)}(\widehat{X}, \mathcal{O}_{\widehat{X}}) = 0,\ 2 \leq i \leq d$$

Thus we obtain (from (4) and (5)) that $H^i_{(0)}(\widehat{X}, \mathcal{O}_{\widehat{X}}) = 0,\ i \le d$. This completes the proof of the Theorem.

As a consequence, we have the following

**Theorem 4.6.0.3** *The Schubert variety $X(\tau)$ in $G_{d,n}$ is arithmetically normal, i.e. the cone $\widehat{X(\tau)}$ is normal.*

*Proof.* First observe that the above Theorem together with Chevalley's result that Schubert varieties are regular in codimension 1 (cf. Theorem A.12.1.10 of Appendix) implies that $X(\tau)$ is normal (in view of Serre Criterion for normality—a ring $A$ is normal if and only if $A$ has $S_2$ and $R_1$). Let $Y = \mathbb{P}(\bigwedge^d V)$. The canonical homomorphism

$$\phi_m : H^0(Y, L^m) \to H^0(X(\tau), L^m)$$

is surjective for $m \gg 0$. Let $S_m = H^0(X(\tau), L^m)$. We have (see [37] for example) that image of $(\phi_m)$ equals $R(\tau)_m$, and $S = \bigoplus_{m \in \mathbb{Z}^+} S_m$ is the integral closure of $R(\tau)$ in its quotient field. Now by Corollary 4.5.0.6, we have $R(\tau)_m = S_m$ for all $m$, and hence we obtain that $R(\tau)$ is normal (note that if a projective variety $X \hookrightarrow \mathbb{P}^n$ is normal, then it is arithmetically normal if and only if the restriction map $H^0(\mathbb{P}^n, L^m) \to H^0(X, L^m)$ is surjective for all $m$, $L$ being the tautological line bundle on $\mathbb{P}^n$). Therefore $X(\tau)$ is arithmetically normal.

### 4.6.1 Factorial Schubert varieties

Recall that an algebraic variety $X$ is said to be factorial at a point $P \in X$, if the stalk $\mathcal{O}_{X,P}$ is factorial, i.e., $\mathcal{O}_{X,P}$ is a unique factorization domain; the variety $X$ is said to be factorial if it is factorial at all $P \in X$. If in addition $X$ is projective, then $X$ is said to be projectively (or also arithmetically) factorial if the cone $\hat{X}$ is factorial at its vertex.

**Theorem 4.6.1.1** *([89]) The following are equivalent for a Schubert variety $X$.*

1. *$X$ is projectively factorial (for the Plücker embedding).*
2. *$X$ is factorial.*
3. *$X$ is factorial at the point cell $e_{id}$.*
4. *$X$ has a unique Schubert divisor.*
5. *$X$ is non–singular.*
6. *$X$ is itself a suitable Grassmannian.*

# 5

# Determinantal varieties

In this chapter, we give a self-contained exposition of the classical determinantal varieties from the view-point of their relationship to Schubert varieties. We follow the same treatment as in [67].

## 5.1 Recollection of facts

Let us fix the integers $1 \leq d < n$ and let $V = K^n$. The *Grassmannian* $G_{d,n}$, and the set $I_{d,n}$ be as in Chapter 4. For $\underline{i} \in I_{d,n}$, let $X_{\underline{i}}$ be the associated Schubert variety in $G_{d,n}$.

Recall the following facts from Chapter 4:

**Fact 1.** $X_{\underline{i}} \subseteq X_{\underline{j}}$ if and only if $\underline{i} \leq \underline{j}$.

**Fact 2.** Bruhat Decomposition: $X_{\underline{j}} = \dot{\bigcup}_{\underline{i} \leq \underline{j}} Be_{\underline{i}}$.

**Fact 3.** $\dim X_{\underline{i}} = \sum_{1 \leq t \leq d} (i_t - t)$.

**Fact 4.** Let $\underline{i}, \underline{j} \in I_{d,n}$. Then $p_{\underline{j}}|_{X_{\underline{i}}} \neq 0 \iff \underline{i} \geq \underline{j}$.

**Fact 5.** Under the set-theoretic bijection between the set of Schubert varieties and the set $I_{d,n}$, the partial order on the set of Schubert varieties given by inclusion induces the partial order $\geq$ on $I_{d,n}$.

Let $R$ be the homogeneous co-ordinate ring of $G_{d,n}$ for the Plücker embedding, and for $\tau \in I_{d,n}$, let $R(\tau)$ be the homogeneous co-ordinate ring of the Schubert variety $X(\tau)$. We have from Chapter 4

**Theorem 5.1.0.1** *Standard monomials on $X(\tau)$ of degree $m$ give a basis for $R(\tau)_m$.*

As seen in Chapter 4, Theorem 5.1.0.1 implies the following qualitative description of a typical quadratic relation on a Schubert variety $X(w)$:

**Proposition 5.1.0.2** *Let $w, \tau, \phi \in I_{d,n}$, $w > \tau, \phi$. Further let $\tau, \phi$ be non-comparable (so that $p_\tau p_\phi$ is a non-standard degree 2 monomial on $X(w)$). Let*

$$p_\tau p_\phi = \sum_{\alpha,\beta} c_{\alpha,\beta} p_\alpha p_\beta, \quad c_{\alpha,\beta} \in K^* \qquad (*)$$

*be the expression for $p_\tau p_\phi$ as a sum of standard monomials on $X(w)$. Then for every $\alpha, \beta$ on the R.H.S. we have, $\alpha \geq$ both $\tau$ and $\phi$, $\beta \leq$ both $\tau$ and $\phi$.*

**Remark 5.1.0.3.** Such a relation as in (*) is called a *straightening relation.*

### 5.1.1 Equations defining Schubert varieties in the Grassmannian

Let $\tau \in I_{d,n}$. Let $\pi_\tau$ be the map $R \to R(\tau)$ (the restriction map, $R$ being the homogeneous co-ordinate ring of $G_{d,n}$ for the Plücker embedding). Let $\ker \pi_\tau = I_\tau$. The following Proposition (cf. Chapter 4) gives a set of ideal generators for $I_\tau$ in terms of Plücker co-ordinates.

**Proposition 5.1.1.1** *$I_\tau$ is generated by $\{p_w, \tau \not\geq w\}$ ( as an ideal in $R$). Further, $Z_\tau$ := {all standard monomials $F$ | $F$ starts with $p_\phi$ for some $\phi \not\leq \tau$} is a basis for $I_\tau$.*

### 5.1.2 Evaluation of Plücker coordinates on the opposite big cell in $G_{d,n}$

Let $P_d$ be the parabolic subgroup of $G$ consisting of all matrices of the form

$$\begin{pmatrix} * & * \\ 0 & * \end{pmatrix}$$

where the 0-matrix is of size $n-d \times d$. Then we have an identification $\varphi_d : G/P_d \cong G_{d,n}$. Denote by $O^-$ the subgroup of $G$ generated by $\{U_\alpha \mid \alpha \in R^- \setminus R_P^-\}$. Then $O^-$ consists of the elements of $G$ of the form

$$\begin{pmatrix} I_d & 0_{d\times(n-d)} \\ A_{(n-d)\times d} & I_{n-d} \end{pmatrix}$$

where $I_d$ (resp. $I_{n-d}$)is the $d \times d$ (resp. $n-d \times n-d$) identity matrix. Further, the restriction of the canonical morphism $\theta_d : G \to G/P_d$ to $O^-$ is an open immersion, and $\theta_d(O^-) = B^- e_{id}$, where $e_{id}$ is the coset $P_d$ of $G/P_d$; also, $\varphi_d(B^- e_{id})$ is the *opposite cell* in $G_{d,n}$. Thus the opposite cell in $G_{d,n}$ gets identified with $O^-$, and in the sequel we shall denote $B^- e_{id}$ by just $O^-$. Note that $O^- \cong M_{n-d,\,d} (\cong \mathbb{A}^{d(n-d)})$.

Consider the morphism $\xi_d : G \to \mathbb{P}(\wedge^d V)$, where $\xi_d = f_d \circ \theta_d$, $f_d$ being the Plücker embedding (cf. chapter 4), and $\theta_d$ (as above) the natural projection $G \to G/P_d$. Then $p_{\underline{j}}(\xi_d(g))$ is simply the minor of $g$ consisting of the first $d$ columns and the rows with indices $j_1, \ldots, j_d$. Let $z \in O^-$, and let $z$ correspond to $A \in M_{n-d,\,d}$. The Plücker coordinate $p_{\underline{j}}$ evaluated at $z$ is simply a certain minor of $A$, which may be explicitly described as follows. Let $\underline{j} = (j_1, \ldots, j_d)$, and let $j_r$ be the largest entry $\leq d$. Let $\{k_1, \ldots, k_{d-r}\}$ be the complement of $\{j_1, \ldots, j_r\}$ in $\{1, \ldots, d\}$. Then this minor of $A$ is given by column indices $k_1, \ldots k_{d-r}$, and row indices $j_{r+1}, \ldots, j_d$ (here the rows of $A$ are indexed as $d+1, \ldots, n$). Conversely, given a minor of $A$, say, with column indices $b_1, \ldots, b_s$, and row indices $i_{d-s+1}, \ldots, i_d$, it is the evaluation of the Plücker coordinate $p_{\underline{i}}$ at $A$, where $\underline{i} = (i_1, \ldots, i_d)$ may be described as follows:

$\{i_1, \ldots, i_{d-s}\}$ is the complement of $\{b_1, \ldots, b_s\}$ in $\{1, \ldots, d\}$, and $i_{d-s+1}, \ldots, i_d$ are simply the row indices (again, the rows of $A$ are indexed as $d+1, \ldots, n$).

**Convention.** If $\underline{j} = (1, \ldots, d)$, then $p_{\underline{j}}$ evaluated at $A$ is 1. We shall consider the element 1 (in $K[M_{n-d,\,d}]$) as the minor of $A$ with row indices (and column indices) given by the empty set.

### 5.1.3 Ideal of the opposite cell in $X(w)$

For a Schubert variety $X(w)$ in $G_{d,n}$, let us denote $B^- e_{\mathrm{id}} \cap X(w)$ by $Y(w)$. We consider $Y(w)$ as a closed subvariety of $O^-$. In view of §5.1.1.1, we obtain that the ideal defining $Y(w)$ in $O^-$ is generated by

$$\{p_{\underline{i}} \mid \underline{i} \in I_{d,n},\ w \not\geq \underline{i}\}.$$

## 5.2 Determinantal varieties

Let $Z = M_{m,n}(K)$, the space of all $m \times n$ matrices with entries in $K$. We shall identify $Z$ with $\mathbb{A}^{mn}$. We have $K[Z] = K[x_{i,j},\ 1 \leq i \leq m,\ 1 \leq j \leq n]$.

### 5.2.1 The variety $D_t$

Let $X = (x_{ij})$, $1 \leq i \leq m$, $1 \leq j \leq n$ be a $m \times n$ matrix of indeterminates. Let $A \subset \{1, \cdots, m\}$, $B \subset \{1, \cdots, n\}$, $\#A = \#B = s$, where $s \leq \min\{m, n\}$. We shall denote by $[A|B]$ the $s$-minor of $X$ with row indices given by $A$, and column indices given by $B$. For $t$, $1 \leq t \leq \min\{m, n\}$, let $I_t(X)$ be the ideal in $K[x_{i,j}]$ generated by $\{[A|B], A \subset \{1, \cdots, m\}, B \subset \{1, \cdots, n\}, \#A = \#B = t\}$. Let $D_t(K)$ (or just $D_t$) be the *determinantal variety*, a closed subvariety of $Z$, with $I_t(X)$ as the defining ideal.

### 5.2.2 Identification of $D_t$ with $Y_\phi$

Let $G = SL_n(K)$. Let $r, d$ be such that $r + d = n$. Let $X$ be a $r \times d$ matrix of indeterminates. As above, let us identify the opposite cell $O^-$ in $G/P_d (\cong G_{d,n})$ as

$$O^- = \left\{ \begin{pmatrix} I_d \\ X \end{pmatrix} \right\}.$$

We have a bijection between {Plücker co-ordinates $p_{\underline{i}}$, $\underline{i} \neq \{1, 2, \cdots, d\}$} and {minors of $X$} (note that if $\underline{i} = \{1, 2, \cdots, d\}$, then $p_{\underline{i}} =$ the constant function 1).

For example, take $r = 3 = d$. We have,

$$O_3^- = \left\{ \begin{pmatrix} I_3 \\ X_{3\times 3} \end{pmatrix} \right\}.$$

We have, $p_{(1,2,4)} = [\{1\}|\{3\}]$, $p_{(2,4,6)} = [\{1,3\}|\{1,3\}]$.
In the sequel, given a Plücker coordinate $p_{\underline{i}}$ (on $O_d^-$), we shall denote the associated minor (of $X_{r,d}$) by $\Delta_{\underline{i}}$.

**The set $Z_t$:** Let $Z_t = \{\underline{i} \in I_{d,n} \mid \Delta_{\underline{i}}$ is a $t$ minor of $X\}$. The partial order on the Plücker co-ordinates induces a partial order on $Z_t$.

**Lemma 5.2.2.1** *Let $\tau$ be the $d$-tuple, $\tau = (1, 2, \cdots, d-t, d+1, d+2, \cdots, d+t)$. Then $\tau \in Z_t$. Further, $\tau$ is the unique smallest element in $Z_t$.*

*Proof.* Clearly $\tau \in Z_t$. Let $\Delta$ be a $t$-minor of $X$, and $p_{\underline{i}}$ the associated Plücker coordinate. Let $\underline{i} = (i_1, \cdots, i_d)$. We have, for $1 \leq k \leq d-t$, $i_k \leq d$, and for $d-t+1 \leq k \leq d$, $i_k > d$. Clearly $\tau$ is the smallest such $d$-tuple.

**Remark 5.2.2.2.** With $\tau$ as in Lemma 5.2.2.1, note that the associated minor $\Delta_\tau$ of $X$ has row and column indices given by $\{d+1, d+2, \cdots, d+t\}$, $\{d+1-t, d+2-t, \cdots, d\}$ respectively, i.e., the right most top corner $t$ minor of $X$, the rows of $X$ being indexed as $d+1, \ldots, n$.

**Remark 5.2.2.3.** With $\tau$ as in Lemma 5.2.2.1, note that $\tau$ is the smallest $d$-tuple $(j_1, \cdots, j_d)$ such that $j_{d-t+1} \geq d+1$.

**Lemma 5.2.2.4** *Let $\tau$ be as in Lemma 5.2.2.1. Let*

$$N_t = \{\underline{i} \in I_{d,n} \mid p_\tau|_{X_{\underline{i}}} = 0\}.$$

*Let $\phi$ be the $d$-tuple, $\phi = (t, t+1, \cdots, d, n+2-t, n+3-t, \cdots, n)$ (note that $\phi$ consists of the two blocks $[t, d]$, $[n+2-t, n]$ of consecutive integers - here, for $i < j$, $[i, j]$ denotes the set $\{i, i+1, \cdots, j\}$). Then $\phi$ is the unique largest element in $N_t$.*

*Proof.* Let $\underline{i} \in N_t$, say, $\underline{i} = (i_1, \cdots, i_d)$. We have, $p_\tau|_{X_{\underline{i}}} = 0$ if and only if $\underline{i} \not\geq \tau$, i.e., if and only if $i_{d-t+1} \not\geq d+1$ (cf. Remark 5.2.2.3), i.e., $i_{d-t+1} \leq d$. Now it is easily checked that $\phi$ is the largest $d$-tuple $(j_1, \cdots, j_d)$ such that $j_{d-t+1} \leq d$.

**Remark 5.2.2.5.** Let $\phi$ be as in Lemma 5.2.2.4. As observed in the proof of Lemma 5.2.2.4, we have $\phi$ is the largest $d$-tuple $(j_1, \cdots, j_d)$ such that $j_{d-t+1} \leq d$.

**Corollary 5.2.2.6** *Let $\underline{i} \in Z_t$. Then $p_{\underline{i}}|_{X_\phi} = 0$, $\phi$ being as in Lemma 5.2.2.4.*

*Proof.* We have (cf. Lemma 5.2.2.4), $p_\tau|_{X_\phi} = 0$, $\tau$ being as in Lemma 5.2.2.1 and hence $\phi \not\geq \tau$ which in turn implies (in view of Lemma 5.2.2.1), $\phi \not\geq \underline{i}$. Hence we obtain $p_{\underline{i}}|_{X_\phi} = 0$.

**Theorem 5.2.2.7** *Let $\phi$ be as in Lemma 5.2.2.4, and let $Y_\phi = \mathcal{O}^- \cap X_\phi$ (the opposite cell in $X_\phi$). Then $D_t \cong Y_\phi$.*

*Proof.* Let $I_\phi$ be the ideal defining $Y_\phi$ in $\mathcal{O}^-$. We have (cf. §5.1.3), $I_\phi$ is generated by $M_t := \{p_{\underline{i}},\ \underline{i} \not\leq \phi\}$. Also, $I(D_t)$ (the ideal defining $D_t$ in $\mathcal{O}^-$) is generated by $\{p_{\underline{i}},\ \underline{i} \in Z_t\}$.

Let $\underline{i} \in Z_t$, say, $\underline{i} = (i_1, \cdots, i_d)$. We have (cf. Lemma 5.2.2.4) $\underline{i} \not\leq \phi$, and hence $p_{\underline{i}} \in M_t$.

Let now $p_{\underline{i}} \in M_t$, say, $\underline{i} = (i_1, \cdots, i_d)$. This implies that $i_{d-t+1} \geq d+1$ (cf. Remark 5.2.2.5), and hence it corresponds to a $s$-minor in $X$, where $s \geq t$. From this it follows that $p_{\underline{i}} \in$ the ideal generated by $\{p_{\underline{i}},\ \underline{i} \in Z_t\}$, i.e., $p_{\underline{i}}$ belongs to $I(D_t)$. Thus we have shown $I_\phi = I(D_t)$ and the result follows from this.

**Corollary 5.2.2.8** *$D_t$ is normal, Cohen-Macaulay of dimension $(t-1)(n-(t-1))$, where note that $n = r + d$.*

*Proof.* The normality and Cohen-Macaulayness of Schubert varieties (cf. Chapter 4) imply corresponding properties for $D_t$ (in view of Theorem 5.2.2.7). Regarding $\dim D_t$, we have $\dim D_t = \dim X_\phi = (t-1)(n-(t-1))$ (recall from Chapter 4 that if $w = (a_1, \cdots, a_d) \in I_{d,n}$, then $\dim X(w) = \sum_{i=1}^{i=d} (a_i - i)$).

### 5.2.3 The bijection $\theta$

Let $n = r + d$. Given $\underline{i} \in I_{d,n}$, let $m$ be such that $i_m \leq d,\ i_{m+1} > d$. Set

$$A_{\underline{i}} = \{n+1-i_d, n+1-i_{d-1}, \cdots, n+1-i_{m+1}\},$$

$$B_{\underline{i}} = \text{ the complement of } \{i_1, i_2 \cdots, i_m\} \text{ in } \{1, 2, \cdots, d\}.$$

Define $\theta : I_{d,n} \setminus \{(12\cdots d)\} \to \{\text{all minors of } X\}$ by setting $\theta(\underline{i}) = [A_{\underline{i}}|B_{\underline{i}}]$. Clearly, $\theta$ is a bijection.

### 5.2.4 The partial order $\succeq$

Define a partial order on the set of all minors of $X$ as follows: Let $[A|B],\ [A'|B']$ be two minors of size $s, s'$ respectively. Let $A = \{a_1, \cdots, a_s\},\ B = \{b_1, \cdots, b_s\},\ A' = \{a'_1, \cdots, a'_s\},\ B = \{b'_1, \cdots, b'_s\}$. We define $[A|B] \succeq [A'|B']$ if $s \leq s',\ a_j \geq a'_j,\ b_j \geq b'_j,\ 1 \leq j \leq s$. Note that the bijection $\theta$ reverses the respective partial orders, i.e., given $\underline{i}, \underline{i}' \in I_{d,n}$, we have, $\underline{i} \leq \underline{i}' \iff \theta(\underline{i}) \succeq \theta(\underline{i}')$. Using this partial order, we define *standard monomials* in $[A, B]$'s:

**Definition 5.2.4.1** *A monomial $[A_1, B_1] \cdots [A_s, B_s]$, $s \in \mathbb{N}$ is standard if $[A_1, B_1] \succeq \cdots \succeq [A_s, B_s]$.*

In view of Theorem 5.1.0.1 and Theorem 5.2.2.7, we obtain

**Theorem 5.2.4.2** *Standard monomials in $[A, B]$'s with $\# A \leq t-1$ form a basis for $K[D_t]$, the algebra of regular functions on $D_t$.*

*Proof.* Theorem 5.2.2.7 identifies $D_t$ as the principal open set $X(\phi)_{p_{id}}$; hence we get an identification

$$K[D_t] \cong K[X(\phi)]_{(p_{id})}$$

where $K[X(\phi)]_{(p_{id})}$ denotes the homogeneous localization; note that $p_{id}$ is a lowest weight vector in $H^0(G_{d,n}, L(\omega_d))$. The fact that standard monomials $\{p_{\tau_1} \cdots p_{\tau_m}, m \in \mathbb{N}, \phi \geq \tau_1\}$ generate $K[X(\phi)]$ implies that

$$\{\frac{p(\tau_1)}{p_{id}} \cdots \frac{p(\tau_m)}{p_{id}}, \ m \in \mathbb{N}, \ \phi \geq \tau_1\}$$

generates $K[D_t]$. Note that the condition that $\phi \geq \tau$ corresponds to the condition that the minor corresponding to $\tau$ has size $\leq t-1$ (note that under the order-reversing bijection $\theta$, $\phi$ corresponds to the pair $((1, \cdots, t-1), (1, \cdots, t-1))$). The linear independence of standard monomials $\{p_{\tau_1} \cdots p_{\tau_m}, m \in \mathbb{N}, \phi \geq \tau_1\}$ clearly implies the linear independence of $\{\frac{p(\tau_1)}{p_{id}} \cdots \frac{p(\tau_m)}{p_{id}}, m \in \mathbb{N}, \phi \geq \tau_1\}$. Further, clearly $p_{\tau_1} \cdots p_{\tau_m}$ is standard if and only if $\frac{p(\tau_1)}{p_{id}} \cdots \frac{p(\tau_m)}{p_{id}}$ is. The result now follows from this.

In particular, taking $t = 1 + min\{m, n\}$, we recover the result due to Doubilet-Rota-Stein ([26], Theorem 2):

**Theorem 5.2.4.3** *Standard monomials in $[A, B]$'s form a basis for $K[Z](\cong K[x_{ij}, 1 \leq i \leq m, 1 \leq j \leq n])$.*

### 5.2.5 Cogeneration of an ideal

As above, let $X$ be a matrix of variables of size $r \times d$, and let $\geq$ be a partial order on the set of all minors of $X$.

**Definition 5.2.5.1** *Given a minor $M$ of $X$, the ideal $I_M$ of $K[x_{ij}, 1 \leq i \leq r, 1 \leq j \leq d]$ generated by $\{M' \mid M'$ a minor of $X$ such that $M' \not\geq M\}$ is called the ideal* cogenerated *by $M$ (for the given partial order).*

**Proposition 5.2.5.2** *Given a minor $M$ of $X$, let $\tau_M$ be the element of $I_{d,n}$ such that $\theta(\tau_M) = M$. Then $I_M = I(Y_{\tau_M})$, where $Y_{\tau_M} = \mathcal{O}_d^- \cap X_{P_d}(\tau_M)$ and $I(Y_{\tau_M})$ is the ideal defining $Y_{\tau_M}$ as a closed subvariety of $\mathcal{O}_d^-$.*

*Proof.* We have (cf. Theorem 5.2.2.7), $I(Y_{\tau_M})$ is generated by $\{p_{\underline{i}} \mid \underline{i} \in I_{d,n}, \mid \tau_M \not\geq \underline{i}\}$. The required result follows from this in view of the bijection $\theta$.

**Corollary 5.2.5.3** *$\{I_M, M$ a minor of $X\}$ is precisely the set of ideals of the opposite cells in Schubert varieties in $G_{d,n}$.*

*Proof.* The result follows in view of Proposition 5.2.5.2 and the bijection $\theta$ (cf. §5.2.3).

**Corollary 5.2.5.4** *Let $M$ be the $(t-1)$-minor of $X$ consisting of the first $(t-1)$-rows of $X$ and the first $(t-1)$-columns of $X$. Then $I_t(X)$ $(=$ the ideal of $D_t)$ equals $I_M$.*

*Proof.* We have (cf. Theorem 5.2.2.7), $I_t(X) = I(Y_\phi)$, $\phi$ being as in Theorem 5.2.2.7. Now, under the bijection $\theta$, we have $\theta(\phi) = [\{1, 2, \cdots, t-1\} \mid \{1, 2, \cdots, t-1\}]$. The result now follows from Proposition 5.2.5.2.

### 5.2.6 The monomial order $\prec$ and Gröbner bases

We introduce a total order on the variables as follows:

$$x_{m1} > x_{m2} > \cdots > x_{mn} > x_{m-1\,1} > x_{m-1\,2} > \cdots > x_{m-1\,n} > \dots$$
$$> x_{11} > x_{12} > \cdots > x_{1n}.$$

This induces a total order, namely the lexicographic order, on the set of monomials in $K[X] = K[x_{11}, \dots, x_{mn}]$, denoted by $\prec$. The largest monomial (with respect to $\prec$) present in a polynomial $f \in K[X]$ is called the *initial term* of $f$, and is denoted by $in(f)$. Note that the initial term (with respect to $\prec$) of a minor of $X$ is equal to the product of its elements on the skew diagonal.

Given an ideal $I \subset K[X]$, denote by $in(I)$ the ideal generated by the initial terms of the elements in $I$. A set $G \subset I$ is called a *Gröbner basis* of $I$ (with respect to the monomial order $\prec$), if $in(I)$ is generated by the initial terms of the elements in $G$. Note that a Gröbner basis of $I$ generates $I$ as an ideal.

We recall the following (cf. [38])

**Theorem 5.2.6.1** *Let $M = [i_1, \dots, i_r | j_1, \dots, j_r]$ be a minor of $X$, and $I$ the ideal of $K[X]$ cogenerated by $M$. For $1 \le s \le r+1$, let $G_s$ be the set of all $s$-minors $[i'_1, \dots, i'_s | j'_1, \dots, j'_s]$ satisfying the conditions*

$$i'_s \le i_r, i'_{s-1} \le i_{r-1}, \dots, i'_2 \le i_{r-s+2}, \qquad (1)$$
$$j'_{s-1} \ge j_{s-1}, \dots, j'_2 \ge j_2, j'_1 \ge j_1$$
$$\text{if } s \le r, \text{ then } i'_1 > i_{r-s+1} \text{ or } j'_s < j_s. \qquad (2)$$

*Then the set $G = \cup_{i=1}^{r+1} G_i$ is a Gröbner basis for the ideal $I$ with respect to the monomial order $\prec$.*

**Theorem 5.2.6.2** *Let $t$, $1 \le t \le \min\{m, n\}$. Let $\mathcal{F}_t = \{M' \mid M'$ is a minor of $X$ of size $t\}$. Then $\mathcal{F}_t$ is a Gröbner basis for $I_t(X)$.*

*Proof.* We have $D_t$ is generated by $\mathcal{F}_t$. Let $M$ be the $(t-1)$-minor $[A|B]$ of $X$, where $A = \{m+2-t, m+3-t, \cdots, m\}$, $B = \{1, 2, \cdots, t-1\}$. Let us write $A = \{i_1, \cdots, i_r\}$, $B = \{j_1, \cdots, j_r\}$. We have $r = t-1$, $i_l = m+l-r$, $j_l = l$, $1 \le l \le r$.

Let $M' \in G$, $G$ being as in Theorem 5.2.6.1, say, $M' = [i'_1, \dots, i'_s | j'_1, \dots, j'_s]$ be a minor of $X$ of size $s$, $s \le r+1\,(= t)$.The inequalities regarding $i$'s and $i'$'s in condition (1) of Theorem 5.2.6.1 are redundant, since $i_l = m+l-r$, $1 \le l \le r$. Similarly, the inequalities regarding $j$'s and $j'$'s in condition (1) of Theorem 5.2.6.1 are again redundant (since $j_l = l$, $1 \le l \le r$; also, condition (2) reduces to the condition that if $s \le r (= t-1)$, then $i'_1 > i_{r-s+1}$ (since $j_s = s$). Therefore, for the above choice of $M$, the conditions (1) and (2) of Theorem 5.2.6.1 are equivalent to

$$\text{if } s \le r\ (= t-1), \text{ then } i'_1 > i_{r-s+1}\ (= i_{t-s}).$$

Let $s \le t-1$. The condition that $i'_1 > i_{t-s}\ (= m+1-s)$ implies $i'_s > m+1$, which is not possible. Hence $G_s = \emptyset$, if $s \le t-1$; thus we obtain $G = G_t$. Now for

$s = t$, as discussed above, condition (1) Theorem 5.2.6.1 is satisfied by any $t$-minor $M'$ while condition (2) is vacuous. Hence we obtain $G_t$ is the set of all $t$-minors, i.e., $G = \mathcal{F}_t$. The result now follows from Theorem 5.2.6.1.

We conclude this chapter with the following

**Theorem 5.2.6.3** *The singular locus of $D_t$ is $D_{t-1}$.*

This theorem will be proved in Chapter 13 (cf. Theorem 13.3.0.3) as an application of SMT.

# 6

# Symplectic Grassmannian

In this chapter we give an exposition of the determinantal varieties in $Sym\ M_n$ (the space of symmetric $n \times n$ matrices) from the view-point of their relationship to Schubert varieties in the symplectic (or also Lagrangian) Grassmannian varieties. After reviewing some basic algebraic group-theoretic results on symplectic groups, we establish the identification of $Sym\ M_n$ with the "opposite cell" in a suitable symplectic Grassmannian, which then yields an identification of a determinantal variety in $Sym\ M_n$ with the "opposite cell" in a suitable Schubert variety in the symplectic Grassmannian (cf. [67]). Using the standard monomial basis for the homogeneous co-ordinate ring of a Schubert variety in the symplectic Grassmannian, we obtain a standard monomial basis for the ring of regular functions on a determinantal variety in $Sym\ M_n$. This basis in fact coincides with that of DeConcini-Procesi (cf. [22]). Thus Schubert-variety-theoretic approach gives a different proof of DeConcini-Procesi's basis for these varieties.

Let $V = K^{2n}$ together with a non-degenerate, skew-symmetric bilinear form $(\cdot,\cdot)$. Let $H = SL(V)$ and $G = Sp(V) = \{A \in SL(V) \mid A$ leaves the form $(\cdot,\cdot)$ invariant $\}$. Taking the matrix of the form (with respect to the standard basis $\{e_1, ..., e_{2n}\}$ of $V$) to be

$$E = \begin{pmatrix} 0 & J \\ -J & 0 \end{pmatrix}$$

where $J$ is the anti-diagonal $(1, \ldots, 1)$ of size $n \times n$, we may realize $Sp(V)$ as the fixed point set of a certain involution $\sigma$ on $SL(V)$, namely $G = H^\sigma$, where $\sigma : H \longrightarrow H$ is given by $\sigma(A) = E({}^tA)^{-1}E^{-1}$. Thus

$$\begin{aligned} G = Sp(2n) &= \{A \in SL(2n) \mid {}^tAEA = E\} \\ &= \{A \in SL(2n) \mid E^{-1}({}^tA)^{-1}E = A\} \\ &= \{A \in SL(2n) \mid E({}^tA)^{-1}E^{-1} = A\} \\ &= H^\sigma. \end{aligned}$$

(Note that $E^{-1} = -E$). Denoting by $T_H$ (resp. $B_H$) the maximal torus in $H$ consisting of diagonal matrices (resp. the Borel subgroup in $H$ consisting of upper triangular matrices) we see easily that $T_H$, $B_H$ are stable under $\sigma$. We set $T_G = T_H^\sigma$, $B_G = B_H^\sigma$. Then it can be seen easily that $T_G$ is a maximal torus in $G$ and $B_G$ is a Borel subgroup in $G$; in particular, $T_G$ consists of diagonal matrices of the form diag $(t_1, \cdots, t_n, t_n^{-1}, \cdots, t_1^{-1})$. We note the following specific facts for this group. See [67] for details.

## 6.1 Some basic facts on $Sp(V)$

Let $N_G$ (resp. $N_H$) denote the normalizer in $G$ (resp. $H$) of $T_G$ (resp. $T_H$). We have, $N_G \subset N_H$; further, $N_H$ is stable under $\sigma$, and we have

$$N_G = N_H^\sigma$$
$$N_G/T_G \hookrightarrow N_H/T_H$$

Thus we obtain

$$W_G \hookrightarrow W_H$$

where $W_G$, $W_H$ denote the Weyl groups of $G$, $H$ respectively (with respect to $T_G$, $T_H$ respectively). Further, $\sigma$ induces an involution on $W_H$:

$$w = (a_1, \cdots, a_{2n}) \in W_H, \sigma(w) = (c_1, \cdots, c_{2n}),\ c_i = 2n + 1 - a_{2n+1-i}$$

and

$$W_G = W_H^\sigma$$

Thus we obtain

$$W_G = \{(a_1 \cdots a_{2n}) \in S_{2n} \mid a_i = 2n + 1 - a_{2n+1-i},\ 1 \leq i \leq 2n\}.$$

(here, $S_{2n}$ is the symmetric group on $2n$ letters). Thus $w = (a_1 \cdots a_{2n}) \in W_G$ is known once $(a_1 \cdots a_n)$ is known. We shall denote an element $(a_1 \cdots a_{2n})$ in $W_G$ by just $(a_1 \cdots a_n)$. For example, $(4231) \in S_4$ represents $(42) \in W_G$, $G = Sp(4)$.
**Root system of type $C_n$:** $\sigma$ induces an involution on $X(T_H)$, the character group of $T_H$:

$$\chi \in X(T_H),\ \sigma(\chi)(D) = \chi(\sigma(D)),\ D \in T_H$$

Let $\epsilon_i$, $1 \leq i \leq 2n$ be the character in $X(T_H)$, $\epsilon_i(D) = d_i$, the $i$-th entry in $D(\in T_H)$. We have

$$\sigma(\epsilon_i) = -\epsilon_{2n+1-i}$$

Now it is easily seen that under the canonical surjective map

$$\varphi : X(T_H) \to X(T_G)$$

we have

$$\varphi(\epsilon_i) = -\varphi(\epsilon_{2n+1-i}),\ 1 \leq i \leq 2n$$

Let $R_H := \{\epsilon_i - \epsilon_j, 1 \leq i, j \leq 2n\}$, the root system of $G$ (relative to $T_H$), and $R_H^+ := \{\epsilon_i - \epsilon_j, 1 \leq i < j \leq 2n\}$, the set of positive roots (relative to $B_H$). We have the following:

1. $\sigma$ leaves $R_H$ (resp. $R_H^+$) stable.
2. For $\alpha, \beta \in R_H, \varphi(\alpha) = \varphi(\beta) \Leftrightarrow \alpha = \sigma(\beta)$.
3. $\varphi$ is equivariant for the canonical action of $W_G$ on $X(T_H), X(T_G)$.
4. $R_H^\sigma = \{\pm(\epsilon_i - \epsilon_{2n+1-i}), 1 \leq i \leq n\}$.

Let $R_G$ (resp. $R_G^+$) the set of roots of $G$ with respect to $T_G$ (resp. the set of positive roots with respect to $B_G$ ). Using the above facts and the explicit nature of the adjoint representation of $G$ on Lie $G$, we deduce that

$$R_G = \varphi(R_H),\ R_G^+ = \varphi(R_H^+)$$

In particular, $R_G$ (resp. $R_G^+$) gets identified with the orbit space of $R_H$ (resp. $R_H^+$) modulo the action of $\sigma$. Thus we obtain the following identification:

$$R_G = \{\pm(\epsilon_i \pm \epsilon_j),\ 1 \leq i < j \leq n\} \cup \{\pm 2\epsilon_i,\ i = 1, ..., n\}$$

$$R_G^+ = \{(\epsilon_i \pm \epsilon_j),\ 1 \leq i < j \leq n\} \cup \{2\epsilon_i,\ i = 1, ..., n\}.$$

The set $S_G$ of simple roots in $R_G^+$ is given by

$$S_G := \{\alpha_i = \epsilon_i - \epsilon_{i+1},\ 1 \leq i \leq n-1\} \cup \{\alpha_n = 2\epsilon_n\}.$$

Let us denote the simple reflections in $W_G$ by $\{s_i,\ 1 \leq i \leq n\}$, namely, $s_i$= reflection with respect to $\epsilon_i - \epsilon_{i+1},\ 1 \leq i \leq n-1$, and $s_n$ = reflection with respect to $2\epsilon_n$. Then we have (cf. [8])

$$s_i = \begin{cases} r_i r_{2n-i}, & \text{if } 1 \leq i \leq n-1 \\ r_n, & \text{if } i = n \end{cases}$$

where $r_i$ denotes the transposition $(i, i+1)$ in $S_{2n},\ 1 \leq i \leq 2n-1$. Continuing the example above, we have, (42) $= s_1 s_2 s_1$.

**Chevalley basis.** For $1 \leq i \leq 2n$, set $i' = 2n+1-i$. The involution $\sigma : SL(2n) \rightarrow SL(2n), A \mapsto E({}^tA)^{-1}E^{-1}$, induces an involution $\sigma : sl(2n) \rightarrow sl(2n), A \mapsto -E({}^tA)E^{-1}(= E({}^tA)E$, since $E^{-1} = -E)$. In particular, we have, for $1 \leq i$, $j \leq 2n$

$$\sigma(E_{ij}) = \begin{cases} -E_{j'i'}, & \text{if } i, j \text{ are both } \leq n \text{ or both } > n \\ E_{j'i'}, & \text{if one of } \{i, j\} \text{ is } \leq n \text{ and the other } > n \end{cases}$$

where $E_{ij}$ is the elementary matrix with 1 at the $(i, j)$-th place and 0 elsewhere. Further

$$\text{Lie } Sp(2n) = \{A \in \text{sl}(2n) \mid E({}^tA)E = A\}.$$

The Chevalley basis $\{H_{\alpha_i} : \alpha_i \in S_G\} \cup \{X_\alpha : \alpha \in R_G\}$ for Lie $Sp(2n)$ may be given as follows:

$$\begin{aligned}
H_{\epsilon_i-\epsilon_{i+1}} &= E_{ii} - E_{i+1,i+1} + E_{(i+1)',(i+1)'} - E_{i'i'} \\
H_{2\epsilon_n} &= E_{nn} - E_{n'n'} \\
X_{\epsilon_j-\epsilon_k} &= E_{jk} - E_{k'j'} \\
X_{\epsilon_j+\epsilon_k} &= E_{jk'} + E_{kj'} \\
X_{2\epsilon_m} &= E_{mm'} \\
X_{-(\epsilon_j-\epsilon_k)} &= E_{kj} - E_{j'k'} \\
X_{-(\epsilon_j+\epsilon_k)} &= E_{k'j} + E_{j'k} \\
X_{-2\epsilon_m} &= E_{m'm}.
\end{aligned}$$

**Length Formula:** For $w \in W_G$, let us denote $l(w, W_H)$ (resp. $l(w, W_G)$), the length of $w$ as an element of $W_H$ (resp. $W_G$). For $w = (a_1, \cdots, a_{2n}) \in W_H$, denote

$$m_w := \#\{i \leq n \mid a_i > n\}$$

**Proposition 6.1.0.1** *For $w = (a_1, \cdots, a_{2n}) \in W_G$, we have*

$$l(w, W_G) = \frac{1}{2}(l(w, W_H) + m_w)$$

*Proof.* Set

$$S_H(w) = \{\beta \in R_H^+ \mid w(\beta) < 0\},\ S_G(w) = \{\beta \in R_G^+ \mid w(\beta) < 0\}$$

We have

$$l(w, W_H) = \#S_H(w),\ l(w, W_G) = \#S_G(w)$$

The canonical map $\varphi : X(T_H) \to X(T_G)$ induces a surjective map $R_H \to R_G$. Further, we have,

$$\texttt{for } \alpha \in R_H, \alpha > 0 \Leftrightarrow \varphi(\alpha) > 0$$

Hence $\varphi$ induces a surjective map

$$S_H(w) \to S_G(w)$$

Further, $\sigma$ leaves $S_H(w)$ stable, and

$$S_H(w)^\sigma = \{\alpha \mid \alpha = \epsilon_i - \epsilon_{2n+1-i}, 1 \leq i \leq n,\ w(\alpha) < 0 \texttt{ in } R_H\}$$

Now if $\alpha = \epsilon_i - \epsilon_{2n+1-i}$, then $w(\alpha) = \epsilon_{a_i} - \epsilon_{a_{2n+1-i}}$. Hence

$$w(\alpha) < 0 \texttt{ in } R_H \Leftrightarrow a_{2n+1-i} < a_i,\ i \leq n$$

But now $a_{2n+1-i} = 2n + 1 - a_i$ (since $w \in W_G$). Hence

$$w(\alpha) < 0 \texttt{ in } R_H \Leftrightarrow n < a_i, 1 \leq i \leq n$$

From this, we obtain

$$\#S_H(w)^\sigma = m(w)$$

Also, we have,

$$\#S_H(w) = 2\#S_G(w) - \#S_H(w)^\sigma$$

The required result now follows from this.

### 6.1.1 Schubert varieties in $G/B_G$

For $w \in W_H$, let $C_H(w)$ be the *Schubert cell* $B_H w B_H (mod\ B_H)$ in $H/B_H$; if $w \in W_G$, we shall denote by $C_G(w)$ the Schubert cell $B_G w B_G (mod\ B_G)$ in $G/B_G$. We observe that $\sigma$ induces a natural involution on $H/B_H$.

**Proposition 6.1.1.1** *Let $w \in W_G$. The Schubert cell $C_H(w)$ is stable under $\sigma$, and $C_H(w)^\sigma = C_G(w)$, $w \in W_G$.*

*Proof.* The first part is clear. Let $B_H^u$ denote the unipotent part of $B_H$. Clearly, $B_H^u$ is $\sigma$-stable (since $B_H$ is $\sigma$-stable). Let $B_1$ be the isotropy subgroup of $B_H^u$ at the point $wH (\in C_H(w))$; then $B_1$ is stable under $\sigma$ (since $\sigma(w) = w$). We have

$$(*) \qquad B_H^u = \prod_{\alpha \in R_H^+} U_\alpha$$

(here, $U_\alpha$ denotes the root subgroup of $H$ associated to $\alpha$). We see easily

$$B_1 = \prod_{\{\alpha \in R_H^+,\ w^{-1}(\alpha) > 0\}} U_\alpha$$

Let

$$B_2 = \prod_{\{\alpha \in R_H^+,\ w^{-1}(\alpha) < 0\}} U_\alpha$$

Then (*) together with the facts that $B_H^u$, $B_1$ are $\sigma$-stable implies that $B_2$ is $\sigma$-stable. Let $x \in C_H(w)$; then $x$ has a presentation $x = bw$ for a unique $b \in B_2$. Hence $x \in C_H(w)^\sigma \Leftrightarrow \sigma(b) = b \Leftrightarrow b \in B_G$; and therefore $x \in C_H(w)^\sigma \Leftrightarrow x \in C_G(w)$.

**Maximal parabolics.** For $1 \leq d \leq n$, we let $P_d$ be the maximal parabolic subgroup of $G$ with $S_G \setminus \{\alpha_d\}$ as the associated set of simple roots. Then it can be seen easily that $W_G^{P_d}$, the set of minimal representatives of $W_G/W_{P_d}$ can be identified with the set of all $(a_1 \cdots a_d)$ satisfying (1) and (2) below:

(1) $1 \leq a_1 < a_2 < \cdots < a_d \leq 2n$
(2) for $1 \leq i \leq 2n$, if $i \in \{a_1, ..., a_d\}$
then $2n + 1 - i \notin \{a_1, ..., a_d\}$

**Bruhat–Chevalley order.** For $w_1 = (a_1 \cdots a_{2n}) \in W_G$ and $w_2 = (b_1 \cdots b_{2n})$, $w_1$, $w_2$ we have $w_2 \geq w_1 \Leftrightarrow \{b_1, ..., b_d\} \uparrow \geq \{a_1, ..., a_d\} \uparrow$ for each $1 \leq d \leq n$ (cf. [102]). Here $\{a_1, \ldots, a_d\}\uparrow$, $\{b_1, \ldots, b_d\}\uparrow$ are the corresponding $d$-tuples arranged in ascending order. Hence for $w \in W_G$, denoting by $w^{(d)}$ the element in $W_G^{P_d}$ which represents the coset $wW_{P_d}$, we have for $w_1, w_2 \in W_G$ and $1 \leq d \leq n$,

$$w_2^{(d)} \geq w_1^{(d)},\ 1 \leq d \leq n \iff \{b_1, \ldots, b_d\} \uparrow \geq \{a_1, \ldots, a_d\} \uparrow .$$

Further, $w_2 \geq w_1 \iff w_2^{(d)} \geq w_1^{(d)}$, $1 \leq d \leq n$. But now, the latter condition is equivalent to $w_2 \geq w_1$ in $W_H$. Thus we obtain that the partial order on $W_G$ is induced by the partial order on $W_H$ (cf. [102]). In particular, for $w_1 = (a_1 \cdots a_d)$, $w_2 = (b_1 \cdots b_d)$, $w_1, w_2 \in W_G^{P_d}$, we have $w_2 \geq w_1 \Leftrightarrow \{b_1, \ldots, b_d\} \uparrow \geq \{a_1, \ldots, a_d\} \uparrow$.

In the sequel, we shall denote an element $(a_1 \cdots a_n)$ in $W_G^{P_d}$ by just $(a_1 \cdots a_d)$. The fact that the partial order on $W_G$ is induced by the partial order on $W_H$ together with Proposition 6.1.1.1 yields the following:

**Proposition 6.1.1.2** *Let $w \in W_G$; let $X_G(w)$ (resp. $X_H(w)$) be the associated Schubert variety in $G/B_G$ (resp. $H/B_H$). Under the canonical inclusion $G/B_G \hookrightarrow H/B_H$, we have $X_G(w) = X_H(w) \cap G/B_G$. Further, the intersection is scheme-theoretic.*

*Proof.* Denote $Z := X_H(w) \cap G/B_G$. We have $X_G(w) \subseteq X_H(w) \cap G/B_G$ (clearly), and is in fact an irreducible component of $Z$ (in view of Proposition 6.1.1.1). Let $Z'$ be another irreducible component of $Z$. Then $Z' = X_G(w')$, for some $w' \in W_G$ ($Z$ being closed and $B_G$-stable in $G/B_G$, is a union of Schubert varieties in $G/B_G$). Now the inclusion $X_G(w') \subset Z$ implies in particular that $C_G(w') (= C_H(w')^\sigma)$ is contained in $Z$. Hence we obtain that $C_H(w')^\sigma \subseteq X_H(w) \cap C_H(w')$; this implies in particular that $C_H(w') \subseteq X_H(w)$ (since $X_H(w) \cap C_H(w')$ is non-empty, and $X_H$ is $B_H$-stable), and hence we obtain that $w' \leq w$ in $W_H$, and hence also in $W_G$ (in view of the fact that the partial order on $W_G$ is induced by the partial order on $W_H$). Thus we get that $Z' (= X_G(w')) \subset X_G(w)$. The first part of the Lemma follows from this. The second part follows from the fact (cf. Chapter 13, Theorem 13.2.1.10) that for any $\tau \in W_G, \tau \leq w$, we have that the tangent space to $X_H(w)$ at the $T$-fixed point $\tau B_H$ is $\sigma$-stable, and its subspace of $\sigma$-fixed points is the tangent space to $X_G(w)$ at the $T$-fixed point $\tau B_G$.

## 6.2 The variety $G/P_n$

Let $Q_n$ denote the maximal parabolic subgroup of $H$ associated to the simple root $\epsilon_n - \epsilon_{n+1}$. As above, let $P_n$ denote the maximal parabolic subgroup of $G$ associated to the simple root $2\epsilon_n$. It is easily seen that $Q_n$ is $\sigma$-stable so that $\sigma$ induces an involution on $H/Q_n$. Let $Z$ be the subgroup of $H$ consisting of matrices of the form

$$\begin{pmatrix} Id_n & 0 \\ Y & Id_n \end{pmatrix}, \ Y \in M_n$$

(here, $M_n$ denotes the $n \times n$ matrices with entries in $K$). The canonical morphism $H \to H/Q_n$ induces a morphism

$$\psi_H : Z \to H/Q_n$$

We have (cf. Chapter 5)

**Fact:** $\psi_H$ is an open immersion, and $\psi_H(Z)$ gets identified with the opposite big cell $O_H^-$ in $H/Q_n$.

**Lemma 6.2.0.1** *$O_H^-$ is $\sigma$- stable, and we have an identification Sym $M_n \cong (O_H^-)^\sigma$.*

*Proof.* For $z = \begin{pmatrix} Id_n & 0 \\ Y & Id_n \end{pmatrix} \in Z$, we have,

$$\sigma(z) = \begin{pmatrix} Id_n & 0 \\ J\,{}^tYJ & Id_n \end{pmatrix}$$

(where recall that $J$ is the anti-diagonal $(1, \cdots, 1)$ of size $n \times n$). Hence, $Z$ is $\sigma$-stable, and the first part of the Lemma follows (in view of Fact above).
We have,

$$Z^\sigma = \{z \in Z \mid J\,{}^tYJ = Y\}$$

Letting $Y = JX$, we have

$$J\,{}^tYJ = Y \Leftrightarrow {}^t X = X$$

The second part of Lemma follows from this.

Consider $H/Q_n$ as the Grassmannian of $n$-dimensional subspaces of $V := K^{2n}$. It is well-known that the set of maximal totally isotropic subspaces (for the skew-symmetric form $(,)$ on $V$) is a closed subvariety of $H/Q_n$ isomorphic to $G/P_n$, the *symplectic* (or also *the Lagrangian*) *Grassmannian.*

**Lemma 6.2.0.2** *$\psi_H$ induces an inclusion $\psi_H : Z^\sigma \hookrightarrow G/P_n$.*

*Proof.* Now $z = \begin{pmatrix} Id_n & 0 \\ Y & Id_n \end{pmatrix}$ considered as a point of $H/Q_n$ (= the Grassmannian of $n$ dimensional subspaces of $K^{2n}$) corresponds to the $n$-dimensional subspace spanned by the columns of $\begin{pmatrix} Id_n \\ Y \end{pmatrix}$. Let us denote this $n$-dimensional subspace by $U_z$. We have,

$$J\,{}^tYJ = Y \Leftrightarrow \left(Id_n\ {}^tY\right)\begin{pmatrix} 0 & J \\ -J & 0 \end{pmatrix}\begin{pmatrix} Id_n \\ Y \end{pmatrix} = 0$$

Hence we obtain that $z \in Z^\sigma \Leftrightarrow U_z$ is a maximal totally isotropic subspace. The required result follows from this.

Let us denote the restriction of the morphism $\psi_H : Z \to H/Q_n, Y \mapsto \begin{pmatrix} Id_n & 0 \\ JY & Id_n \end{pmatrix}$ to *Sym* $M_n$ by $\psi_G$. Let $O_G^-$ the opposite big cell in $G/P_n$.

### 6.2.1 Identification of *Sym* $M_n$ with $O_G^-$

Let $L_n$ denote the tautological line bundle on $H/Q_n (= G_{n,2n})$ for the Plücker embedding. Then $H^0(H/Q_n, L_n)$ is the dual Weyl $H$-module with highest weight $\omega_n$.

**Proposition 6.2.1.1** *Let $Id = (1, \cdots, n) (\in I_{n,2n})$; denote $f := p_{Id}$, the Plücker co-ordinate corresponding to $Id$. We have*

1. *$f$ is a lowest weight vector in $H^0(H/Q_n, L_n)$.*
2. *$O_H^-$ is the principal open set $(H/Q_n)_f$.*

*Proof.* (1). We have

$$H^0(H/Q_n, L_n) = (\Lambda^n(K^{2n}))^*$$

Further, for $\mathbf{i} = (i_1, \cdots, i_n) \in I_{n,2n}$,

$$p_{\mathbf{i}} = e_{i_1}^* \wedge \cdots \wedge e_{i_n}^*$$

(here $\{e_t^*, 1 \leq t \leq 2n\}$ denotes the basis of $V^*$ dual to the standard basis $\{e_t, 1 \leq t \leq 2n\}$ of $V = K^{2n}$). Hence it follows that $p_{\mathbf{i}}$ is a $T_H$-weight vector of weight $-(\epsilon_{i_1} + \cdots + \epsilon_{i_n})$. Now (1) follows from this.
(2) follows from the fact that the opposite big cell is precisely the set of points in $H/Q_n$ where a lowest weight vector (in $H^0(H/Q_n, L_n)$) does not vanish.

### 6.2.2 Canonical dual pair

Let $Y \in M_n$. Given two $s$-tuples $A := (a_1, \cdots, a_s)$, $B := (b_1, \cdots, b_s)$, in $I_{s,n}$, $s \leq n$, let us denote by $p(A, B)(Y)$, the $s$-minor of $Y$ with row and column indices given by $A, B$ respectively. Consider the identification

$$(*) \qquad M_n \cong \left\{ \begin{pmatrix} Id_n \\ JY \end{pmatrix}, Y \in M_n \right\}$$

where $J$ is as above. Let $\mathbf{i} := (i_1, \cdots, i_n) \in I_{n,2n}$, and let $p_{\mathbf{i}}$ be the associated Plücker co-ordinate on $G_{n,2n}(= H/Q_n)$. Let $f_{\mathbf{i}}$ denote the restriction of $p_{\mathbf{i}}$ to $M_n$ (under the identification (*)). Then $f_{\mathbf{i}} = p(\mathbf{i}(A), \mathbf{i}(B))$, where $\mathbf{i}(A), \mathbf{i}(B)$ are given as follows:

Let $r$ be such that $i_r \leq n, i_{r+1} > n$; let $s = n - r$. Then $\mathbf{i}(A)$ is the $s$-tuple given by $(2n + 1 - i_n, \cdots, 2n + 1 - i_{r+1})$, while $\mathbf{i}(B)$ is the $s$-tuple given by the complement of $(i_1, \cdots, i_r)$ in $(1, \cdots, n)$. We refer to $(\mathbf{i}(A), \mathbf{i}(B))$ as the *canonical dual pair* associated to $\mathbf{i}$.

### 6.2.3 The bijection $\theta$

Define a partial order $\geq$ on the set of minors of $Y \in M_n$ as follows: Given $A, B, A', B'$ where $A, B \in I_{s,n}$, and $A', B' \in I_{s',n}$, $p(A, B) > p(A', B')$ if $s \leq s'$, and $a_t \geq a_t', b_t \geq b_t', 1 \leq t \leq s$. Then the map $\mathbf{i} \mapsto (\mathbf{i}(A), \mathbf{i}(B))$ defines an order-reversing bijection $\theta$ between $I_{n,2n}$ and minors of $Y$. In the above bijection, the $n$-tuple $(1, \cdots, n)$ will correspond to the constant function 1 (corresponding to the minor with the set of row (resp. column) indices being the empty set).

### 6.2.4 The dual Weyl $G$-module with highest weight $\omega_n$

Let $L'_n$ be the restriction of $L_n$ to $G/P_n$. For $\mathbf{i} \in I_{n,2n}$, let $p'_{\mathbf{i}}$ be the restriction of $p_{\mathbf{i}}$ to $G/P_n$; note that $p'_{\mathbf{i}} \in H^0(G/P_n, L'_n)$. Let $p(\mathbf{i}(A), \mathbf{i}(B))$ be the restriction of $p_{\mathbf{i}}$ to $M_n$, $M_n$ being identified with the opposite cell $O^-_H$ as above (cf. (*)); let $p'(\mathbf{i}(A), \mathbf{i}(B))$ be the restriction of $p(\mathbf{i}(A), \mathbf{i}(B))$ to $Sym\, M_n$. Let $f$ be a lowest weight vector in $H^0(H/Q_n, L_n)$, and $f'$ the restriction of $f$ to $G/P_n$.

**Proposition 6.2.4.1** *$H^0(G/P_n, L'_n)$ is the dual Weyl $G$-module with highest weight $\omega_n$.*

*Proof.* We first observe that $p'_{\mathbf{i}}, \mathbf{i} \in I_{n,2n}$ is non-zero; this follows from the fact that $p'(\mathbf{i}(A), \mathbf{i}(B))$ is non-zero on $Sym\, M_n$ ($Sym\, M_n = Z^\sigma$ being identified with an open subset of $G/P_n$). Now $f$ being a lowest weight vector in $H^0(H/Q_n, L_n)$ (cf. Proposition 6.2.1.1), the one-dimensional span $Kf$ is $B^-_H$-stable, where $B^-_H$ is the Borel subgroup of $H$ opposite to $B_H$. Also, one sees easily that $B^-_H$ is $\sigma$-stable, and $(B^-_H)^\sigma (= G \cap B^-_H)$ is precisely $B^-_G$, the Borel subgroup of $G$ opposite to $B_G$. This fact together with the fact that $Kf$ is stabilized by $B^-_G$ implies that $f'$ is a lowest weight vector in $H^0(G/P_n, L'_n)$; further,

$$\texttt{weight of } f' (= \texttt{weight of } f) = -(\epsilon_1 + \cdots + \epsilon_n) = -\omega_n$$

Hence we obtain that a highest weight vector in $H^0(G/P_n, L'_n)$ has weight $w_0(G)(-\omega_n)$ (note that the Weyl involution $i = -w_0(G)$ is $Id$ on $X(T_G)$—here, $w_0(G)$ is the element of largest length in $W_G$). Hence we obtain that $L'_n$ is the ample generator of Pic $G/P_n$ so that $H^0(G/P_n, L'_n)$ is the fundamental representation of $G$ with highest weight $\omega_n$. The result now follows from this.

As an immediate consequence we obtain

**Theorem 6.2.4.2** *We have an identification $Z^\sigma \cong O^-_G$.*

*Proof.* Under the identification $\psi : Z \cong O^-_H$, we have that $Z \cong (H/Q_n)_f$ (cf. Proposition 6.2.1.1,(2)); thus $Z$ is precisely the set of points in $H/Q_n$ where the lowest weight vector $f (\in H^0(H/Q_n, L_n))$ does not vanish. Hence we obtain that

$$(G/P_n)_{f'} = Z \cap G/P_n$$

Let $z \in Z$, say $z = \begin{pmatrix} Id \\ JY \end{pmatrix}$. Then $z \in G/P_n$ if and only if the $n$-dimensional subspace spanned by the columns of $\begin{pmatrix} Id \\ JY \end{pmatrix}$ is totally isotropic, i.e, if and only if

$$\left(Id_n \ {}^tY\right) \begin{pmatrix} 0 & J \\ -J & 0 \end{pmatrix} \begin{pmatrix} Id_n \\ Y \end{pmatrix} = 0$$

i.e, if and only if

$$J\,{}^tYJ = Y$$

Hence, letting $Y = JX$, we obtain that

$$O_G^-(= (G/P_n)_{f'}) = Z \cap G/P_n = Sym\ M_n = Z^\sigma$$

This implies the required result.

**Corollary 6.2.4.3** $O_G^- = (O_H^-)^\sigma = Sym\ M_n$.
*(Thus the opposite cell in $H/Q_n$ induces the opposite cell in $G/P_n$.)*

*Proof.* This follows from the above Theorem, the identification $Z \cong O_H^-$, and Lemma 6.2.0.1.

**Corollary 6.2.4.4** $dim\, G/P_n = \frac{1}{2}n(n+1)$.

### 6.2.5 Identification of $D_t(Sym\ M_n)$ with $Y_{P_n}(\varphi)$

Fix an integer $t$ where $1 \le t \le n$. Let $D_t(Sym\ M_n)$ denote the subscheme of $Sym\ M_n$ defined by the vanishing of $t$ minors in $Sym\ M_n$. In this subsection, we shall describe an identification of $D_t(Sym\ M_n)$ with the opposite cell $Y_{P_n}(\varphi)$ in a suitable Schubert variety $X_{P_n}(\varphi)$.

As a consequence of Proposition 6.1.1.2, we have the following:

**Proposition 6.2.5.1** *Let $w \in W^{P_n}$. Let $X_{P_n}(w)$ (resp. $X_{Q_n}(w)$) be the associated Schubert variety in $G/P_n$ (resp. $H/Q_n$). Then under the canonical inclusion $G/P_n \subset H/Q_n$, we have, $X_{P_n}(w) = X_{Q_n}(w) \cap G/P_n$ (scheme-theoretically).*

Fix a $t$ as above. Let $\varphi \in W^{P_n}$ be defined by

$$\varphi = (t, t+1, \cdots, n, 2n+2-t, 2n+3-t, \cdots, 2n)$$

(note that $\varphi$ consists of two blocks $[t, n]$, $[2n+2-t, 2n]$ of consecutive integers—here, for $i < j$, $[i, j]$ denotes the set $\{i, i+1, \cdots, j\}$). Let $Y_{P_n}(\varphi) := X_{P_n}(\varphi) \cap O_G^-$, the *opposite cell in* $X_G(\varphi)$.

**Theorem 6.2.5.2** *The isomorphism $Sym\ M_n \cong O_G^-$ induces an isomorphism*

$$D_t(Sym\ M_n) \cong Y_{P_n}(\varphi)$$

*Proof.* We have scheme-theoretic identifications

$$D_t(Sym\ M_n) \cong D_t(M_n) \cap Sym\ M_n$$
$$Sym\ M_n \cong O_G^-$$
$$Y_{P_n}(\varphi) \cong Y_{Q_n}(\varphi) \cap O_G^-$$

On the other hand, we have an isomorphism (cf. Chapter 5)

$$D_t(M_n) \cong Y_{Q_n}(\varphi)$$

The required result now follows.

**Corollary 6.2.5.3** *$D_t(Sym\, M_n)$ is normal, Cohen-Macaulay of dimension $n(t-1) - \frac{1}{2}(t-1)(t-2)$.*

*Proof.* The normality and Cohen-Macaulayness follow since Schubert varieties in $G/P_n$ are normal and Cohen-Macaulay (cf. §A.12 of Appendix). The assertion on the dimension follows in view of Proposition 6.1.0.1 (note that for $w \in W^P$, we have, $\dim X_P(w) = l(w)$).

### 6.2.6 Admissible pairs and canonical pairs

Let $(\tau, \varphi)$ be an admissible pair in $W^{P_n}$ (see Appendix A.2.4 for the definition of admissible pairs). We have ([56]) that $(\tau, \varphi)$ is an admissible pair with $X_{P_n}(\varphi)$ being a divisor in $X_{P_n}(\tau)$ if and only if there is a pair $i$, $(i+1)'$ present in $\varphi$ (as an element of $I_{n,2n}$ - also recall that for $1 \leq j \leq 2n$, $j' = 2n+1-j$), for some $i \leq n$, and $\tau$ is obtained from $\varphi$ by replacing $i, (i+1)'$ by $(i+1), i'$. Let us refer to this as a *elementary double move*. Hence, $(\tau, \varphi)$ is an admissible pair if and only if either $\tau = \varphi$(in which case it is a trivial admissible pair) or $\tau$ is obtained from $\varphi$ by a sequence of elementary double moves. From this we obtain the following

**Lemma 6.2.6.1** *Let $\tau, \varphi \in W^{P_n}$. Then $(\tau, \varphi)$ is a non-trivial admissible pair if and only if the following hold:*
*(i) $\tau > \varphi$ (as elements of $I_{n,2n}$)*
*(ii) $m_\tau > 0$*
*(iii) $m_\tau = m_\varphi$*
*where recall that for $\theta := (a_1, \cdots, a_{2n}) \in I_{n,2n}$, $m_\theta = \#\{i \leq n \mid a_i > n\}$.*

### 6.2.7 Canonical pairs

We now define an injection of $I_{n,2n} (= W^{Q_n})$ into $W^{P_n} \times W^{P_n}$. We first associate to $\mathbf{i} \in I_{n,2n}$, a pair $(\mathbf{i}(\alpha), \mathbf{i}(\beta))$ to be called *the canonical pair* as follows: This is defined similar to the canonical dual pair $(\mathbf{i}(A), \mathbf{i}(B))$ (cf. §6.2.1). In §6.2.1, we used the identification

$$(*) \qquad M_n(K) \cong \left\{ \begin{pmatrix} Id_n \\ JY \end{pmatrix}, Y \in M_n(K) \right\}$$

where $J$ is the anti-diagonal $(1, \cdots, 1)$ of size $n \times n$. If $\mathbf{i} \in I_{n,2n}$, then the restriction to $M_n$ of the Plücker co-ordinate $p_{\mathbf{i}}$ (under the identification (*)) evaluated at $Y$ corresponds to the minor of $Y$ with row and column indices given by $\mathbf{i}(A), \mathbf{i}(B)$ respectively. To define the canonical pair for an $\mathbf{i} \in I_{n,2n}$ we use the identification:

$$(**) \qquad M_n(K) \cong \left\{ \begin{pmatrix} Y \\ J \end{pmatrix}, Y \in M_n(K) \right\}$$

For $\mathbf{i} \in I_{n,2n}$, the pair $(\mathbf{i}(\alpha), \mathbf{i}(\beta))$ is defined to be the set of row and column indices of the minor of $Y$ which is the evaluation at $Y$ of the restriction to $M_n(K)$ of the Plücker co-ordinate $p_{\mathbf{i}}$ (under the identification (**)). Let $\mathbf{i} = (i_1, \cdots, i_n)$. Then $\mathbf{i}(\alpha), \mathbf{i}(\beta)$ have the following description:

Let $r$ be such that $i_r \leq n, i_{r+1} > n$. For $1 \leq j \leq 2n$, let $j' = 2n + 1 - j$. Then $\mathbf{i}(\alpha)$ is the $r$-tuple $(i_1, \cdots, i_r)$, while $\mathbf{i}(\beta)$ is the $r$-tuple which is the complement in $(1, \cdots, n)$ of $i'_{r+1}, i'_{r+2}, \cdots, i'_n$. Note that the canonical pair $(\mathbf{i}(\alpha), \mathbf{i}(\beta))$, and the canonical dual pair $(\mathbf{i}(A), \mathbf{i}(B))$ are related as follows:

$$\mathbf{i}(A) = \texttt{complement of } \mathbf{i}(\beta) \texttt{ in } (1, \cdots, n),$$
$$\mathbf{i}(B) = \texttt{complement of } \mathbf{i}(\alpha) \texttt{ in } (1, \cdots, n)$$

Further we have the map

$$\mathbf{i} \mapsto (\mathbf{i}(\alpha), \mathbf{i}(\beta))$$

is an order-preserving bijection between $I_{n,2n}$ and minors of $Y$; while the map

$$\mathbf{i} \mapsto (\mathbf{i}(A), \mathbf{i}(B))$$

is an order-reversing bijection between $I_{n,2n}$ and minors of $Y$ (the partial order among the minors of $Y$ being as in §6.2.3). In the above bijections, the $n$-tuple $(1, \cdots, n)$ will correspond to the constant function 1 (corresponding to the minor with the set of row (resp. column) indices being the empty set).

### 6.2.8 The inclusion $\eta : I_{n,2n} \hookrightarrow W^{P_n} \times W^{P_n}$

First observe that for $\tau := (a_1, \cdots, a_r, b_1, \cdots, b_s) \in W^{P_n}$, where $a_1 < \cdots < a_r \leq n < b_1 < \cdots < b_s$, we have, $\{b_1, \cdots, b_s\}$ is just the complement in $\{n + 1, n+2, \cdots, 2n\}$ of $\{a'_1, \cdots, a'_r\}$ (arranged in ascending order). Hence, $\tau$ is completely determined by $(a_1, \cdots, a_r)$. Denoting

$$\mathcal{I} := \bigcup_{1 \leq r \leq n} I_{r,n}$$

the map $\tau \mapsto (a_1, \cdots, a_r)$ defines a bijection

$$\upsilon : W^{P_n} \overset{\texttt{bij}}{\to} \mathcal{I}$$

Set

$$\mathcal{A} := \{(I, J) \in \bigcup_{r \leq n} I_{r,n} \times I_{r,n}, \mid I \geq J\}$$

Then the bijection $\upsilon$ induces a bijection:

$$\rho : \{\texttt{admissible pairs in } W^{P_n}\} \overset{\texttt{bij}}{\to} \mathcal{A}$$

(cf. Lemma 6.2.6.1; note that for $I = J$, we obtain $W^{P_n}$ (being identified with trivial admissible pairs)). We shall refer to the elements of $\mathcal{A}$ as *admissible elements* in $\mathcal{I} \times \mathcal{I}$. Using the bijection $\upsilon$, we define $\eta$ to be the injection

$$\eta : I_{n,2n} \hookrightarrow W^{P_n} \times W^{P_n}, \mathbf{i} \mapsto (\mathbf{i}(\alpha), \mathbf{i}(\beta))$$

Identifying $W^{P_n}$ with the diagonal set in $W^{P_n} \times W^{P_n}$, we have that $\eta(\mathbf{i}) \in W^{P_n}$ if and only if $\mathbf{i}(\alpha) = \mathbf{i}(\beta)$. Further, if $\mathbf{i}(\alpha) \neq \mathbf{i}(\beta)$, then $\eta(\mathbf{i})$ is an admissible pair if and only if the following hold:
(i) $\mathbf{i}(\alpha) \geq \mathbf{i}(\beta))$
(ii) $\#\mathbf{i}(\alpha)(=\#\mathbf{i}(\beta)) < n$.

We shall refer to an $\mathbf{i} \in I_{n,2n}$ such that $\eta(\mathbf{i}) \in \mathcal{A}$ as *an admissible element of* $I_{n,2n}$. Let us denote

$$\mathcal{A}_{n,2n} := \{\texttt{admissible elements in } I_{n,2n}\}$$

Then we have a bijection

$$\mathcal{A}_{n,2n} \overset{\texttt{bij}}{\to} \{\texttt{admissible pairs in } W^{P_n}\}$$

The comparison order $\geq$ among admissible pairs in $W^{P_n}$ (namely, $(\tau, \varphi) \geq (\tau', \varphi') \Leftrightarrow \varphi \geq \tau'$) induces a comparison order on $\mathcal{A}_{n,2n}$:

$$\mathbf{i} \geq \mathbf{j}, \;\; \texttt{if} \;\; \mathbf{i}(\beta) \geq \mathbf{j}(\alpha)$$

(here, an $s$-tuple $(a_1, \cdots, a_s) \geq$ an $t$-tuple $(b_1, \cdots, b_t)$, if $s \leq t$ and $a_r \geq b_r$, $1 \leq r \leq s$).

### 6.2.9 A standard monomial basis for $D_t(Sym\ M_n)$

Fix $1 \leq t \leq n$. For $A := (a_1, \cdots, a_s)$, $B := (b_1, \cdots, b_s)$, in $I_{s,n}$, $s \leq n$, let $p'(A, B)$ denote the restriction of $p(A, B)(\in K[M_n])$ to $Sym\ M_n$; if $\mathbf{i}$ is the element of $I_{n,2n}$ corresponding to $(A, B)$ under the bijection $\theta$ (cf. §6.2.3), we shall denote $p'(A, B)$ also by $p'(\mathbf{i})$.

**Definition 6.2.9.1** *A monomial $p'(\mathbf{i}_1) \cdots p'(\mathbf{i}_r)$, $\mathbf{i}_t \in \mathcal{A}_{n,2n}$ is standard if $\mathbf{i}_1 \geq \cdots \geq \mathbf{i}_r$, i.e., if $\mathbf{i}_1(\alpha) \geq \mathbf{i}_1(\beta) \geq \mathbf{i}_2(\alpha) \geq \cdots \geq \mathbf{i}_r(\beta)$.*
*Note that for any $t \leq r$, we have, $\mathbf{i}_t(\alpha) \geq \mathbf{i}_t(\beta)$ (since $\mathbf{i}_t \in \mathcal{A}_{n,2n}$).*

Using the identification

$$M_n \cong \left\{ \begin{pmatrix} Y \\ J \end{pmatrix}, Y \in M_n \right\}$$

and the isomorphism (cf. Theorem 6.2.5.2)

$$D_t(Sym\ M_n) \cong Y_G(\varphi)$$

we obtain

**Theorem 6.2.9.2** *$K[D_t(Sym\ M_n)]$ has a basis consisting of standard monomials of the form $p'(A_1, B_1) \cdots p'(A_r, B_r)$, $A_1 \geq B_1 \geq A_2 \geq \cdots \geq B_r$, $r \in \mathbb{N}$, $A_1 \leq (t, t+1, \cdots, n)$.*

*Proof.* We have following facts:

• $K[X_{P_n}(\varphi)]$ has a basis consisting of standard monomials on $X_{P_n}(\varphi)$ of the form $p(\tau_1, \delta_1)\cdots p(\tau_r, \delta_r), r \in \mathbb{N}, \varphi \geq \tau_1 \geq \delta_1 \geq \tau_2 \geq \cdots \geq \delta_r, (\tau_i, \delta_i)$ being admissible pairs (cf. Chapter 8, Theorem 8.2.0.7).

• $\mathbf{i} \mapsto (\mathbf{i}(\alpha), \mathbf{i}(\beta))$ is an order-preserving bijection between $I_{n,2n}$ and minors of (a generic matrix) $Y \in M_n$. Further, under this bijection, the admissible elements of $I_{n,2n}$ are mapped bijectively onto admissible pairs in $W^{P_n}$.

• Under the bijection $\upsilon : W^{P_n} \overset{\text{bij}}{\to} \mathcal{I}$, we have, $\upsilon(\varphi) = ([t, n], [2n + 2 - t, 2n])$ (cf. Theorem 6.2.5.2).

• The canonical pair corresponding to $\varphi(= ([t, n], [(t-1)', 1']))$ is $((t, t+1, \cdots, n), (t, t+1, \cdots, n))$.

These facts together with Theorem 6.2.5.2 identify $K[D_t(Sym\, M_n)]$ with $K[X_G(\varphi)]_{(p_{id})}$ (the homogeneous localization of $K[X_{P_n}(\varphi)]$ at $p_{id}$). Now, the fact that $p_{id} \leq p_{\tau,\delta}$ (for all admissible pairs $(\tau, \delta)$) implies that $p(\tau_1, \delta_1)\cdots p(\tau_r, \delta_r)$, $r \in \mathbb{N}$ is standard if and only if $\frac{p(\tau_1,\delta_1)}{p_{id}} \cdots \frac{p(\tau_r,\delta_r)}{p_{id}}$ is. Also, the fact that the standard monomials

$$\{p(\tau_1, \delta_1)\cdots p(\tau_r, \delta_r), \varphi \geq \tau_1, r \in \mathbb{N}\}$$

generate $K[X_{P_n}(\varphi)]$ implies that the standard monomials

$$\{\frac{p(\tau_1, \delta_1)}{p_{id}} \cdots \frac{p(\tau_r, \delta_r)}{p_{id}}, \varphi \geq \tau_1, r \in \mathbb{N}\}$$

generate $K[D_t(Sym\, M_n)](= K[X_G(\varphi)]_{(p_{id})})$; further, the linear independence of the standard monomials $\{p(\tau_1, \delta_1)\cdots p(\tau_r, \delta_r), \varphi \geq \tau_1, r \in \mathbb{N}\}$ clearly implies the linear independence of the standard monomials $\{\frac{p(\tau_1,\delta_1)}{p_{id}} \cdots \frac{p(\tau_r,\delta_r)}{p_{id}}, \varphi \geq \tau_1, r \in \mathbb{N}\}$. The result now follows from this.

### 6.2.10 De Concini-Procesi's basis for $D_t(Sym\, M_n)$

In arriving at Theorem 6.2.9.2, we have used the order-preserving bijection between $I_{n,2n}$ and minors of $Y \in M_n(K)$ given by the canonical pairs of elements of $I_{n,2n}$. If instead we use the order-reversing bijection between $I_{n,2n}$ and minors of $Y \in M_n(K)$ given by the canonical dual pairs of elements of $I_{n,2n}$, then we obtain De Concini-Procesi's basis for $D_t(Sym\, M_n)$ (cf. [22]) described below.

First, we note (or recall) the following:

(i) The canonical dual pair of $\mathbf{i} := ([t, n], [(t-1)', 1'])(= \varphi)$ is given by

$$\mathbf{i}(A) = (1, \cdots, t-1),\ \mathbf{i}(B) = (1, \cdots, t-1)$$

(ii) Under the identification

$$M_n \cong \left\{ \begin{pmatrix} Id_n \\ JY \end{pmatrix}, Y \in M_n \right\}$$

we have

$$p'_{\mathbf{i}}(Y) = p(\mathbf{i}(A), \mathbf{i}(B))(Y)$$

(iii) Defining $\eta'$ to be the injection

$$\eta' : I_{n,2n} \hookrightarrow W^{P_n} \times W^{P_n}, \mathbf{i} \mapsto (\mathbf{i}(A), \mathbf{i}(B))$$

we have that $\eta'(\mathbf{i}) \in W^{P_n}$ if and only if $\mathbf{i}(A) = \mathbf{i}(B)$. Further, if $i(A) \neq i(B)$, then $\eta'(\mathbf{i})$ is an admissible pair if and only if the following hold:
(1) $\mathbf{i}(A) \leq \mathbf{i}(B))$
(2) $\#\mathbf{i}(\alpha)(= \#\mathbf{i}(\beta)) < n$.

We shall refer to such $\mathbf{i}$'s as *dual admissible elements of* $I_{n,2n}$, and denote

$$\mathcal{A}'_{n,2n} := \{\texttt{dual admissible elements in } I_{n,2n}\}$$

(iv) The comparison order $\geq$ among admissible pairs in $W^{P_n}$ induces a comparison order on $\mathcal{A}'_{n,2n}$:

$$\mathbf{i} \leq \mathbf{j}, \ \texttt{if } \mathbf{i}(B) \leq \mathbf{j}(A)$$

**Definition 6.2.10.1** *A monomial $p'(\mathbf{i}_1)\cdots p'(\mathbf{i}_r), \mathbf{i}_t \in \mathcal{A}'_{n,2n}$, is standard if $\mathbf{i}_1 \leq \cdots \leq \mathbf{i}_r$, i.e., if $\mathbf{i}_1(A) \leq \mathbf{i}_1(B) \leq \mathbf{i}_2(A) \leq \cdots \leq \mathbf{i}_r(B)$.*
*Note that for any $t \leq r$, we have, $\mathbf{i}_t(A) \leq \mathbf{i}_t(B)$ (since $\mathbf{i}_t \in \mathcal{A}'_{n,2n}$).*

**Theorem 6.2.10.2** *$K[D_t(Sym\ M_n)]$ has a basis consisting of standard monomials of the form $p'(A_1, B_1)\cdots p'(A_r, B_r)$, $A_1 \leq B_1 \leq A_2 \leq \cdots \leq B_r, r \in \mathbb{N}, A_1 \geq (1, \cdots, t-1)$.*
*(Note that the condition that $(1, \cdots, t-1) \leq A \leq B$ amounts to the condition that $\#A(= \#B) \leq t-1$, i.e., $p'(A, B)$'s are minors of size $\leq t-1$.)*

*Proof.* Proof is similar to that of Theorem 6.2.9.2. As in the proof of Theorem 6.2.9.2, we have the following facts:

• $K[X_{P_n}(\varphi)]$ has a basis consisting of standard monomials on $X_{P_n}(\varphi)$ of the form $p(\tau_1, \delta_1)\cdots p(\tau_r, \delta_r), r \in \mathbb{N}, \varphi \geq \tau_1 \geq \delta_1 \geq \tau_2 \geq \cdots \geq \delta_r, (\tau_i, \delta_i)$ being admissible pairs.

• $\mathbf{i} \mapsto (\mathbf{i}(A), \mathbf{i}(B))$ is an order-reversing bijection between $I_{n,2n}$ and minors of (a generic matrix) $Y \in M_n$. Further, under this bijection, the dual admissible elements of $I_{n,2n}$ are mapped bijectively onto admissible pairs in $W^{P_n}$.

• Under the bijection $\upsilon : W^{P_n} \overset{\texttt{bij}}{\to} \mathcal{I}$, we have, $\upsilon(\varphi) = [t, n](= (t, t+1, \cdots, n))$ (cf. Theorem 6.2.5.2).

• The canonical dual pair corresponding to $\varphi(= ([t, n], [(t-1)', 1']))$ is $((1, \cdots, t-1), (1, \cdots, t-1))$.

These facts together with Theorem 6.2.5.2 imply the required result.

Taking $t = n+1$, we have $D_t(Sym\ M_n)$ equals $Sym\ M_n$, and we obtain

**Theorem 6.2.10.3** *$K[Sym\ M_n]$ has a basis consisting of standard monomials of the form $p'(A_1, B_1)\cdots p'(A_r, B_r)$, $A_1 \leq B_1 \leq A_2 \leq \cdots \leq B_r, r \in \mathbb{N}$.*

# 7

# Orthogonal Grassmannian

In this chapter we give an exposition of determinantal varieties in $Sk\, M_n$ (the space of skew symmetric $n \times n$ matrices) from the view-point of their relationship to Schubert varieties in the orthogonal Grassmannian varieties (cf. [67]). After reviewing some basic algebraic group-theoretic results on even special orthogonal groups, we establish the identification of $Sk\, M_n$ with the "opposite cell" in a suitable orthogonal Grassmannian, which then yields an identification of a determinantal variety in $Sk\, M_n$ with the "opposite cell" in a suitable Schubert variety in the orthogonal Grassmannian. Using the standard monomial basis for the homogeneous co-ordinate ring of a Schubert variety in the orthogonal Grassmannian, we obtain a standard monomial basis for the ring of regular functions on a determinantal variety in $Sk\, M_n$. This basis in fact coincides with that of DeConcini-Procesi (cf. [22]). Thus the Schubert-variety-theoretic approach gives a different proof of DeConcini-Procesi's basis for these varieties.

## 7.1 The even orthogonal group $SO(2n)$

Let $K$ be an algebraically closed field of characteristic not equal to 2. Let $V = K^{2n}$ together with a non-degenerate symmetric bilinear form $(\cdot, \cdot)$. Taking the matrix of the form $(\cdot, \cdot)$ (with respect to the standard basis $\{e_1, \ldots, e_{2n}\}$ of $V$) to be $E$, the anti-diagonal $(1, \ldots, 1)$ of size $2n \times 2n$. We may realize $G = SO(V)$ as the fixed point set $SL(V)^\sigma$, where $\sigma : SL(V) \to SL(V)$ is given by $\sigma(A) = E({}^tA)^{-1}E$. Set $H = SL(V)$.

Denoting by $T_H$ (resp. $B_H$) the maximal torus in $H$ consisting of diagonal matrices (resp. the Borel subgroup in $H$ consisting of upper triangular matrices) we see easily that $T_H$, $B_H$ are stable under $\sigma$. We set $T_G = T_H^\sigma$, $B_G = B_H^\sigma$. Then it follows that $T_G$ is a maximal torus in $G$ and $B_G$ is a Borel subgroup in $G$.

**Some basic facts on $SO(V)$:** Let $N_G$ (resp. $N_H$) denote the normalizer in $G$ (resp. $H$) of $T_G$ (resp. $T_H$). We have, $N_G \subset N_H$; further, $N_H$ is stable under $\sigma$, and we have

$$N_G = N_H^\sigma$$
$$N_G/T_G \hookrightarrow N_H/T_H$$

Thus we obtain

$$W_G \hookrightarrow W_H$$

where $W_G, W_H$ denote the Weyl groups of $G, H$ respectively (with respect to $T_G, T_H$ respectively). Further, $\sigma$ induces an involution on $W_H$:

$$w = (a_1, \cdots, a_{2n}) \in W_H, \sigma(w) = (c_1, \cdots, c_{2n}),\ c_i = 2n + 1 - a_{2n+1-i}$$

and

$$W_G = \{w \in W_H^\sigma \mid w \texttt{ is an even permutation}\}$$

Thus we obtain

$$W_G = \{(a_1 \cdots a_{2n}) \in S_{2n} \mid a_i = 2n + 1 - a_{2n+1-i},\ 1 \le i \le 2n,\ \textit{and } m_w \textit{ is even}\}.$$

where $m_w = \#\{i \le n \mid a_i > n\}$. Thus $w = (a_1 \cdots a_{2n}) \in W_G$ is known once $(a_1 \cdots a_n)$ is known.

**Root system of type $D_n$:** $\sigma$ induces an involution on $X(T_H)$, the character group of $T_H$:

$$\chi \in X(T_H),\ \sigma(\chi)(D) = \chi(\sigma(D)),\ D \in T_H$$

Let $\epsilon_i$, $1 \le i \le 2n$ be the character in $X(T_H)$, $\epsilon_i(D) = d_i$, the $i$-th entry in $D(\in T_H)$. We have

$$\sigma(\epsilon_i) = -\epsilon_{2n+1-i}$$

Now it is easily seen that the under the canonical surjective map

$$\varphi : X(T_H) \to X(T_G)$$

we have

$$\varphi(\epsilon_i) = -\varphi(\epsilon_{2n+1-i}),\ 1 \le i \le 2n$$

Let $R_H := \{\epsilon_i - \epsilon_j, 1 \le i, j \le 2n\}$, the root system of $G$ (relative to $T_H$), and $R_H^+ := \{\epsilon_i - \epsilon_j, 1 \le i < j \le 2n\}$, the set of positive roots (relative to $B_H$). We have the following:

1. $\sigma$ leaves $R_H$ (resp. $R_H^+$) stable.
2. For $\alpha, \beta \in R_H, \varphi(\alpha) = \varphi(\beta) \Leftrightarrow \alpha = \sigma(\beta)$
3. $\varphi$ is equivariant for the canonical action of $W_G$ on $X(T_H), X(T_G)$.
4. $R_H^\sigma = \{\pm(\epsilon_i - \epsilon_{2n+1-i}), 1 \le i \le n\}$

Let $R_G$ (resp. $R_G^+$) the set of roots of $G$ with respect to $T_G$ (resp. the set of positive roots with respect to $B_G$ ). Using the above facts and the explicit nature of the adjoint representation of $G$ on Lie $G$, we deduce that

$$R_G = \varphi(R_H \setminus R_H^\sigma),\ R_G^+ = \varphi(R_H^+ \setminus (R_H^+)^\sigma)$$

In particular, $R_G$ (resp. $R_G^+$) gets identified with the orbit space of $R_H$ (resp. $R_H^+$) modulo the action of $\sigma$ minus the fixed point set under $\sigma$. Thus we obtain the following identification:

$$R_G = \{\pm(\epsilon_i \pm \epsilon_j),\ 1 \leq i < j \leq n\}$$

$$R_G^+ = \{(\epsilon_i \pm \epsilon_j),\ 1 \leq i < j \leq n\}.$$

The simple roots in $R_G^+$ are given by

$$\{\alpha_i = \epsilon_i - \epsilon_{i+1},\ 1 \leq i \leq n-1\} \cup \{\alpha_n = \epsilon_{n-1} + \epsilon_n\}.$$

Let us denote the simple reflections in $W_G$ by $\{s_i,\ 1 \leq i \leq n\}$, namely, $s_i$= reflection with respect to $\epsilon_i - \epsilon_{i+1},\ 1 \leq i \leq n-1$, and $s_n$ = reflection with respect to $\epsilon_{n-1}+\epsilon_n$. Then we have (cf. [8])

$$s_i = \begin{cases} r_i r_{2n-i}, & \text{if } 1 \leq i \leq n-1 \\ r_{n-1} r_n r_{n-1} r_n r_{n+1} r_n, & \text{if } i = n \end{cases}$$

where $r_i$ denotes the transposition $(i, i+1)$ in $S_{2n},\ 1 \leq i \leq 2n-1$.

**Chevalley basis.** Recall, for $1 \leq k \leq 2n$, we set $k' = 2n + 1 - k$. The involution $\sigma : SL(2n) \to SL(2n), A \mapsto E({}^tA)^{-1}E$, induces an involution $\sigma : sl(2n) \to sl(2n), A \mapsto -E({}^tA)E$. In particular, we have, for $1 \leq i, j \leq 2n$, $\sigma(E_{ij}) = -E_{j'i'}$, where $E_{ij}$ is the elementary matrix with 1 at the $(i, j)$-th place and 0 elsewhere. Further,

$$\text{Lie } SO(2n) = \{A \in \text{sl}(2n) \mid E({}^tA)E = -A\}.$$

The Chevalley basis for Lie $SO(2n)$ may be given as follows:

$$\begin{aligned} H_{\epsilon_i - \epsilon_{i+1}} &= E_{ii} - E_{i+1,i+1} + E_{(i+1)',(i+1)'} - E_{i'i'} \\ H_{\epsilon_{n-1}+\epsilon_n} &= E_{n-1,n-1} + E_{n,n} - E_{n',n'} - E_{(n-1)',(n-1)'} \\ X_{\epsilon_j - \epsilon_k} &= E_{jk} - E_{k'j'} \\ X_{\epsilon_j + \epsilon_k} &= E_{jk'} - E_{kj'} \\ X_{-(\epsilon_j - \epsilon_k)} &= E_{kj} - E_{j'k'} \\ X_{-(\epsilon_j + \epsilon_k)} &= E_{k'j} - E_{j'k}. \end{aligned}$$

**Length Formula:** For $w \in W_G$, let us denote $l(w, W_H)$ (resp. $l(w, W_G)$), the length of $w$ as an element of $W_H$ (resp. $W_G$). For $w = (a_1, \cdots, a_{2n}) \in W_H$, denote

$$m_w := \#\{i \leq n \mid a_i > n\}$$

**Proposition 7.1.0.1** *For $w = (a_1, \cdots, a_{2n}) \in W_G$, we have $l(w, W_G) = \frac{1}{2}(l(w, W_H) - m_w)$.*

*Proof.* Set

$$S_H(w) = \{\beta \in R_H^+ \mid w(\beta) < 0\},\ S_G(w) = \{\beta \in R_G^+ \mid w(\beta) < 0\}$$

We have

$$l(w, W_H) = \#S_H(w),\ l(w, W_G) = \#S_G(w)$$

The canonical map $\varphi : X(T_H) \to X(T_G)$ induces a surjective map $\varphi : R_H \setminus R_H^\sigma \to R_G$. Further, we have,

$$\texttt{for}\ \alpha \in R_H, \alpha > 0 \Leftrightarrow \varphi(\alpha) > 0$$

Hence $\varphi$ induces a surjective map

$$S_H(w) \setminus S_H(w)^\sigma \longrightarrow S_G(w)$$

(note that $\sigma$ leaves $S_H(w)$ stable), and

$$S_H(w)^\sigma = \{\alpha \mid \alpha = \epsilon_i - \epsilon_{2n+1-i}, 1 \le i \le n,\ w(\alpha) < 0\ \texttt{in}\ R_H\}$$

If $\alpha = \epsilon_i - \epsilon_{2n+1-i}$, then $w(\alpha) = \epsilon_{a_i} - \epsilon_{a_{2n+1-i}}$. Hence

$$w(\alpha) < 0\ \texttt{in}\ R_H \Leftrightarrow a_{2n+1-i} < a_i,\ i \le n$$

But now $a_{2n+1-i} = 2n + 1 - a_i$ (since $w \in W_G$). Hence

$$w(\alpha) < 0\ \texttt{in}\ R_H \Leftrightarrow n < a_i, 1 \le i \le n$$

From this, we obtain

$$\#S_H(w)^\sigma = m(w)$$

Also, we have,

$$\#S_H(w) = 2\#S_G(w) + \#S_H(w)^\sigma$$

The required result now follows from this.

### 7.1.1 Schubert varieties in $G/B_G$

For $w \in W_H$, let $C_H(w)$ be the *Schubert cell* $B_H w B_H (mod\ B_H)$ in $H/B_H$; if $w \in W_G$, we shall denote by $C_G(w)$ the Schubert cell $B_G w B_G (mod\ B_G)$ in $G/B_G$. We observe that $\sigma$ induces a natural involution on $H/B_H$.

**Proposition 7.1.1.1** *Let $w \in W_G$. The Schubert cell $C_H(w)$ is stable under $\sigma$, and $C_H(w)^\sigma = C_G(w),\ w \in W_G$.*

*Proof.* The first part is clear. Let $B_H^u$ denote the unipotent part of $B_H$. Clearly, $B_H^u$ is $\sigma$-stable (since $B_H$ is $\sigma$-stable). Let $B_1$ be the isotropy subgroup of $B_H^u$ at the point

$wH(\in C_H(w))$; then $B_1$ is stable under $\sigma$ (since $\sigma(w)=w$). We have

$$(*)\qquad B_H^u=\prod_{\alpha\in R_H^+}U_\alpha$$

(here, $U_\alpha$ denotes the root subgroup of $H$ associated to $\alpha$). We see easily

$$B_1=\prod_{\{\alpha\in R_H^+,\ w^{-1}(\alpha)>0\}}U_\alpha$$

Let

$$B_2=\prod_{\{\alpha\in R_H^+,\ w^{-1}(\alpha)<0\}}U_\alpha$$

Then (*) together with the facts that $B_H^u$, $B_1$ are $\sigma$-stable implies that $B_2$ is $\sigma$-stable. Let $x\in C_H(w)$; then $x$ has a presentation $x=bw$ for a unique $b\in B_2$. Hence $x\in C_H(w)^\sigma\Leftrightarrow\sigma(b)=b\Leftrightarrow b\in B_G$; and therefore $x\in C_H(w)^\sigma\Leftrightarrow x\in C_G(w)$.

**Maximal parabolics.** For $1\le d\le n$, we let $P_d$ be the maximal parabolic subgroup of $G$ with $S\setminus\{\alpha_d\}$ as the associated set of simple roots. Then it can be seen easily that $W_G^{P_d}$, $d\ne n-1$, can be identified with

$$\left\{(a_1\cdots a_d)\ \middle|\ \begin{array}{l}(1)1\le a_1<a_2<\cdots<a_d\le 2n\\ (2)\text{for }\ 1\le i\le 2n,\ \text{ if }\ i\in\{a_1,...,a_d\}\\ \qquad\text{then }\ 2n+1-i\notin\{a_1,...,a_d\}\end{array}\right\}.$$

For $d=n-1$, if $w\in W_G^{P_d}$, then

$$w\equiv wu_i\ (\mathrm{mod}\ W_{P_{n-1}}),\ 0\le i\le n, i\ne n-1,$$

where

$$u_i=\begin{cases}s_{\alpha_n}, & \text{if } i=n\\ id, & \text{if i=0}\\ s_{\alpha_i}s_{\alpha_{i+1}}\cdots s_{\alpha_{n-2}}s_{\alpha_n}, & \text{if } 1\le i\le n-2.\end{cases}$$

Note that the set $\{wu_i,\ 0\le i\le n, i\ne n-1\}$ is totally ordered under the Bruhat order; note also that given $w\in W$, there are $n$ different $n-1$-tuples representing the coset $wW_{P_d}$, namely, the tuples given respectively by the first $n-1$ entries in $wu_i$, $0\le i\le n, i\ne n-1$. For $w\in W$, denoting by $w^{(d)}$ the element of $W_G^{P_d}$ representing $wW_{P_d}$, we can have $w_1^{(n-1)}=w_2^{(n-1)}$, with $\{a_1,\dots,a_{n-1}\}\uparrow$ and $\{b_1,\dots,b_{n-1}\}\uparrow$ being different (here, $\{i_1,\dots,i_s\}\uparrow$ denotes the entries $i_1,\dots,i_s$ arranged in ascending order). For $w\in W$, say $w=(a_1\cdots a_{2n})$, we see easily that

$$w^{(d)}=\{a_1,\dots,a_d\}\uparrow,\ 1\le d\le n,\ d\ne n-1$$

and

$$w^{(n-1)}=\text{the least (under }\ge\text{) in the totally ordered set } Y$$

where

$$Y = \{(y_1^{(i)}, \dots, y_{n-1}^{(i)}) \uparrow 0 \le i \le n,\ i \ne n-1\}.$$

$y_1^{(i)}, \dots, y_{n-1}^{(i)}$ being the first $(n-1)$ entries in $wu_i$, $0 \le i \le n$, $i \ne n-1$. (Here, the partial order $\ge$ is the usual partial order, namely, $(i_1, \dots, i_{n-1}) \ge (j_1, \dots, j_{n-1})$, if $i_t \ge j_t$, $1 \le t \le n-1$, where $(i_1, \dots, i_{n-1})$, $(j_1, \dots, j_{n-1})$ are two increasing sequences of $(n-1)$-tuples.)

**Bruhat–Chevalley order.** For $1 \le i \le 2n$, let $i' = 2n+1-i$, and $|i| = \min\{i, i'\}$. We shall denote the Bruhat–Chevalley order on $W_G$ by $\succeq$. Given $w_1 = (a_1 \cdots a_{2n})$, $w_2 = (b_1 \cdots b_{2n})$, $w_1, w_2 \in W_G$, we have $w_2 \succeq w_1$ if and only if the following two conditions hold [102].

1. For $1 \le d \le n$, we have $\{b_1, \dots, b_d\} \uparrow \ge \{a_1, \dots, a_d\} \uparrow$, for all $d$.
2. Let $(c_1, \dots, c_d)$ (resp. $(e_1, \dots, e_d)$) be the increasing sequence $\{a_1, \dots, a_d\} \uparrow$ (resp. $\{b_1, \dots, b_d\} \uparrow$). Suppose for some $r$, $1 \le r \le d$, and some $i$, $0 \le i \le d-r$, $\{|c_{i+1}|, \dots, |c_{i+r}|\} = \{|e_{i+1}|, \dots, |e_{i+r}|\} = \{n+1-r, \dots, n\}$ as sets (order doesn't matter). Then $\#\{j : i+1 \le j \le i+r \text{ and } c_j > n\}$, and $\#\{j : i+1 \le j \le i+r \text{ and } e_j > n\}$ should both be even or both be odd.

**Remark 7.1.1.2.** Thus the Bruhat order $\succeq$ on $W_G$ is *not* induced from the Bruhat order on $W_H$.

Following the terminology in [102], we shall refer to $\{|c_{i+1}|, \dots, |c_{i+r}|\}$ and $\{|e_{i+1}|, \dots, |e_{i+r}|\}$ as *analogous parts* if they satisfy the hypothesis in Condition 2 above, and if $\{c_1, \dots, c_d\}$ and $\{e_1, \dots, e_d\}$ have analogous parts of the same parity, we say they are **D**-compatible.

**Remark 7.1.1.3.** (a) Let $(c_1, \dots, c_d)$, $(e_1, \dots, e_d) \in W_G^{P_d}$, where $(c_1, \dots, c_d) \succeq \{e_1, \dots, e_d)$. Suppose $(c_1, \dots, c_d)$, $(e_1, \dots, e_d)$ have analogous parts. Then it is easily seen that the condition (2) is equivalent to the condition that $\#\{j,\ 1 \le j \le d \mid c_j > n\}$ and $\#\{j,\ 1 \le j \le d \text{ and } e_j > n\}$ are both even or both odd.

As a consequence, we have the following

**Proposition 7.1.1.4** *The Bruhat-Chevalley order $\succeq$ on $W^{P_n}$ is induced by the Bruhat-Chevalley order $\succeq$ on $W^{Q_n}$, $P_n$ (resp. $Q_n$) being the maximal parabolic subgroup of $G$ (resp. $H = SL_{2n}(K)$) associated to the simple root $\epsilon_{n-1} + \epsilon_n$ (resp. $\epsilon_n - \epsilon_{n+1}$).*

*Proof.* Let $w_1, w_2 \in W_G^{P_n}$, and let $w_1 \ge w_2$ (as elements of $W_H{}^{Q_n}$). Let

$$w_1 = (c_1, \dots, c_n),\ w_2 = (e_1, \dots, e_n)$$

Then we have

$$(c_1, \dots, c_n) \ge (e_1, \dots, e_n)$$

(since $w_1 \ge w_2$ in $W_H{}^{Q_n}$). Condition (2) above is trivially satisfied (cf. Remark 7.1.1.3). The result now follows.

## 7.2 The variety $G/P_n$

As above, let $Q_n$ denote the maximal parabolic subgroup of $H$ associated to the simple root $\epsilon_n - \epsilon_{n+1}$, and $P_n$ denote the maximal parabolic subgroup of $G$ associated to the simple root $\epsilon_{n-1} + \epsilon_n$. It is easily seen that $Q_n$ is $\sigma$-stable so that $\sigma$ induces an involution on $H/Q_n$. Let $Z$ be the subgroup of $H$ consisting of matrices of the form

$$\begin{pmatrix} Id_n & 0 \\ Y & Id_n \end{pmatrix}, \; Y \in M_n(K)$$

(here, $M_n(K)$ denotes the $n \times n$ matrices with entries in $K$). The canonical morphism $H \to H/Q_n$ induces a morphism

$$\psi_H : Z \to H/Q_n$$

Recall (cf. Chapter 5)

**Fact:** $\psi_H$ is an open immersion, and $\psi_H(Z)$ gets identified with the opposite big cell $O_H^-$ in $H/Q_n$.

**Lemma 7.2.0.1** *$O_H^-$ is $\sigma$- stable, and we have an identification $Sk\, M_n \cong (O_H^-)^\sigma$, where $Sk\, M_n$ is the space of skew symmetric $n \times n$ matrices.*

*Proof.* For $z = \begin{pmatrix} Id_n & 0 \\ Y & Id_n \end{pmatrix} \in Z$, we have,

$$\sigma(z) = \begin{pmatrix} Id_n & 0 \\ -J\,{}^tYJ & Id_n \end{pmatrix}$$

(where recall that $J$ is the anti-diagonal $(1, \cdots, 1)$ of size $n \times n$). Hence, $Z$ is $\sigma$-stable, and the first part of the Lemma follows (in view of Fact above).
We have,

$$(\dagger) \qquad Z^\sigma = \{z \in Z \mid -J\,{}^tYJ = Y\}$$

Letting $Y = JX$, we have

$$-J\,{}^tYJ = Y \Leftrightarrow {}^tX = -X$$

The second part of Lemma follows from this.

Consider $H/Q_n$ as the Grassmannian of $n$-dimensional subspaces of $V := K^{2n}$. It is well-known that the set of maximal totally isotropic subspaces (for the symmetric form ( , ) on $V$) is a closed subvariety of $H/Q_n$ isomorphic to $G/P_n$, *the orthogonal Grassmannian.*

**Lemma 7.2.0.2** *$\psi_H$ induces an inclusion $\psi_H : Z^\sigma \hookrightarrow G/P_n$*

*Proof.* Now $z = \begin{pmatrix} Id_n & 0 \\ Y & Id_n \end{pmatrix}$ considered as a point of $H/Q_n$ corresponds to the $n$-dimensional subspace spanned by the columns of $\begin{pmatrix} Id_n \\ Y \end{pmatrix}$. Let us denote this $n$-dimensional subspace by $U_z$. We have,

$$-J\,{}^tYJ = Y \Leftrightarrow \left(Id_n\ {}^tY\right)\begin{pmatrix} 0 & J \\ J & 0 \end{pmatrix}\begin{pmatrix} Id_n \\ Y \end{pmatrix} = 0$$

Hence we obtain in view of (†) in the proof of Lemma 7.2.0.1 that $z \in Z^\sigma \Leftrightarrow U_z$ is a maximal totally isotropic subspace. The required result follows from this.

Let us denote the restriction of the morphism $\psi_H : Z \to H/Q_n, Y \mapsto \begin{pmatrix} Id_n & 0 \\ JY & Id_n \end{pmatrix}$ to $Sk\,M_n$ by $\psi_G$. Let $O_G^-$ the opposite big cell in $G/P_n$.

### 7.2.1 Identification of $Sk\,M_n$ with $O_G^-$

Let $L_n$ denote the tautological line bundle on $H/Q_n(= G_{n,2n})$ for the Plücker embedding. Then $H^0(H/Q_n, L_n)$ is the dual Weyl $H$-module with highest weight $\omega_n$. Recall (cf. Chapter 5)

**Proposition 7.2.1.1** *Let $Id = (1, \cdots, n)(\in I_{n,2n})$; denote $f := p_{Id}$. We have*

1. *$f$ is a lowest weight vector in $H^0(H/Q_n, L_n)$.*
2. *$O_H^-$ is the principal open set $(H/Q_n)_f$.*

### 7.2.2 Canonical dual pair

Let $Y \in M_n$. Given two $s$-tuples $A := (a_1, \cdots, a_s)$, $B := (b_1, \cdots, b_s)$, in $I_{s,n}$, $s \leq n$, let us denote by $p(A, B)(Y)$, the $s$-minor of $Y$ with row and column indices given by $A, B$ respectively. Consider the identification

$$(*) \qquad M_n(K) \cong \left\{ \begin{pmatrix} Id_n \\ JY \end{pmatrix}, Y \in M_n(K) \right\} \cong O_H^-$$

where $J$ is as above. Let $\mathbf{i} := (i_1, \cdots, i_n) \in I_{n,2n}$, and let $p_{\mathbf{i}}$ be the associated Plücker co-ordinate on $G_{n,2n}(= H/Q_n)$. Let $f_{\mathbf{i}}$ denote the restriction of $p_{\mathbf{i}}$ to $M_n$ (under the identification (*)). Then $f_{\mathbf{i}}(Y) = p(\mathbf{i}(A), \mathbf{i}(B))$, where $\mathbf{i}(A), \mathbf{i}(B)$ are given as follows:

Let $r$ be such that $i_r \leq n, i_{r+1} > n$; let $s = n - r$. Then $\mathbf{i}(A)$ is the $s$-tuple given by $(2n+1-i_n, \cdots, 2n+1-i_{r+1})$, while $\mathbf{i}(B)$ is the $s$-tuple given by the complement of $(i_1, \cdots, i_r)$ in $(1, \cdots, n)$. Recall (cf. Chapter 6) that $(\mathbf{i}(A), \mathbf{i}(B))$ is the canonical dual pair associated to $\mathbf{i}$.

### 7.2.3 The bijection $\theta$

Define a partial order $\geq$ on the set of minors of $Y \in M_n$ as follows: Given $A, B, A', B'$ where $A, B \in I_{s,n}$, and $A', B' \in I_{s',n}$, $p(A, B) > p(A', B')$ if $s \leq s'$, and $a_t \geq a'_t, b_t \geq b'_t, 1 \leq t \leq s$. Then the map $\mathbf{i} \mapsto (\mathbf{i}(A), \mathbf{i}(B))$ defines an order-reversing bijection $\theta$ between $I_{n,2n}$ and minors of $M_n$. In the above bijection, the $n$-tuple $(1, \cdots, n)$ will correspond to the constant function 1 (corresponding to the minor with the set of row (resp. column) indices being the empty set).

### 7.2.4 The dual Weyl $G$-module with highest weight $\omega_n$

Let $L'_n$ be the restriction of $L_n$ to $G/P_n$. For $\mathbf{i} \in I_{n,2n}$, let $p'_\mathbf{i}$ be the restriction of $p_\mathbf{i}$ to $G/P_n$; note that $p'_\mathbf{i} \in H^0(G/P_n, L'_n)$. Let $p(\mathbf{i}(A), \mathbf{i}(B))$ be the restriction of $p_\mathbf{i}$ to $M_n$, $M_n$ being identified with the opposite cell $O_H^-$ as above (cf. (*)); let $p'(\mathbf{i}(A), \mathbf{i}(B))$ be the restriction of $p(\mathbf{i}(A), \mathbf{i}(B))$ to $Sk\, M_n$. Let $f = p_{Id}$, a lowest weight vector in $H^0(H/Q_n, L_n)$, and $f'$ the restriction of $f$ to $G/P_n$.

Let $Y \in Sk\, M_n$. Consider $\mathbf{i} \in I_{n,2n}$ such that $\mathbf{i}(A) = \mathbf{i}(B)$. Then $p'(\mathbf{i}(A), \mathbf{i}(A))(Y)$ is a principal minor of the skew symmetric matrix $Y$. Hence $p'(\mathbf{i}(A), \mathbf{i}(A))(Y)$ is a square; denoting $q_\mathbf{i}(Y)$, the corresponding Pfaffian, we obtain a regular function

$$q_\mathbf{i} : Sk\, M_n \to K$$

Let $r = \#\mathbf{i}(A)$; then we have that $q_\mathbf{i}$ is non-zero if and only if $r$ is even (since the determinant of a skew symmetric $r \times r$ matrix is zero, if $r$ is odd).

**Proposition 7.2.4.1** *Let notation be as above. We have*

1. *For $\mathbf{i} \in I_{n,2n}$ such that $\mathbf{i}(A) = \mathbf{i}(B)$, we have a regular function $q_\mathbf{i} : Sk\, M_n \to K$ such that $q_\mathbf{i}^2 = p'(\mathbf{i}(A), \mathbf{i}(A))$; further, $q_\mathbf{i}$ is non-zero if and only if $\#\mathbf{i}(A)$ is even.*
2. *Let $\mathcal{L}$ be the ample generator of Pic $G/P_n$. Then $\mathcal{L}^2 = L'_n$.*

*Proof.* (1) follows from the discussion above.
(2). Let $f$ be a lowest weight vector in $H^0(H/Q_n, L_n)$; then the one-dimensional span $Kf$ is $B_H^-$-stable, where $B_H^-$ is the Borel subgroup of $H$ opposite to $B_H$. Also, one sees easily that $B_H^-$ is $\sigma$-stable, and $(B_H^-)^\sigma (= G \cap B_H^-)$ is precisely $B_G^-$, the Borel subgroup of $G$ opposite to $B_G$. This fact together with the fact that $Kf$ is stabilized by $B_G^-$ implies that $f'$ is a lowest weight vector in $H^0(G/P_n, L'_n)$; further,

$$\texttt{weight of } f' (= \texttt{weight of } f) = -(\epsilon_1 + \cdots + \epsilon_n) = -2\omega_n$$

(note that $\omega_n = \frac{1}{2}(\epsilon_1 + \cdots + \epsilon_n)$). Hence we obtain that a highest weight vector in $H^0(G/P_n, L'_n)$ has weight $w_0(G)(-2\omega_n)$ (here, $w_0(G)$ denotes the element of largest length in $W_G$). On the other hand, $H^0(G/P, \mathcal{L})$ is the dual Weyl module with highest weight $i(\omega_n)$(note that $i = -w_0(G)$, the Weyl involution). These two facts together with (1) imply (2).

As an immediate consequence we obtain

**Theorem 7.2.4.2** *We have an identification* $Z^\sigma \cong O_G^-$.

*Proof.* Under the identification $\psi : Z \cong O_H^-$, we have that $Z \cong (H/Q_n)_f$; thus $Z$ is precisely the set of points in $H/Q_n$ where the lowest weight vector $f (\in H^0(H/Q_n, L_n))$ does not vanish. Hence we obtain that

$$(G/P_n)_{f'} = Z \cap G/P_n$$

Let $z \in Z$, say $z = \begin{pmatrix} Id \\ JY \end{pmatrix}$. Then $z \in G/P_n$ if and only if the $n$-dimensional subspace spanned by the columns of $\begin{pmatrix} Id \\ JY \end{pmatrix}$ is totally isotropic, i.e, if and only if

$$\begin{pmatrix} Id_n & {}^tY \end{pmatrix} \begin{pmatrix} 0 & J \\ J & 0 \end{pmatrix} \begin{pmatrix} Id_n \\ Y \end{pmatrix} = 0$$

i.e, if and only if

$$-J\,{}^tYJ = Y$$

Hence, letting $Y = JX$, we obtain that

$$O_G^-(= (G/P_n)_{f'}) = Z \cap G/P_n = Sk\,M_n = Z^\sigma$$

This implies the required result.

**Corollary 7.2.4.3** $O_G^- = (O_H^-)^\sigma = Sk\,M_n$
*(Thus the opposite cell in $H/Q_n$ induces the opposite cell in $G/P_n$.)*

*Proof.* This follows from the above Theorem, the identification $Z \cong O_H^-$, and Lemma 7.2.0.1.

**Corollary 7.2.4.4** *dim* $G/P_n = \frac{1}{2}n(n-1)$.

### 7.2.5 Identification of $D_t(Sk\,M_n)$ with $Y_G(\varphi)$

Fix an odd integer $t$ where $1 \leq t \leq n$. Let $D_t(Sk\,M_n)$ denote the subscheme of $Sk\,M_n$ defined by the vanishing of $t$ minors in $Sk\,M_n$. In this subsection, we shall describe an identification of $D_t(Sk\,M_n)$ with the opposite cell $Y_{P_n}(\varphi)$ in a suitable Schubert variety $X_{P_n}(\varphi)$ in $G/P_n$. Unlike the symplectic group case where we have that for $w \in W_{Sp(2n)}$, $X_{Sp(2n)}(w) = X_H(w) \cap Sp(2n)/B$, scheme-theoretically (cf. Chapter 6, Proposition 6.1.1.2), for $G = SO(2n)$, the Schubert variety $X_G(w)$, $w \in W_G$ need not equal $X_H(w) \cap G/B_G$; in fact, the intersection $X_H(w) \cap G/B_G$ need not even be irreducible (for details, see [65]). Nevertheless it is true that for $w \in W^{P_n}$,

$$X_{P_n}(w) = X_{Q_n}(w) \cap G/P_n \ \texttt{(scheme-theoretically)}$$

as given by the following

**Proposition 7.2.5.1** *Let $w \in W^{P_n}$. Let $X_{P_n}(w)$ (resp. $X_{Q_n}(w)$) be the associated Schubert variety in $G/P_n$ (resp. $H/Q_n$). Then under the canonical inclusion $G/P_n \subset H/Q_n$, we have, $X_{P_n}(w) = X_{Q_n}(w) \cap G/P_n$ (scheme-theoretically).*

*Proof.* Denote $Z := X_{Q_n}(w) \cap G/P_n$. We have $X_{P_n}(w) \subseteq X_{Q_n}(w) \cap G/P_n$ (clearly), and is in fact an irreducible component of $Z$ (in view of Proposition 7.1.1.1). Let $Z'$ be another irreducible component of $Z$. Then $Z' = X_{P_n}(w')$, for some $w' \in W^{P_n}$ ($Z$ being closed and $B_G$-stable in $G/P_n$, is a union of Schubert varieties in $G/P_n$). Now the inclusion $X_{P_n}(w') \subset Z$ implies in particular that $C_{P_n}(w')(= C_{Q_n}(w')^{\sigma})$ is contained in $Z$ (here, $C_{P_n}(w')$ (resp. $C_{Q_n}(w')$) is the Schubert cell in $G/P_n$ (resp. $H/Q_n$) associated to $w'$). Hence we obtain that $C_{Q_n}(w')^{\sigma} \subseteq X_{Q_n}(w) \cap C_{Q_n}(w')$; this implies in particular that $C_{Q_n}(w') \subseteq X_{Q_n}(w)$ (since $X_{Q_n}(w) \cap C_{Q_n}(w')$ is non-empty, and $X_{Q_n}$ is $B_H$-stable), and hence we obtain that $w' \leq w$ in $W^{Q_n}$, and hence also in $W^{P_n}$ also (in view of the fact that the partial order on $W^{P_n}$ is induced by the partial order on $W^{Q_n}$ (cf. Proposition 7.1.1.4)). Thus we get that $Z'(= X_{P_n}(w')) \subset X_{P_n}(w)$. Hence we obtain that

$$(*) \qquad X_{P_n}(w) = X_{Q_n}(w) \cap G/P_n \ (\texttt{set-theoretically})$$

The assertion that (*) is in fact a scheme-theoretic equality follows from the fact that for any $\tau \in W^{P_n}, \tau \leq w$, the tangent space to $X_{Q_n}(w)$ at the $T$-fixed point $\tau Q_n$ is $\sigma$-stable, and its subspace of $\sigma$-fixed points is the tangent space to $X_{P_n}(w)$ at the $T$-fixed point $\tau P_n$ (cf. Chapter 13, Theorem 13.2.1.8).

Fix an odd integer $t \leq n$. Let $\varphi \in W^{P_n}$ be defined by

$$\varphi = (t, t+1, \cdots, n, 2n+2-t, 2n+3-t, \cdots, 2n)$$

(note that $\varphi$ consists of two blocks $[t, n]$, $[2n+2-t, 2n]$ of consecutive integers - here, for $i < j$, $[i, j]$ denotes the set $\{i, i+1, \cdots, j\}$; also note that $m_{\varphi}(= t-1)$ is even). Let $Y_{P_n}(\varphi) := X_{P_n}(\varphi) \cap O_G^-$, the *opposite cell in $X_G(\varphi)$*.

**Theorem 7.2.5.2** *The isomorphism $Sk\, M_n \cong O_G^-$ induces an isomorphism*

$$D_t(Sk\, M_n) \cong Y_{P_n}(\varphi)$$

*Proof.* We have scheme-theoretic identifications

$$D_t(Sk\, M_n) \cong D_t(M_n) \cap Sk\, M_n$$
$$Sk\, M_n \cong O_G^-$$
$$Y_{P_n}(\varphi) \cong Y_{Q_n}(\varphi) \cap O_G^-$$

On the other hand, we have an isomorphism (cf. Chapter 5)

$$D_t(M_n) \cong Y_{Q_n}(\varphi)$$

The required result now follows.

**Corollary 7.2.5.3** *$D_t(Sk\, M_n)$ is normal, Cohen-Macaulay of dimension $n(t-1)-\frac{1}{2}t(t-1)$.*

*Proof.* *The normality and Cohen-Macaulayness follow since Schubert varieties in $G/P_n$ are normal and Cohen-Macaulay (cf. §A.12 of Appendix). The assertion on the dimension follows in view of Proposition 7.1.0.1 (note that for $w \in W^P$, we have, dim $X_P(w) = l(w)$).*

**The bijection $\eta$:**
First observe that for $\tau := (a_1, \cdots, a_r, b_1, \cdots, b_s) \in W^{P_n}$, where $a_1 < \cdots < a_r \leq n < b_1 < \cdots < b_s$ ($s$ being even), we have, $\{b_1, \cdots, b_s\}$ is just the complement in $\{n+1, n+2, \cdots, 2n\}$ of $\{a'_1, \cdots, a'_r\}$ (arranged in ascending order). Hence, $\tau$ is completely determined by $(a_1, \cdots, a_r)$. Denoting

$$\mathcal{I}_{\mathtt{even}} := \bigcup_{\{r|n-r \ \mathtt{is\ even}\}} I_{r,n}$$

the map $\tau \mapsto (a_1, \cdots, a_r)$ defines a bijection

$$\upsilon : W^{P_n} \stackrel{\mathtt{bij}}{\to} \mathcal{I}_{\mathtt{even}}$$

Set

$$\mathcal{A} := \{\mathtt{the\ diagonal\ of}\ \mathcal{I}_{\mathtt{even}} \times \mathcal{I}_{\mathtt{even}}\}$$

Then we have an obvious bijection:

$$\rho : W^{P_n} \stackrel{\mathtt{bij}}{\to} \mathcal{A}$$

We shall refer to the elements of $\mathcal{A}$ as *admissible elements* in $\mathcal{I}_{\mathtt{even}} \times \mathcal{I}_{\mathtt{even}}$. Set

$$\mathcal{A}_{n,2n} := \left\{ \mathbf{i} \in I_{n,2n} \;\middle|\; \begin{array}{c} \mathbf{i}(A) = \mathbf{i}(B) \\ \#\mathbf{i}(A)\ \mathtt{is\ even} \end{array} \right\}$$

(here, recall that $(\mathbf{i}(A), \mathbf{i}(B))$ is the dual canonical pair associated to $\mathbf{i}$). Using the bijection $\upsilon$, we define $\eta$ to be the bijection

$$\eta : \mathcal{A}_{n,2n} \stackrel{\mathtt{bij}}{\to} \mathcal{A},\ \mathbf{i} \mapsto (\mathbf{i}(A), \mathbf{i}(A))$$

This defines an order-reversing bijection

$$\mathcal{A}_{n,2n} \stackrel{\mathtt{bij}}{\to} W^{P_n}$$

### 7.2.6 A standard monomial basis for $D_t(Sk\, M_n)$

Fix an odd integer $t$, $1 \leq t \leq n$. For $A := (a_1, \cdots, a_s)$, $s$ being even, let $q(A)$ denote Pfaffian of the restriction of $p(A, A)(\in K[M_n])$ to $Sk\, M_n$; if $\mathbf{i}$ is the element of $I_{n,2n}$ with $(A, A)$ as the associated dual canonical pair, then we shall denote $q(A)$ also by $q(\mathbf{i})$.

**Definition 7.2.6.1** *A monomial* $q(\mathbf{i}_1)\cdots q(\mathbf{i}_r)$, $\mathbf{i}_t \in \mathcal{A}_{n,2n}$ *is standard if* $\mathbf{i}_1 \geq \cdots \geq \mathbf{i}_r$, *i.e., if* $\mathbf{i}_1(A) \leq \mathbf{i}_2(A) \leq \cdots \leq \mathbf{i}_r(A)$.

**Theorem 7.2.6.2** *Let notation be as above.* $K[D_t(Sk\, M_n)]$ *has a basis consisting of standard monomials of the form* $q(A_1)\cdots q(A_r)$, $A_1 \leq \cdots \leq A_r$, $r \in \mathbb{N}$, $A_1 \geq (1, \cdots, t-1)$, $A_i \in \mathcal{A}_{n,2n}$.

*(Note that the condition that* $(1, \cdots, t-1) \leq A$ *amounts to the condition that* $\#A \leq t-1$, *i.e.,* $q(A)$*'s are Pfaffians of principal minors of size* $\leq t-1$.*)*

*Proof.* We have the following facts:

• $K[X_{P_n}(\varphi)]$ has a basis consisting of standard monomials on $X_{P_n}(\varphi)$ of the form $p(\tau_1)\cdots p(\tau_r)$, $r \in \mathbb{N}$, $\varphi \geq \tau_1 \geq \cdots \geq \tau_r$, $\tau_i \in W^{P_n}$ (cf. Chapter 8, Theorem 8.1.0.2; note that $\omega_n$ is minuscule).

• $\mathbf{i} \mapsto (\mathbf{i}(A), \mathbf{i}(A))$ is an order-reversing bijection between $\mathcal{A}_{n,2n}$ and Pfaffians of even-sized principal minors of (a generic matrix) $Y \in Sk\, M_n$.

• We have a bijection, $\mathcal{A}_{n,2n} \overset{\mathtt{bij}}{\to} W^{P_n}$.

• Under the bijection $\upsilon : W^{P_n} \overset{\mathtt{bij}}{\to} \mathcal{I}_{\mathtt{even}}$, we have, $\upsilon(\varphi) = [t, n](=(t, t+1, \cdots, n))$ (cf. Theorem 6.2.5.2).

• The canonical dual pair corresponding to $\varphi(=([t, n], [(t-1)', 1']))$ is $((1, \cdots, t-1), (1, \cdots, t-1))$.

These facts together with Theorem 7.2.5.2 imply the required result.

**Remark 7.2.6.3.** The basis described in Theorem 7.2.6.2 is precisely De Concini-Procesi's basis for $K[D_t(Sk\, M_n)]$ (cf. [22]). Thus our approach gives a different proof of DeConcini-Procesi's basis for $K[D_t(Sk\, M_n)]$.

Taking $t > n$, we have $D_t(Sk\, M_n)$ equals $Sk\, M_n$, and we obtain

**Theorem 7.2.6.4** $K[Sk\, M_n]$ *has a basis consisting of standard monomials of the form* $q(A_1)\cdots q(A_r)$, $A_1 \leq \cdots \leq A_r$, $r \in \mathbb{N}$, $A_s \in \mathcal{A}_{n,2n}$, $1 \leq s \leq r$.

# 8

# The standard monomial theoretic basis

Standard monomial theory (SMT) for Schubert varieties in the Grassmannian was presented in sections 4.3, 4.4, 4.5 of Chapter 4. Some of its applications, including the proof of normality and Cohen-Macaulayness of Schubert varieties, were presented in section 4.6 of Chapter 4. As we will see in Chapter 10, this is already good enough for deriving the first and second fundamental theorems of classical invariant theory (CIT) for the general linear group (cf. Example **A** of Chapter 1 and also Theorem 10.2.2.2, Chapter 10). The reason is that the ring generated by the basic invariants for the action of the general linear group is the ring of functions on a determinantal variety, and this variety is the opposite cell of a certain Schubert variety in the Grassmannian—the first of these facts appears in Theorem 10.3.1.1 of Chapter 10, and the second was already proved in Theorem 5.2.2.7 of Chapter 5.

The rings generated by the basic invariants in the case of the symplectic and orthogonal group actions (Examples **B** and **C** respectively of Chapter 1) are rings of functions on skew-symmetric and symmetric determinantal varieties respectively—these facts will be proved respectively in Theorems 10.5.0.1 and 10.4.0.2 of Chapter 10. And, as was already seen in Theorem 6.2.5.2 of Chapter 6 and Theorem 7.2.5.2 of Chapter 7 respectively, these varieties are the opposite cells of certain Schubert varieties in the orthogonal and symplectic Grassmannians respectively. In order to derive the results of CIT in these cases in a fashion analogous to the case of the general linear group, we need an SMT for Schubert varieties in these Grassmannians. As pointed out in Chapter 1, this line of thought was a motivating factor in the development of SMT.

SMT for the orthogonal Grassmannian is due originally to Seshadri [111], and for the symplectic Grassmannian to Lakshmibai-Musili-Seshadri [61, 62] and independently De Concini [19]. We state the theorems respectively in the two sections below. The proofs are postponed to the appendix, for giving them here would divert the flow of the book. The reader is encouraged to accept these statements without proof and go on.

## 8.1 SMT for the even orthogonal Grassmannian

The purpose of this section is to state the main theorem of SMT for the even orthogonal Grassmannian. The preparatory material below is dealt with more systematically in Chapter 7. See also Example A.3.2 of the Appendix.

Let $K$ be an algebraically closed field of characteristic not equal to 2. Fix a vector space $V$ of finite dimension over $K$ and a non-degenerate symmetric form $\langle\ ,\ \rangle$ on $V$. We will assume the dimension of $V$ to be even, say $2n$, for that is true in the case of the invariant theoretic application we have in mind. A linear subspace of $V$ is said to be *isotropic* if the form $\langle\ ,\ \rangle$ vanishes identically on it. It is elementary to see that an isotropic subspace of $V$ has dimension at most $n$ and that every isotropic subspace is contained in one of dimension $n$. Denote by $G_n(V)$ the Grassmannian of $n$-dimensional subspaces of $V$ and by $\mathfrak{I}'$ the set of all $n$-dimensional isotropic subspaces of $V$. Then $\mathfrak{I}'$ is a closed subvariety of $G_n(V)$.

The orthogonal group $\mathrm{O}(V)$ of linear automorphisms of $V$ preserving $\langle\ ,\ \rangle$ acts transitively on $\mathfrak{I}'$—this follows from Witt's theorem that an isometry between subspaces can be lifted to one of the whole vector space. However, the special orthogonal group $\mathrm{SO}(V)$ does not act transitively on $\mathfrak{I}'$. There are two connected components of $\mathfrak{I}'$. We take $\mathfrak{I}$ to be one of the these components (to be specific, the one containing the point corresponding to the span of $e_1, \ldots, e_n$ in the notation below), and call it the *even orthogonal Grassmannian*. Then $\mathrm{SO}(V)$ acts transitively on $\mathfrak{I}$. Using this we can identify $\mathfrak{I}$ as the quotient of $\mathrm{SO}(V)$ by the stabilizer of any point (for example, the span of $e_1, \ldots, e_n$ in the notation to be defined below)—one needs also to check that the differential at the identity of the orbit map defined by the point is surjective. These stabilizers are (conjugate) maximal parabolic subgroups of $\mathrm{SO}(V)$. Thus, after fixing a Borel subgroup $B$ of $\mathrm{SO}(V)$, we can identify $\mathfrak{I}$ as the quotient of $\mathrm{SO}(V)$ by a certain maximal parabolic subgroup $P$ containing $B$:

$$\mathfrak{I} \cong G/P,\ G = SO(V).$$

Further, the fundamental weight $\omega$ associated to $P$ is minuscule (see §A.2.3 of the Appendix for details and also for the definition of a minuscule weight). The *Schubert varieties* of $\mathfrak{I}$ are defined with respect to a Borel subgroup of $\mathrm{SO}(V)$—they are the Borel orbit closures of $\mathfrak{I}$ with the canonical reduced scheme structure.

We now make some choices that are convenient for the study of Schubert varieties. For $j$ an integer such that $1 \leq j \leq 2n$, set $j^* := 2n+1-j$. Fix a basis $e_1, \ldots, e_{2n}$ of $V$ such that

$$\langle e_i, e_j \rangle = \begin{cases} 1 & \text{if } i = j^* \\ 0 & \text{otherwise} \end{cases}$$

The elements of $\mathrm{SO}(V)$ that are diagonal with respect to the basis $e_1, \ldots, e_{2n}$ form a maximal torus $T$ of $\mathrm{SO}(V)$. Similarly the elements of $\mathrm{SO}(V)$ that are upper triangular with respect to $e_1, \ldots, e_{2n}$ form a Borel subgroup $B$ of $\mathrm{SO}(V)$—a linear transformation is *upper triangular* with respect to $e_1, \ldots, e_{2n}$ if for each $j$, $1 \leq j \leq 2n$, the image of $e_j$ under the transformation is a linear combination of $e_i$ with $i \leq j$.

To index the Schubert varieties, we choose a natural combinatorial indexing set for the $T$-fixed points. Denote by $I(n, 2n)$ the set of all subsets of distinct entries of $\{1, \ldots, 2n\}$ of cardinality $n$. An element $v$ of $I(n, 2n)$ may be written as $v = (v_1, \ldots, v_n)$ where $1 \leq v_1 < \ldots < v_n \leq 2n$ and $v = \{v_1, \ldots, v_n\}$. Given $v = (v_1, \ldots, v_n)$ and $w = (w_1, \ldots, w_n)$ in $I(n, 2n)$, we say $v \leq w$ if $v_1 \leq w_1, \ldots, v_n \leq w_n$. Clearly, $\leq$ defines a partial order on $I(n, 2n)$. Let $J(n)$ denote the set of elements $v$ of $I(n, 2n)$ with the following properties: (1) exactly one of $j, j^*$ belongs to $v$ for every $j$, $1 \leq j \leq n$; and (2) the number of elements in $v$ that are bigger than $n$ is even (possibly 0). The partial order on $J(n)$ induced from the partial order $\leq$ on $I(n, 2n)$ is also denoted $\leq$.

The $T$-fixed points of $\mathfrak{I}$ are parametrized by $J(n)$: for $v = (v_1, \ldots, v_n)$ in $J(n)$, the corresponding $T$-fixed point, denoted $e^v$, is the span of $e_{v_1}, \ldots, e_{v_n}$. These points lie in different $B$-orbits and the union of their $B$-orbits is all of $\mathfrak{I}$. Schubert varieties are thus indexed by the $T$-fixed points and so in turn by $J(n)$. Given $w$ in $J(n)$, we denote by $X(w)$ the closure of the $B$-orbit of the $T$-fixed point $e^w$. We have the Bruhat ($B$-orbit) decomposition:

$$X(w) = \coprod_{v \leq w} Be^v.$$

In other words, the orbit $Be^v$ belongs to the closure of the orbit $Be^w$ if and only if $v \leq w$.

Let $\mathfrak{I} \subseteq G_n(V) \hookrightarrow \mathbb{P}(\wedge^n V)$ be the Plücker embedding (as described in Chapter 4). The pull-back to $\mathfrak{I}$ of the line bundle $\mathcal{O}(1)$ on $\mathbb{P}(\wedge^n V)$ is the square of the ample generator of the Picard group of $\mathfrak{I}$—this follows from Proposition 7.2.4.1,(2) of Chapter 7 (or also the calculation done in Example A.3.2 of the Appendix). Letting $L$ denote the ample generator, we want to describe the homogeneous coordinate rings of $\mathfrak{I}$ and its Schubert subvarieties in the embedding defined by $L$–we have in fact that $L$ is very ample (cf. Chapter 3, §3.3.2, Summary (4)).

For $\theta$ in $I(n, 2n)$, let $p_\theta$ denote the corresponding Plücker coordinate. Consider the affine patch $\mathbb{A}$ of $\mathbb{P}(\wedge^n V)$ given by $p_\epsilon = 1$, where $\epsilon := (1, \ldots, n)$. The intersection $\mathbb{A} \cap G_n(V)$ of this patch with the Grassmannian is an affine space. Indeed the $n$-plane corresponding to an arbitrary point $z$ of $\mathbb{A} \cap G_n(V)$ has a basis consisting of column vectors of a matrix of the form

$$C = \begin{pmatrix} I \\ A \end{pmatrix}$$

where $I$ is the identity matrix of size $n \times n$ and $A$ is an arbitrary matrix of size $n \times n$. The association $z \mapsto A$ is bijective. The restriction of a Plücker coordinate $p_\theta$ to $\mathbb{A} \cap G_n(V)$ is given by the determinant of a submatrix of size $n \times n$ of $C$, the entries of $\theta$ determining the rows to be chosen from $C$ to form the submatrix.

As can be readily verified, a point $z$ of $\mathbb{A} \cap G_n(V)$ belongs to $\mathfrak{I}'$ if and only if the corresponding matrix $A = (a_{ij})$ is *skew-symmetric with respect to the anti-diagonal*: $a_{ij} + a_{j^*i^*} = 0$, where the columns and rows of $A$ are numbered $1, \ldots, n$ and $n + 1, \ldots, 2n$ respectively. For example, if $n = 4$, then a matrix that is skew-symmetric with respect to the anti-diagonal looks like this:

$$\begin{pmatrix} -d & -c & -b & 0 \\ -g & -f & 0 & b \\ -i & 0 & f & c \\ 0 & i & g & d \end{pmatrix}$$

Since the set of these matrices is connected and contains the point that is spanned by $e_1, \ldots, e_n$, it follows that this affine patch of $\mathfrak{I}'$ intersects only $\mathfrak{I}$. In other words, $p_\epsilon$ vanishes everywhere on $\mathfrak{I}' \setminus \mathfrak{I}$.

Computing $p_\theta / p_\epsilon$ as a function on the affine patch $p_\epsilon \neq 0$, we see that it is the determinant of a skew-symmetric matrix of even size, and therefore a square—the square root, which is determined up to sign, is called the *Pfaffian*. This suggests that $p_\theta$ itself is a square: more precisely that there exists a section $q_\theta$ of the line bundle $L$ on $\mathfrak{I}$ such that $q_\theta^2 = p_\theta$. A weight calculation confirms this to be the case.[1] In fact, we have that $q_\theta$'s are precisely the extremal weight vectors (cf. Chapter 3, § 3.3.1) in $H^0(G/P, L)$. The $q_\theta$ are also called *Pfaffians*.

**Definition 8.1.0.1** A *standard monomial* in $J(n)$ is a totally ordered sequence $v_1 \geq \ldots \geq v_t$ (with repetitions allowed) of elements of $J(n)$. Such a standard monomial is said to be *$w$-dominated* for $w \in J(n)$ if $w \geq v_1$. To a standard monomial $v_1 \geq \ldots \geq v_t$ in $J(n)$ we associate the product $q_{v_1} \cdots q_{v_t}$, where the $q_v$ are sections of the line bundle $L$ defined above. Such a product is also called a *standard monomial* and it is said to be *dominated by $w$* for $w \in J(n)$ if the underlying monomial in $J(n)$ is dominated by $w$.

We can now state the main theorem of standard monomial theory for $\mathfrak{I}$ and its Schubert subvarieties.

**Theorem 8.1.0.2** (Seshadri [111]) *Standard monomials $q_{v_1} \cdots q_{v_r}$ of degree $r$ form a basis for the space of forms of degree $r$ in the homogeneous coordinate ring of $\mathfrak{I}$ in the embedding defined by the ample generator $L$ of the Picard group. More generally, for $w \in J(n)$, the $w$-dominated standard monomials of degree $r$ form a basis for the space of forms of degree $r$ in the homogeneous coordinate ring of the Schubert subvariety $X(w)$ of $\mathfrak{I}$.*

*Proof.* The theorem will be derived as a special case of the more general theorem proved in the Appendix (see Theorem A.4.0.1 of the Appendix).

More generally, we have results similar to Theorem 8.1.0.2 for Schubert varieties in a minuscule $G/P$ (cf. (Seshadri [111])): A *minuscule fundamental weight* $\omega$ has the property (in fact, this is a defining property) that the weights (for the $T$-action) in $H^0(G/P, L)$ form one orbit for the action of the Weyl group. Hence the extremal

[1] With notation as in Example A.3.2 of the Appendix, the weight of $p_\theta$, for $\theta$ in $I(n, 2n)$, is $-(\varepsilon_{\theta_1} + \cdots + \varepsilon_{\theta_n})$—of course, while trying to calculate the restriction to $SO(2n)$, we would have to set $\varepsilon_j + \varepsilon_{j*} = 0$, that is, $\varepsilon_{n+1} = -\varepsilon_n, \ldots, \varepsilon_{2n} = -\varepsilon_1$. For $\theta$ in $I(n, 2n)$, if $\mu_1, \mu_2$ are weights of $H^0(G/P, L)$ such that $\mu_1 + \mu_2$ equals the weight of $p_\theta$, then $\mu_1 = \mu_2 = -(\varepsilon_{\theta_1} + \cdots + \varepsilon_{\theta_n})/2$.

weight vectors (cf. Chapter 3, § 3.3.1) in $H^0(G/P, L)$ form a basis for $H^0(G/P, L)$; further, for $w \in W^P$, the standard monomials $q_{v_1} \cdots q_{v_r}$, $w \geq v_1 \geq \cdots \geq v_r$ of degree $r$ form a basis for $H^0(X(w), L^r)$.

## 8.2 SMT for the symplectic Grassmannian

The purpose of this section is to state the main theorem of SMT for the symplectic Grassmannian. The preparatory material below is dealt with again and more systematically in Chapter 6. See also Example A.3.3 of the Appendix.

Let $K$ be an algebraically closed field of arbitrary characteristic. Fix a vector space $V$ of finite dimension over $K$ and fix a non-degenerate alternating form $\langle\ ,\ \rangle$ on $V$. Then the dimension of $V$ is necessarily even, say $2n$. A linear subspace of $V$ is said to be *isotropic* if the form $\langle\ ,\ \rangle$ vanishes identically on it. It is elementary to see that an isotropic subspace of $V$ has dimension at most $n$ and that every isotropic subspace is contained in one of dimension $n$. Denote by $G_n(V)$ the Grassmannian of $n$-dimensional subspaces of $V$ and by $\mathfrak{M}_n(V)$ the set of all $n$-dimensional isotropic subspaces of $V$. Then $\mathfrak{M}_n(V)$ is a closed subvariety of $G_n(V)$, and is called the *variety of maximal isotropic subspaces* or the *symplectic Grassmannian* or also *the Lagrangian Grassmannian.*

The group $\mathrm{Sp}(V)$ of linear automorphisms of $V$ preserving $\langle\ ,\ \rangle$ acts transitively on $\mathfrak{M}_n(V)$—this follows from Witt's theorem that an isometry between subspaces can be lifted to one of the whole vector space. Using this we can identify $\mathfrak{M}_n(V)$ as the quotient of $\mathrm{Sp}(V)$ by the stabilizer of any point (for example, the span of $e_1, \ldots, e_n$ in the notation to be defined below)—one needs also to check that the differential at the identity of the orbit map defined by the point is surjective. These stabilizers are (conjugate) maximal parabolic subgroups of $\mathrm{Sp}(V)$. Thus, after fixing a Borel subgroup $B$ of $\mathrm{Sp}(V)$, we can identify $\mathfrak{M}_n(V)$ as the quotient of $\mathrm{Sp}(V)$ by a certain maximal parabolic subgroup $P$ containing $B$:

$$\mathfrak{M}_n(V) \cong G/P,\ G = Sp(V)$$

The *Schubert varieties* of $\mathfrak{M}_n(V)$ are defined with respect to a Borel subgroup of $\mathrm{Sp}(V)$—they are the Borel orbit closures of $\mathfrak{M}_n(V)$ with the canonical reduced scheme structure.

We now make some choices that are convenient for the study of Schubert varieties. For $j$ an integer such that $1 \leq j \leq 2n$, set $j^* := 2n + 1 - j$. Fix a basis $e_1, \ldots, e_{2n}$ of $V$ such that

$$\langle e_i, e_j \rangle = \begin{cases} 1 & \text{if } i = j^* \text{ and } i < j \\ -1 & \text{if } i = j^* \text{ and } i > j \\ 0 & \text{otherwise} \end{cases}$$

The elements of $\mathrm{Sp}(V)$ that are diagonal with respect to the basis $e_1, \ldots, e_{2n}$ form a maximal torus $T$ of $\mathrm{Sp}(V)$. Similarly the elements of $\mathrm{Sp}(V)$ that are upper triangular with respect to $e_1, \ldots, e_{2n}$ form a Borel subgroup $B$ of $\mathrm{Sp}(V)$—a linear transformation is *upper triangular* with respect to $e_1, \ldots, e_{2n}$ if for each $j$, $1 \leq j \leq 2n$, the image of $e_j$ under the transformation is a linear combination of $e_i$ with $i \leq j$.

To index the Schubert varieties, we choose a natural combinatorial indexing set for the $T$-fixed points. Denote by $I(n, 2n)$ the set of all subsets of distinct entries of $\{1, \ldots, 2n\}$ of cardinality $n$. An element $v$ of $I(n, 2n)$ may be written as $v = (v_1, \ldots, v_n)$ where $1 \leq v_1 < \ldots < v_n \leq 2n$ and $v = \{v_1, \ldots, v_n\}$. Given $v = (v_1, \ldots, v_n)$ and $w = (w_1, \ldots, w_n)$ in $I(n, 2n)$, we say $v \leq w$ if $v_1 \leq w_1, \ldots, v_n \leq w_n$. Clearly, $\leq$ defines a partial order on $I(n, 2n)$. Let $I(n)$ denote the set of elements $v$ of $I(n, 2n)$ with the property that exactly one of $j$, $j^*$ belongs to $v$ for every $j$, $1 \leq j \leq n$. The partial order on $I(n)$ induced from the partial order $\leq$ on $I(n, 2n)$ is also denoted $\leq$.

The $T$-fixed points of $\mathfrak{M}_n(V)$ are parametrized by $I(n)$: for $v = (v_1, \ldots, v_n)$ in $I(n)$, the corresponding $T$-fixed point, denoted by $e^v$, is the span of $e_{v_1}, \ldots, e_{v_n}$. These points lie in different $B$-orbits and the union of their $B$-orbits is all of $\mathfrak{M}_n(V)$. Schubert varieties are thus indexed by the $T$-fixed points and so in turn by $I(n)$. Given $w$ in $I(n)$, we denote by $X(w)$ the closure of the $B$-orbit of the $T$-fixed point $e^w$. We have the Bruhat ($B$-orbit) decomposition:

$$X(w) = \coprod_{v \leq w} Be^v.$$

In other words, the orbit $Be^v$ belongs to the closure of the orbit $Be^w$ if and only if $v \leq w$.

Let $\mathfrak{M}_n(V) \subseteq G_n(V) \hookrightarrow \mathbb{P}(\wedge^n V)$ be the Plücker embedding (as described in Chapter 4). The pull-back to $\mathfrak{M}_n(V)$ of the line bundle $\mathcal{O}(1)$ on $\mathbb{P}(\wedge^n V)$ is the ample generator of the Picard group of $\mathfrak{M}_n(V)$ (cf. Chapter 6, Proposition 6.2.4.1). We want to describe the homogeneous coordinate rings of $\mathfrak{M}_n(V)$ and its Schubert subvarieties in this embedding.

For $\theta$ in $I(n, 2n)$, let $p_\theta$ denote the corresponding Plücker coordinate. Consider the affine patch $\mathbb{A}$ of $\mathbb{P}(\wedge^n V)$ given by $p_\epsilon = 1$, where $\epsilon := (1, \ldots, n)$. The intersection $\mathbb{A} \cap G_n(V)$ of this patch with the Grassmannian is an affine space. Indeed the $n$-plane corresponding to an arbitrary point $z$ of $\mathbb{A} \cap G_n(V)$ has a basis consisting of column vectors of a matrix of the form

$$C = \begin{pmatrix} I \\ A \end{pmatrix}$$

where $I$ is the identity matrix of size $n \times n$ and $A$ is an arbitrary matrix of size $n \times n$. The association $z \mapsto A$ is bijective. The restriction of a Plücker coordinate $p_\theta$ to $\mathbb{A} \cap G_n(V)$ is given by the determinant of a submatrix of size $n \times n$ of $C$, the entries of $\theta$ determining the rows to be chosen from $C$ to form the submatrix.

As can be readily verified, a point $z$ of $\mathbb{A} \cap G_n(V)$ belongs to $\mathfrak{M}_n(V)$ if and only if the corresponding matrix $A = (a_{ij})$ is *symmetric with respect to the anti-diagonal*: $a_{ij} = a_{j^* i^*}$, where the columns and rows of $A$ are numbered $1, \ldots, n$ and $n + 1, \ldots, 2n$ respectively. For example, if $n = 4$, then a matrix that is symmetric with respect to the anti-diagonal looks like this:

$$\begin{pmatrix} d & c & b & a \\ g & f & e & b \\ i & h & f & c \\ j & i & g & d \end{pmatrix}$$

**Notation 8.2.0.1** For $u = (u_1, \ldots, u_n)$ in $I(n, 2n)$, set $u^* := (u_n^*, \ldots, u_1^*)$. The association $u \mapsto u^*$ is an order reversing involution of $I(n, 2n)$. There is another order reversing involution on $I(n, 2n)$, namely $u \mapsto \{1, \ldots, 2n\} \setminus u$. These two involutions commute with each other. Composing them, we obtain an order preserving involution on $I(n, 2n)$: $u \mapsto u^\# := \{1, \ldots, 2n\} \setminus u^*$. Note that $u \in I(n)$ if and only if $u = u^\#$.

**Lemma 8.2.0.2** *Given any $\theta \in I(n, 2n)$, we have $p_\theta = \pm p_{\theta^\#}$ on $\mathfrak{M}_n(V)$. The sign factor can easily be determined explicitly but we will have no use for it*[2].

*Proof.* Since $\mathfrak{M}_n(V)$ is irreducible and its intersection $\mathbb{A} \cap \mathfrak{M}_n(V)$ with the affine patch is non-empty, it is enough to check that $p_\theta = \pm p_{\theta^\#}$ on $\mathbb{A} \cap \mathfrak{M}_n(V)$, and this follows from the symmetry property just mentioned of the matrix $A$.

The relations $p_\theta = p_{\theta^\#}$ do not span the space of all linear relations among the $p_\theta$—see Example 8.1 below. In order to describe a nice parametrizing set for a basis for the space of linear forms in the homogeneous coordinate ring of $\mathfrak{M}_n(V)$—in fact for describing bases for spaces of forms of any given degree—we make the following definition.

**Definition 8.2.0.3**[3] Denote by $\epsilon$ the element $(1, \ldots, n)$ of $I(n)$. The *$\epsilon$-degree* of an element $x$ of $I(n)$ is the cardinality of $x \setminus \{1, \ldots, n\}$ or equivalently that of $\{1, \ldots, n\} \setminus x$. More generally, given any $v \in I(n)$, the *$v$-degree* of an element $x$ of

[2] The sign factor is 1 or $-1$ according as

$$\{(\theta_1 - 1) + \cdots + (\theta_k - k)\} + \{(\theta_{k+1} - (d+1)) + \cdots + (\theta_d - (2d - k))\}$$

is even or odd and $k$ is the cardinality of $[d] \cap \theta$.

To prove this, let $\theta = (\theta_1, \ldots, \theta_k, \theta_{k+1}, \ldots, \theta_d)$ and $\varphi = (\varphi_1, \ldots, \varphi_{d-k}, \varphi_{d-k+1}, \ldots, \varphi_d)$ be the complement of $\theta$ in $\{1, \ldots, 2d\}$. Then, as can be easily seen, $p_\theta$ is $(-1)^{\{(\theta_1-1)+\cdots+(\theta_k-k)\}}$ times the determinant of the $d-k \times d-k$ submatrix of the matrix $A$ above obtained by taking the entries in rows $\theta_{k+1}, \ldots, \theta_d$ and columns $\varphi_1, \ldots, \varphi_{d-k}$.

We have $\theta^\# = (\varphi_d^*, \ldots, \varphi_{d-k+1}^*, \varphi_{d-k}^*, \ldots, \varphi_1^*)$. Since $u \mapsto u^*$ commutes with taking complements, it follows that the complement in $\{1, \ldots, d\}$ of $(\varphi_d^*, \ldots, \varphi_{d-k+1}^*)$ is $(\theta_d^*, \ldots, \theta_{k+1}^*)$. By applying to $p_{\theta^\#}$ the observation in the last paragraph about $p_\theta$, we see that $p_{\theta^\#}$ is $(-1)^{\{(\varphi_d^*-1)+\cdots+(\varphi_{d-k+1}^*-k)\}}$ times the determinant submatrix of size $d-k \times d-k$ of the matrix $A$ above obtained by taking the entries in rows $\varphi_{d-k}^*, \ldots, \varphi_1^*$ and columns $\theta_d^*, \ldots, \theta_{k+1}^*$.

It follows from the symmetry of $A$ about the anti-diagonal that the two submatrices of $A$ considered above have the same determinant. And it is easy to see that $(\varphi_d^* - 1) + \cdots + (\varphi_{d-k+1}^* - k)$ equals $(\theta_{k+1} - (d+1)) + \cdots + (\theta_d - (2d - k))$.

[3] Admissible pairs as defined here are a special case of the *admissible pairs* to be defined in the Appendix.

$I(n)$ is the cardinality of $x \setminus v$ or equivalently that of $v \setminus x$. An ordered pair $\Lambda = (x, y)$ of elements of $I(n)$ is called an *admissible pair* if $x \geq y$ and the $\epsilon$-degrees of $x$ and $y$ are equal.

Given admissible pairs $\Lambda = (x, y)$ and $\Lambda' = (x', y')$, we say $\Lambda \geq \Lambda'$ if $y \geq x'$, that is, if $x \geq y \geq x' \geq y'$. An ordered sequence $(\Lambda_1, \ldots, \Lambda_t)$ of admissible pairs is called a *standard tableau* if $\Lambda_i \geq \Lambda_{i+1}$ for $1 \leq i < t$. We often write $\Lambda_1 \geq \ldots \geq \Lambda_t$ to denote the standard tableau $(\Lambda_1, \ldots, \Lambda_t)$. Given $w \in I(n)$, we say that a standard tableau $\Lambda_1 \geq \ldots \geq \Lambda_t$ is *$w$-dominated* if $w \geq x_1$, where $\Lambda_1 = (x_1, y_1)$.

**Proposition 8.2.0.4** *There is a bijective map $(x, y) \mapsto \theta$ from the set of admissible pairs $(x, y)$ to elements $\theta$ of $I(n, 2n)$ such that*

$$\theta \cap [n] \geq \theta^{\#} \cap [n],$$

*where $[n] := \{1, \ldots, n\}$.*

*Proof.* Let $[n]^{\mathrm{c}} := \{n + 1, \ldots, 2n\}$. Given an admissible pair $(x, y)$, set

$$\theta = (x \cap [n]) \cup \left(y \cap [n]^{\mathrm{c}}\right)$$

That $\theta$ satisfies the required condition is readily verified. For the map in the other direction, set

$$x = (\theta \cap [n]) \cup \left(\theta^{\#} \cap [n]^{\mathrm{c}}\right) \quad \text{and} \quad y = \left(\theta^{\#} \cap [n]\right) \cup \left(\theta \cap [n]^{\mathrm{c}}\right)$$

It is easy to verify that $(x, y)$ is an admissible pair.

**Remark 8.2.0.5.** An element $\theta$ of $I(n, 2n)$ can be specified by giving $\theta \cap [n]$ and $\theta^{\#} \cap [n]$. The ordered pairs $(\theta \cap [n], \theta^{\#} \cap [n])$ associated to the admissible pair $(x, y)$ by the above proposition were called admissible elements in Chapter 6.

*Example 8.1.* The relations $p_\theta = p_{\theta^{\#}}$ do not span the linear space of relations. For example, let

$$n = 4, \quad \theta_1 = (1, 2, 7, 8), \quad \theta_2 = (1, 4, 5, 8) \quad \text{and} \quad \theta_3 = (1, 3, 6, 8).$$

Then

$$\theta_1^{\#} = (3, 4, 5, 6), \quad \theta_2^{\#} = (2, 3, 6, 7) \quad \text{and} \quad \theta_3^{\#} = (2, 4, 5, 7).$$

From the form displayed above of a matrix that is symmetric with respect to the anti-diagonal, we get

$$p_{\theta_1} = p_{\theta_1^{\#}} = df - cg, \quad p_{\theta_2} = p_{\theta_2^{\#}} = cg - bi \quad \text{and} \quad p_{\theta_3} = p_{\theta_3^{\#}} = bi - df,$$

and consequently, $p_{\theta_1} + p_{\theta_2} + p_{\theta_3} = 0$.

The association of the previous proposition is not a bijection if the condition on $\theta$ is dropped. In the case $n = 4$ for example there are 42 admissible pairs $(x, y)$ and 43 pairs $(\theta, \theta^{\#})$. If we try the procedure in the proof above for recovering $(x, y)$ on an arbitrary $(\theta, \theta^{\#})$, the two resulting elements may not be comparable. Taking for example $\theta = (2, 3, 6, 7)$, $\theta^{\#} = (1, 4, 5, 8)$, we recover $(2, 3, 5, 8)$ and $(1, 4, 6, 7)$.

**Definition 8.2.0.6** Given an admissible pair $\Lambda = (x, y)$, we define the associated Plücker co-ordinate $p_\Lambda$ to be $p_\theta$ where $(x, y) \mapsto (\theta, \theta^\#)$ by the association of Proposition 8.2.0.4. Observe that $p_\theta = p_{\theta^\#}$ by Lemma 8.2.0.2. To formal products of admissible pairs we associate the product of the associated Plücker coordinates. In particular, to a standard tableau $\Lambda_1 \geq \ldots \geq \Lambda_t$ we associate $p_{\Lambda_1} \cdots p_{\Lambda_t}$. Such monomials associated to standard tableaux are called *standard monomials*. Given $w$ in $I(n)$, we say that a standard monomial is *$w$-dominated* if the corresponding standard tableau is $w$-dominated.

We can now state the main theorem of standard monomial theory for $\mathfrak{M}_n(V)$ and its Schubert subvarieties.

**Theorem 8.2.0.7** (De Concini [19], Lakshmibai-Musili-Seshadri [61,62]) *Standard monomials $p_{\Lambda_1} \cdots p_{\Lambda_t}$ of degree $r$ form a basis for the space of forms of degree $r$ in the homogeneous coordinate ring of $\mathfrak{M}_n(V)$ in the Plücker embedding. More generally, for $w \in I(n)$, the $w$-dominated standard monomials of degree $r$ form a basis for the space of forms of degree $r$ in the homogeneous coordinate ring of the Schubert subvariety $X(w)$ of $\mathfrak{M}_n(V)$.*

*Proof.* The theorem will be derived as a special case of the more general theorem to be proved in the Appendix (see Theorem A.4.0.1 of the Appendix).

# 9

# Review of GIT

In this chapter, we do a brief review of GIT - Geometric Invariant Theory. Our treatment follows that of [96]. For further details, we refer the reader to [87] & [110]. We fix a base field $K$ which we assume to be algebraically closed of arbitrary characteristic. In this chapter, we shall work just with the closed points of an algebraic variety.

## 9.1 $G$-spaces

Consider an algebraic variety $X$ together with an algebraic group action, i.e., we are given an algebraic group $G$ and a morphism $G \times X \to X$. For $x \in X$, we shall denote the orbit through $x$ by $O(x)$. We shall denote the fixed points set by $X^G$, namely, $X^G = \{x \in X \mid g \cdot x = x, \forall g \in G\}$ Defining "quotients" (which are again algebraic varieties) is the central theme in GIT. It turns out that actions by "reductive groups" do admit some natural notion of "quotients". We shall define such quotients for actions by reductive Groups. We first start with recalling the definition of reductive groups (and definitions of related notions). After pointing out (with an example) that the orbit space (whose underlying set is the set of orbits) need not have a variety structure, we recall the definitions of various quotients, and then explore these in the case of affine and projective varieties.

Throughout $G$ will denote an affine algebraic group defined over $K$.

### 9.1.1 Reductive groups

We first recall the definitions of reductive groups, linearly reductive groups, geometrically reductive groups, and completely reducible groups. For generalities on reductive groups, we refer the reader to Borel [7] & Humphreys [43].

**Definition 9.1.1.1** *A* $G$-module *is a finite-dimensional* $K$*-vector space* $V$ *together with an algebraic group morphism* $\rho : G \to GL(V)$(= *the group of* $K$*-linear automorphisms of* $V$*, also denoted* $\mathrm{Aut}_K V$ *or just* $\mathrm{Aut}\, V$*). We refer to* $\rho$ *as* "a rational representation" *of* $G$.

**Remark 9.1.1.2.** Note that a rational representation $\rho : G \to GL(V)$, determines a morphism $G \times V \to V$, $(g, v) \mapsto \rho(g)(v)$, and hence an action of $G$ on $V$. We refer to this as a *linear action* (of $G$ on $V$).

**Definition 9.1.1.3** *A $G$-module $V$ is* irreducible, *if it has no proper, non-zero $G$-stable subspaces.*

**Definition 9.1.1.4** *A $G$-module is $V$ is* completely reducible, *if it can be written as $V = \bigoplus_i V_i$, where each $V_i$ is an irreducible $G$-module.*

**Definition 9.1.1.5** *An algebraic group $G$ is* completely reducible *if every (finite dimensional) $G$-module $V$ is completely reducible.*

**Definition 9.1.1.6** *An algebraic group $G$ is* linearly reductive *if given a finite dimensional $G$-module $V$, and a non-zero $v \in V^G$, there exists a $G$-invariant linear form $F \in V^*$ such that $F(v) \neq 0$.*

**Fact:** $G$ is completely reducible if and only if $G$ is linearly reductive.
For a proof, see Nagata [92]

**Definition 9.1.1.7** *An algebraic group $G$ is* geometrically reductive *if given a finite dimensional $G$-module $V$, and a non-zero $v \in V^G$, there exists a $G$-invariant form $F \in Sym(V^*)$ with $deg(F) > 0$ such that $F(v) \neq 0$.*

Before defining reductive groups, we first recall the radical of an algebraic group $G$. Denote by $G_u$ the unipotent part of $G$.

Recall that a *Borel subgroup* of an algebraic group is a maximal connected solvable subgroup. The *radical* of an algebraic group $G$, denoted $R(G)$, is the connected component through $e$ (the identity element in $G$) of the intersection of all Borel subgroups of $G$, i.e.

$$R(G) := (\cap_{B \in \mathcal{B}} \mathrm{B})^\circ$$

$\mathcal{B}$ being the set of all Borel subgroups in $G$.

**Definition 9.1.1.8** *$G$ is* reductive *if $R(G)_u = (e)$ (the trivial subgroup).*

(**Note:** This is equivalent to saying that $R(G)$ is a *torus*, i.e, a direct product of copies of $K^*$.)

**Theorem 9.1.1.9** *(H. Weyl) In characteristic 0, a reductive group is linearly reductive.*

For a proof see [29].

**Remark 9.1.1.10.** In positive characteristics, a reductive group need not be linearly reductive as explained in the example below:

**Example:** Let $char(K) = p > 0$, and $G = GL(V) = GL(n)$ where $V$ is a vector space over $K$ of dimension $n$ and $p|n$. Let $X = M_n(K)$, the space of $n \times n$ matrices with entries in $K$, which we identify with $End(V)$. Let $G$ act on $X$ by conjugation. We then have that $v := Id_V \in X$ is $G$-invariant.

**Fact**: Under the action of $GL(n)$ on $M_n(K)$ by conjugation, a $G$-invariant polynomial form is a symmetric function of the eigenvalues of a matrix (for details, see §9.3.1 below). Therefore if there exists a linear $G$-invariant form $F$, then it has to be a multiple of the sum of the eigenvalues, that is, $F = \lambda tr(A)$, for some $\lambda \in K$ (here, tr$(A)$ is the trace of $A$ = sum of the diagonal entries of $A$). Therefore for $v = Id$, $F(v) = \lambda n = 0$ (since $p|n$). Therefore $G$ cannot be linearly reductive.

Let $G$ be an algebraic group acting (say, on the right) on an algebraic variety $X$. Let $Y := X/G$ be the orbit space of $X$ under the action of $G$; a point in $Y$ is a $G$-orbit, i.e., $O(x)$, for some $x \in X$.

In general, $Y$ need not have a variety structure; if $Y$ has a variety structure, then it forces each orbit to be closed, which may not happen in general.

As an example, consider

$$X = \{\text{upper triangular matrices in } M_2(K)\}$$
$$G = \{\text{invertible upper triangular matrices in } M_2(K)\}$$

Let $G$ act on $X$ by conjugation. Consider $A = \begin{pmatrix} 1 & 1 \\ 0 & 1 \end{pmatrix} \in X$ and $g = \begin{pmatrix} a & b \\ 0 & d \end{pmatrix} \in G$. Then

$$gAg^{-1} = \begin{pmatrix} a & b \\ 0 & d \end{pmatrix} A \frac{1}{ad} \begin{pmatrix} d & -b \\ 0 & a \end{pmatrix} = \begin{pmatrix} 1 & a/d \\ 0 & 1 \end{pmatrix}$$

where $a/d \in K^*$. Thus $G \cdot A = O(A) = \left\{ \begin{pmatrix} 1 & t \\ 0 & 1 \end{pmatrix}, t \in K^* \right\}$.

Now $\overline{O(A)} = O(A) \cup \left\{ \begin{pmatrix} 1 & 0 \\ 0 & 1 \end{pmatrix} \right\}$, so $O(A)$ is not closed!

Let $X = Spec\, R$ where $R$ is a finitely generated $K$-algebra. Each $f \in R$ corresponds to a map $f : X \to K$. A group action $G \circlearrowleft X$ (say, on the left) induces an action $G \circlearrowleft R$ defined by $g \cdot f(x) := f(g^{-1}x)$ for $g \in G$, $f \in R$, and $x \in X$. Let

$$R^G = \{f \in R \mid g \cdot f = f, \forall g \in G\}.$$

If $R^G$ is a finitely generated $K$-algebra, then we can use the variety $Spec\, R^G$ for defining a quotient. But $R^G$ need not be a finitely generated $K$-algebra in general:

**Hilbert's Fourteenth Problem:** Given a finitely generate $K$-algebra $R$ and an algebraic group action $G \circlearrowleft R$, is $R^G$ also a finitely generated $K$-algebra?

Nagata answered this in the negative in 1958 (cf. [91, 94]) by providing a counterexample. However, if $G$ is *reductive* (such as is the case with $GL_n$, $SL_n$, $Sp_{2n}$, $O_n$, or any of the exceptional groups), then we shall see (thanks to Mumford's conjecture) below that for the action of $G$ on an affine variety $Spec\, R$, $R^G$ is finitely generated.

**Mumford's conjecture:** If $G$ is reductive, then $G$ is geometrically reductive (in any characteristic).

Mumford's conjecture was proved in 1975 by Haboush [36]. In fact, we have,

**Theorem 9.1.1.11** *The following are equivalent:*

1. *$G$ is reductive.*
2. *$G$ is geometrically reductive.*
3. *If $G \circlearrowleft R$ where $R$ is a finitely generated $K$-algebra, then $R^G$ is finitely generated.*

The implication (1) ⇒ (2) was proved by Haboush (cf. [36]).

The implication (2) ⇒ (3) was proved by Nagata (cf. [93]).

The implication (3) ⇒ (1) was proved by Popov (cf. [98]).

The proof of the above Theorem may be found in GIT [87, Appendix 1].

## 9.2 Affine quotients

We start with a fundamental result concerning geometrically reductive groups.

**Proposition 9.2.0.1** *Let $G$ be a geometrically reductive algebraic group acting on an affine variety $X = Spec\, R$. Given two disjoint $G$-stable closed subsets $W_1, W_2 \subset X$, there exists $f \in R^G$ such that $F(W_1) = 0$ and $F(W_2) = 1$.*

*Proof.* Since $W_1 \cap W_2 = \emptyset$, we have $I_1 + I_2 = R$, $I_j$ being the defining ideal of $W_j$, $j = 1, 2$. Hence there exist $f_j \in I_j$, $j = 1, 2$ such that $f_1 + f_2 = 1$. Then $h := f_1$ has the property that $h(W_1) = 0$ and $h(W_2) = 1$.

Let $V =$ the span of $\{g \cdot h, g \in G\}$. Note that $V$ is the smallest $G$-submodule of $R$ containing $h$. Then $V$ is $G$-stable and finite dimensional (recall the well-known fact that any finite dimensional subspace $U$ of $R$ is contained in a finite dimensional $G$-stable subspace of $R$ (for a proof, see Borel [7] for example); in the present case, $U = Kh$). Let $\{h_1 = h, \cdots, h_n\}$ be a basis for $V$. Write

$$g \cdot h_i = \sum_j a_{ij}(g)h_j,\ a_{ij} \in K[G]$$

This gives rise to

$$G \to Aut\, V,\ g \mapsto (a_{ij}(g))_{1\le i,j\le n}$$

The morphism

$$\varphi : X \to V(= K^n), x \mapsto (h_1(x), \cdots, h_n(x))$$

is a $G$-morphism, i.e., $\varphi(g \cdot x) = g \cdot \varphi(x)$.

Further $\varphi(W_1) = 0, \varphi(W_2) = (h_1(W_2), \cdots, h_n(W_2)) = v$, say (note that writing $h_i = \sum_j c_{ij} g_{ij} h$, we have, $h_i(W_1) = 0, h_i(W_2) =$ a constant, since $W_1, W_2$ are $G$-stable, and $h(W_1) = 0, h(W_2) = 1$). Write $v = (h_1(x), \cdots, h_n(x))$, where $x$ is any point in $W_2$. Note that $v \neq 0$ (since $h_1(W_2) = 1$); further, for $g \in G$, we have,

$$\begin{aligned} g \cdot v &= (g \cdot h_1(x), \cdots, g \cdot h_n(x)) \\ &= (h_1(g^{-1} \cdot x), \cdots, h_n(g^{-1} \cdot x)) \\ &= (h_1(x), \cdots, h_n(x)) \\ &= v \end{aligned}$$

Thus $v \in V^G$. Hence the geometric reductivity of $G$ implies that there exists a $F$ in $K[x_1, \cdots, x_n]$ such that $F(0) = 0$, $F(v) = 1$. Then $f := F \circ \varphi$ has the required properties.

### 9.2.1 Affine actions

Let $G$ be a reductive group acting on an affine variety $X = Spec R$ where $R$ is a finitely generated $K$-algebra. Let $Y := Spec R^G$. Let $\varphi : X \to Y$ be the morphism induced by the inclusion $R^G \hookrightarrow R$. We first list some properties of the morphism $\varphi$.

**Proposition 9.2.1.1** *1. $\varphi$ is surjective.*

*2. $\varphi$ is $G$-invariant, i.e., $\varphi$ is constant on orbits:*

$$\varphi(g \cdot x) = \varphi(x),\ g \in G,\ x \in X.$$

*3. $\varphi$ is an affine morphism, i.e. the inverse image of an open affine set is an open affine set.*

*4. If $W \subset X$ is $G$-stable and closed, then $\varphi(W)$ is closed.*

*5. If $W_1, W_2$ are two disjoint $G$-stable closed subsets of $X$, then $\varphi(W_1) \cap \varphi(W_2) = \emptyset$.*

*6. Given open $U \subset Y$, the map*

$$\varphi^* : K[U] \to K[\varphi^{-1}(U)]^G$$

*is an isomorphism.*

*Proof.* (1) Let $y$ be a (closed) point in $Y$. Let $\mathbf{m}_y$ be the maximal ideal in $K[Y](= R^G)$ corresponding to $y$. Let $f_1, \cdots, f_r$ be generators for $\mathbf{m}_y$. Then we have that $\sum f_i R \neq R$ (in view of [96], Lemma 3.4.2, we have that for any $f \in (\sum f_i R) \cap R^G$, $f^m \in \sum f_i R^G$). Hence there exists a maximal ideal $\mathfrak{m}$ containing $\sum f_i R$. Let $x$ be the point corresponding to $\mathfrak{m}$. Then $f_i(x) = 0, \forall i$, and it follows that $\varphi(x) = y$. Thus we obtain that $\varphi$ is surjective.

(2) To show $\varphi(g \cdot x) = \varphi(x),\ g \in G,\ x \in X$, we should show that

$$(*) \qquad f(\varphi(g \cdot x)) = f(\varphi(x)),\ f \in K[Y](= R^G),\ g \in G,\ x \in X.$$

Now for any $f \in R^G, g \in G, x \in X$, we have

$$\begin{gathered} f(gx) = f(\varphi(g \cdot x)), f(x) = f(\varphi(x)) \\ f(g \cdot x) = g^{-1} \cdot f(x) = f(x) \end{gathered}$$

from which (*), and hence the $G$-invariance of $\varphi$ follows.

Assertion (3) follows trivially from the fact that a morphism $f : X \to Y$ is an affine morphism if and only if there exists an open affine cover $Y = \cup V_i$ such that $f^{-1}(V_i)$ is open affine.
(4) Assume if possible that $\varphi(W)$ is not closed in $Y$. Let $y \in \overline{\varphi(W)} \setminus \varphi(W)$. Then $\varphi^{-1}(y)$, $W$ are two $G$-stable, disjoint, closed subsets of $X$. Hence by Proposition 9.2.0.1, there exists a $f \in R^G$ such that $f(W) = 0$, $f(\varphi^{-1}(y)) = 1$. Hence we obtain $f(\overline{\varphi(W)}) = 0$, $f(y) = 1$, a contradiction (note that $y \in \overline{\varphi(W)}$). Hence our assumption is wrong and (4) follows.
(5) Since $W_1, W_2$ are two $G$-stable, disjoint, closed subsets of $X$, we have by Proposition 9.2.0.1 that there exists a $f \in R^G$ such that $f(W_1) = 0$, $f(W_2) = 1$. Hence writing $W_1, W_2$ as unions of $G$-orbits and using the $G$-invariance of $\varphi$, we obtain $f(\varphi(W_1)) = 0$, $f(\varphi(W_2)) = 1$. Assertion (5) follows from this.
(6) It suffices to prove assertion (6) for principal open subsets, that is, for $Y_f := \{y \in Y | f(y) \neq 0\}$, where $f \in K[Y]$ is a regular function (since the collection $\{Y_f,\ f \in K[Y]\}$ is a base for the Zariski topology on $Y$). Let $X_f := \varphi^{-1}(Y_f)$. Take $U = Y_f$, then $K[U] = K[X]^G_f$. On the other hand, $\varphi^{-1}(U) = X_f$ and $K[X_f] = K[X]_f$. It is seen easily that the localization of the $G$-invariants equals the $G$-invariants of the localization, $(K[X]^G)_f = (K[X]_f)^G$. This proves (6).

As an important consequence, we obtain that $\varphi$ has the universal mapping property:

**Proposition 9.2.1.2** ***(Universal mapping property)*** *With $X, Y, \varphi$ as above, given an algebraic variety $Z$, and a $G$-invariant morphism $F : X \to Z$, there exists a unique morphism $\chi : Y \to Z$ such that the diagram*

$$\begin{array}{ccc} X & & \\ \downarrow{\scriptstyle\varphi} & \searrow^{F} & \\ Y & \xrightarrow{\chi} & Z \end{array}$$

*is commutative.*

*Proof.* If possible, let $\chi_1, \chi_2$ be two such morphisms. Let $y \in Y$, say, $y = \varphi(x)$, for some $x \in X$ (cf. Proposition 9.2.1.1, (1)). For $i = 1, 2$, we have

$$\chi_i(y) = \chi_i(\varphi(x)) = F(x).$$

Hence we obtain $\chi_1 = \chi_2$, and the uniqueness of $\chi$ follows.
We shall now prove the existence of $\chi$. We first
**Claim 1:** For $y \in Y$, $F$ is constant on $\varphi^{-1}(y)$.
If possible, assume that $z_1, z_2 \in F(\varphi^{-1}(y))$, $z_1 \neq z_2$. Write $z_i = F(x_i), i = 1, 2$, where $x_1, x_2$ are two (distinct) points of $\varphi^{-1}(y)$. Let

$$W_1 = F^{-1}(z_1),\ W_2 = F^{-1}(z_2)$$

Then $W_1, W_2$ are two disjoint, $G$-stable, closed subsets of $X$. Hence by Proposition 9.2.1.1, (5), we obtain that $\varphi(W_1) \cap \varphi(W_2) = \emptyset$; but $y \in \varphi(W_1) \cap \varphi(W_2)$ (note that

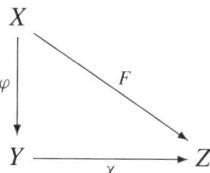

$y = \varphi(x_i) \in \varphi(W_i), i = 1, 2)$), a contradiction. Hence our assumption is wrong, and Claim 1 follows.

Take an affine cover $Z := \cup V_i$ for $Z$. Let $U_i = \{y \in Y \mid F(\varphi^{-1}(y)) \in V_i\}$. Denote by $\varphi_i, F_i$, the restriction of $\varphi$, $F$ respectively, to $\varphi^{-1}(U_i)$. Define $\chi_i : U_i \to V_i, y \mapsto F(\varphi^{-1}(y))$ (note that $\chi_i$ is well-defined in view of Claim 1). Thus we have the following commutative diagram:

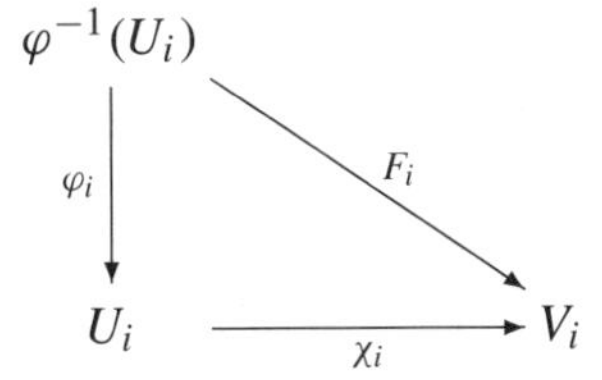


**Claim 2:** The map $\chi_i : U_i \to V_i$ is in fact a morphism of varieties.

The comorphism $F_i^*$ maps $K[V_i]$ inside $K[F_i^{-1}V_i]^G$:

$$(*) \qquad F_i^* : K[V_i] \to K[F_i^{-1}V_i]^G$$

(since $F_i$ is $G$-invariant). Also, by Proposition 9.2.1.1, (6), the comorphism $\varphi_i^*$ maps $K[U_i]$ isomorphically onto $K[\varphi_i^{-1}(U_i)]^G$:

$$(**) \qquad \varphi_i^* : K[U_i] \cong K[\varphi_i^{-1}(U_i)]^G$$

Now

$$K[\varphi_i^{-1}(U_i)]^G = K[\varphi_i^{-1}\chi_i^{-1}(V_i)]^G = K[F_i^{-1}(V_i)]^G$$

This together with (**) implies

$$(***) \qquad K[U_i] \cong K[F_i^{-1}V_i]^G$$

Hence from (*), (***), we obtain that $\chi* : K[V_i] \to K[U_i]$ is a $K$-algebra homomorphism. Claim 2 follows from this (in view of the well-known fact that if $U, V$ are two varieties with $V$ affine, then the map $Mor(U, V) \to Hom_{K\text{-alg}}(K[V], K[U])$, $f \mapsto f^*$ is a bijection. See for example [86] for a proof.) Clearly, the $\chi_i, \chi_j$ agree on $U_i \cap U_j$, and hence may be glued to define a morphism $\chi : Y \to Z$. This completes the proof of Proposition 9.2.1.2.

## 9.3 Categorical quotients

**For the remaining sections of this chapter, we shall assume $G$ to be a reductive group.**

Let $X = Spec R$ be an affine variety on which $G$ acts. Let $Y := Spec R^G$ (note that $R^G$ is a finitely generated $K$-algebra). Let $\varphi : X \to Y$ be the morphism induced by the inclusion $R^G \hookrightarrow R$.

Consider the following two properties of $\varphi$ (seen in the previous section):

(i) $\varphi$ is $G$-invariant, i.e, constant on $G$-orbits.
(ii) $\varphi$ has the universal mapping property. (As above, given $f : X \to Z$ which is constant on $G$-orbits, there exists a unique $g : Y \to Z$ such that $f = g \circ \varphi$:

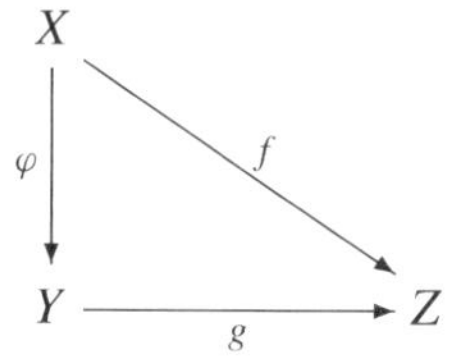


We make an abstraction of (i) and (ii), and define categorical quotients:

**Definition 9.3.0.1** *Let $X$ be an algebraic variety with an action by an algebraic group $G$. A pair $(Y, \varphi)$ is called a* categorical quotient *if*

*(i) $Y$ is an algebraic variety.*
*(ii) $\varphi$ is morphism of varieties from $X$ to $Y$.*
*(iii) $\varphi$ is $G$-invariant, i.e., $\varphi$ is constant on orbits*
*(iv) $\varphi$ has the universal mapping property.*

**Remark 9.3.0.2.** It may be seen easily that if a categorical quotient exists, then it is unique up to isomorphism.

**Notation:** We shall denote the categorical quotient (of $X$ by $G$) by $X/\!\!/G$ ( if $G$ acts on the right; a similar notation if $G$ acts on the left).

Immediately from the definition, we have
**Fact:** If $X$ is affine, then $X/\!\!/G = Spec(K[X]^G)$.

### 9.3.1 Examples

Let $X = M_n(K) = Spec\ R$ where $R = K[x_{11}, \ldots, x_{nn}]$. Let $G = GL_n(K)$, and let $G$ act on $X$ by conjugation.

**Description of $R^G$:** Let $A \in X$. Up to conjugation, $A$ is determined by its Jordan blocks. For $f \in R^G$, $f$ being constant on conjugacy classes, $f$ is completely known by specifying its values on Jordan blocks. In the case of $A$ being a diagonal matrix, its Jordan blocks are given by the diagonal entries. Further, denoting by $D_n$ the set of diagonal matrices, we have $\bigcup_j gD_ng^{-1}$ is dense in $X$. Hence, $f \in R^G$ is uniquely determined by knowing its values on diagonal matrices. Thus we obtain an injective homomorphism

$$R^G \hookrightarrow K[t_1, \cdots, t_n],\ f \mapsto f|_{D_n}.$$

Since two diagonal matrices with permuted diagonal entries are conjugate, $f$ must be a symmetric function in the $t_i$'s, and we obtain an identification of $R^G$ with the space of symmetric polynomials in the $t_i$'s. Now a symmetric polynomial in the $t_i$'s is just a polynomial in the elementary symmetric functions $e_j$'s in the $t_i$'s, and the $e_j$'s are simply the coefficients of the characteristic polynomial.

Thus we obtain

$$R^G \cong K[e_1, \cdots, e_n]$$

where $e_i$ is the elementary symmetric polynomial of degree $i$ in the eigenvalues of $A$, i.e., the coefficients of the characteristic polynomial $x^n + a_1x^{n-1} + \cdots + a_{n-1}x + a_n$ of $A$. Hence, $X/\!\!/G = \mathbb{A}^n$, and the map

$$X \to \mathbb{A}^n,\ A \mapsto (a_1, \cdots, a_n)$$

is the corresponding (categorical) quotient map.

Let $X = M_{n,d}(K) = Spec\ R = \mathbb{A}^{dn}$, where $d \leq n$.
Let $G = SL_d(K)$, and let $G$ act on $X$ by multiplication on the right. Consider the $d$-minors of a matrix $A \in X$. Let $\underline{i} = (i_1, \ldots, i_d)$ be an ordered sequence of indices such that $1 \leq i_1 < \ldots < i_d \leq n$. Let $I_{d,n}$ denote the set of all such sequences. For each $\underline{i} \in I_{d,n}$, let $A_{\underline{i}}$ denote the $d \times d$ submatrix of $A$ with rows given by the indices $\{i_1, \ldots, i_d\}$, and define the map $p_{\underline{i}} : X \to K$ by $A \overset{p_{\underline{i}}}{\to} det(A_{\underline{i}})$. Then $p_{\underline{i}} \in R^G$ for all $\underline{i} \in I_{d,n}$. In fact, we have,

$$R^G = K[p_{\underline{i}}, \underline{i} \in I_{d,n}]$$

We shall see later (in Chapter 10) that the categorical quotient $X/\!\!/G = \widehat{G_{d,n}} = Spec(R^G)$, the cone over the Grassmannian.

## 9.4 Good quotients

We now make an abstraction of Proposition 9.2.1.1 to define a good quotient.

**Definition 9.4.0.1** *Given an algebraic variety $X$ and an algebraic group $G$ acting on $X$, a* good quotient *is defined as a pair $(Y, \varphi)$ satisfying the following:*

*(i)* $\varphi : X \to Y$ *is a surjective morphism.*
*(ii)* $\varphi$ *is $G$-invariant, i.e., $\varphi$ is constant on orbits:*

$$\varphi(g \cdot x) = \varphi(x),\ g \in G,\ x \in X.$$

*(iii)* $\varphi$ *is an affine morphism, i.e. the inverse image of an open affine set is again an open affine set.*
*(iv) If $W \subset X$ is $G$-stable and closed, then $\varphi(W)$ is closed.*
*(v) If $W_1, W_2$ are two disjoint $G$-stable closed subsets of $X$, then $\varphi(W_1) \cap \varphi(W_2) = \emptyset$.*
*(vi) Given open $U \subset Y$, the map*

$$\varphi^* : K[U] \to K[\varphi^{-1}(U)]^G$$

*is an isomorphism.*

*In addition, if the underlying set of $Y$ is the orbit space $X/G$, then $(Y, \varphi)$ is called a* geometric quotient.

The following facts are immediate from the definition.
**Facts**:

(1) A good quotient is a categorical quotient.
(2) For $X = Spec\ R$, $X/\!\!/G = Spec(R^G)$ is a good quotient.
(3) Consider any closed subgroup $H \subset G$ (any algebraic group). Then the coset space $G/H$ is a geometric quotient for the action of $H$ on $G$ by right multiplication. In [107], Rosenlicht describes the variety structure of $G/H$.
(4) If the action of $G$ on $X$ is free, i.e., we have an inclusion $G \hookrightarrow Aut\ X$ and a geometric quotient exists, then $X$ is a principal fiber bundle over $Y$ with group $G$; recall (cf. [87]) that $X$ is a principal fiber bundle over $Y$ with group $G$ if there exists a morphism $\varphi : X \to Y$ such that
(i) $\varphi$ is a flat morphism of finite type.
(ii) The canonical map $\psi : G \times X \to X \times X$ is an isomorphism.
(5) Just as $\mathbb{P}^n$ may be realized as a (geometric) quotient of $\mathbb{C}^{n+1} \setminus \{0\}$ for the natural action of $\mathbb{C}^*$, any toric variety $Z$ (see [30] for the definition of a toric variety) may be realized as a categorical quotient $X/\!\!/G$ for a suitable $X$ open in an affine space and a suitable group $G$; if in addition, $Z$ is smooth, then it is a good quotient (cf. [18]).

**Further properties of good & geometric quotients:** We list below some properties of good & geometric quotients which are immediate from the definition.

**Proposition 9.4.0.2** *(1) If $(Y, \varphi)$ is a good (resp. geometric) quotient of $X$ by $G$, then for any open $U \subset Y$, $res(\varphi) : \varphi^{-1}(U) \to U$ is a good (resp. geometric) quotient of $\varphi^{-1}(U)$ by $G$ (here, $res(\varphi)$ denotes the restriction).*
*(2) Suppose $Y$ has an affine cover, i.e. $Y = \bigcup U_i$ where each $U_i$ is affine, and suppose each $res(\varphi) : \varphi^{-1}(U_i) \to U_i$ is a good (resp. geometric) quotient of $\varphi^{-1}(U_i)$ by $G$. Then $(Y, \varphi)$ is a good (resp. geometric) quotient of $X$ by $G$.*
*(3) Let $(Y, \varphi)$ be a good (resp. geometric) quotient of $X$ by $G$. For $x_1, x_2 \in X$, $\varphi(x_1) = \varphi(x_2) \Leftrightarrow \overline{O(x_1)} \cap \overline{O(x_2)} \neq \emptyset$.*

*(This follows from (2) and (5) of Proposition 9.2.1.1)*

*(4) Each fiber contains a unique closed orbit.*

*(Clearly, each fiber cannot have two or more closed orbits by (5) of Proposition 9.2.1.1; also, each fiber $\varphi_y$ being $G$-stable has a closed $G$-orbit (by "closed orbit lemma" stated below) which is in fact closed in $X$, since $\varphi_y$ is closed in $X$.*

***Lemma 9.4.0.3 Closed orbit Lemma:*** *Let an algebraic group $G$ act on an algebraic variety $X$. Then each orbit is a smooth variety which is open in its closure in $X$. Its boundary is a union of orbits of strictly smaller dimensions. In particular, the orbits of minimal dimension are closed.*

*See Borel [7] for a proof of the closed orbit Lemma.)*

(5) *If the action of G on X is closed, i.e., each orbit is closed, then $(Y, \varphi)$is a geometric quotient.*

*(If the action of G on X is closed, then Y is clearly the orbit space.)*

### 9.4.1 Some results on good quotients

In this subsection, we shall prove some results on good quotients to be used for our discussion in the following sections.

We first recall (the well-known) *upper semi-continuity* Theorem:

**Theorem 9.4.1.1** *Let $f : X \to Y$ be a morphism of finite type of algebraic varieties $X, Y$. Then for any integer $n$, $\{x \in X \mid dim(f^{-1}(f(x))) \geq n\}$ is a closed subset of $X$.*
*(**An equivalent formulation:** For any integer $m$, $\{x \in X \mid dim(f^{-1}(f(x))) \leq m\}$ is an open subset of $X$.)*

Let $X$ be an algebraic variety together with an action by an algebraic group $G$. As a first application of "upper semi-continuity" we have

**Lemma 9.4.1.2** *(1) For any integer $n$, $\{x \in X \mid dim\, G_x \geq n\}$ is a closed subset of $X$.*
*(2) For any integer $m$, $\{x \in X \mid dim\, O(x) \geq m\}$ is an open subset of $X$.*

*Proof.* Consider the morphism

$$\phi : X \times G \to X \times X, \quad (x, g) \longmapsto (x \cdot g, x).$$

(Here, we have supposed the action of $G$ on the right). We see that the fiber of $\phi$ through $(x, g)$ can be identified with the isotropy subgroup of $G$ at $x$. Applying Theorem 9.4.1.1 (rather, its equivalent formulation) to $\phi$, we obtain that the set $V$ of points $(x, g)$ such that $\dim G_x < n$, is open in $X \times G$. Now the canonical projection $X \times G \to X$, $(x, g) \longmapsto x$ is open. Hence, $\{x \in X \mid dim\, G_x < n\}$ is open in $X$; assertion (1) follows from this.
Assertion (2) follows from the fact $dim\, G_x + dim\, O(x) = dim\, G$.

**Lemma 9.4.1.3** *Let $d$ be the maximum of the orbit dimensions.*
*Let $X_{\max} = \{x \in X \mid dim\, O(x) = d\}$. Then $X_{\max}$ is a $G$-stable open subset of $X$.*

*Proof.* The $G$-stability of $X_{\max}$ is clear. The assertion that $X_{\max}$ is open follows from Lemma 9.4.1.2, (2) (note that by the maximality of $d$, we have $X_{\max} = \{x \in X \mid dim\, O(x) \geq d\}$).

**Lemma 9.4.1.4** *Let $(Y, \varphi)$ be a good quotient of $X$ by $G$. Let $x \in X$ be such that $O(x)$ is a closed orbit of dimension $d$ (here, $d$ is as in Lemma 9.4.1.3). Let $q : X_{\max} \to Y$ denote the restriction of $\varphi$ to $X_{\max}$. Then $O(x)$ is the fiber of $q$ at $q(x)$.*

*Proof.* Let $x' \in X_{\max}$ be such that $q(x') = q(x)$. We are required to show that $x'$ belongs to $O(x)$. Assume if possible that $O(x) \neq O(x')$. We have that $\varphi(x') = \varphi(x)$ (since $q(x') = q(x)$), and hence $\overline{O(x)} \cap \overline{O(x')} \neq \emptyset$ (by Definition 9.4.0.1,(v)). But by hypothesis, $\overline{O(x)} = O(x)$. Hence we obtain that $O(x) \subseteq \overline{O(x')}$; this together with the assumption that $O(x) \neq O(x')$, and the "closed orbit Lemma" implies that $\dim O(x) < d$ (since $\dim O(x') = d$), which is a contradiction. Hence our assumption is wrong and the result follows.

**Lemma 9.4.1.5** *Let $(Y, \varphi)$ be a good quotient of $X$ by $G$.*
*Let $X_1 = \{x \in X_{\max} \mid \dim(\text{fiber at } q(x)) = d\}$, $q$ being the restriction of $\varphi$ to $X_{\max}$, and $d$ being as in Lemma 9.4.1.3. Then $X_1$ is a $G$-stable open subset of $X$.*

*Proof.* The $G$-stability of $X_1$ is clear. Now $G$-orbits in $X$ map to points in $Y$, so that this is also the case for $q : X_{\max} \to Y$. Thus we have that the fibers of $q$ are of dimension $\geq d$ ($d$ being as in Lemma 9.4.1.3). Hence we obtain by Theorem 9.4.1.1 applied to $q$ (we take $m = d$ in the equivalent formulation of Theorem 9.4.1.1) that $X_1$ is an open subset of $X_{\max}$ and hence of $X$ (since $X_{\max}$ is an open subset of $X$ (cf. Lemma 9.4.1.3)).

**Corollary 9.4.1.6** *Let $(Y, \varphi)$ be a good quotient of $X$ by $G$. Let $d$ be as in Lemma 9.4.1.3. Suppose that there exists a closed orbit of dimension $d$. Then $\dim Y = \dim X - d$.*

*Proof.* Let $x \in X$ be such that $O(x)$ is closed. Let $X_1$ be as in Lemma 9.4.1.5. By Lemma 9.4.1.4, the orbit $O(x)$ is the fiber at $q(x)$. Thus the fiber at $q(x)$ has dimension $d$, and hence $x \in X_1$; in particular, $X_1$ is non-empty. Now denoting the restriction of $q$ to $X_1$ also by just $q$, we have, $q : X_1 \to Y$ is dominant and equidimensional with fiber dimension $= d$ (that $q$ is dominant follows from the fact that $\varphi$ is surjective, in particular dominant, and $X_1$ is a non-empty open subset of $X$). The required result follows from this.

**Lemma 9.4.1.7** *Let $(Y, \varphi)$ be a good quotient of $X$ by $G$.*
*Let $X'_{\max} = \{x \in X_{\max} \mid O(x) \text{ is closed}\}$. Then $X'_{\max}$ is a $G$-stable open subset of $X$.*

*Proof.* First observe that in view of Lemma 9.4.1.4, we have, $X'_{\max} \subseteq X_1$. The $G$-stability of $X'_{\max}$ is clear. Now $X \setminus X_{\max}$ being a $G$-stable closed subset of $X$, we have, $\varphi(X \setminus X_{\max})$ is a closed subset of $Y$. Hence $Y' := Y \setminus \varphi(X \setminus X_{\max})$ is an open subset of $Y$.

**Claim:** $X'_{\max} = \varphi^{-1}(Y')$.

**The inclusion $\subseteq$:**

Let $x \in X'_{\max} (\subseteq X_{\max})$. Then $\varphi(x) \notin \varphi(X \setminus X_{\max}) (= Y \setminus Y')$. Hence $\varphi(x) \in Y'$.

**The inclusion $\supseteq$:**

Let $x \in X$ be such that $\varphi(x) \in Y'$. We need to show that $x$ is in $X'_{\max}$. Let us assume if possible that $x \notin X'_{\max}$. Then either $x \in X \setminus X_{\max}$, or $x \in X_{\max}$, but $O(x)$ is not closed.

In the former case, $\varphi(x) \in \varphi(X \setminus X_{\max})(= Y \setminus Y')$. Hence $\varphi(x) \notin Y'$, a contradiction (note that by hypothesis, $\varphi(x) \in Y'$).

In the latter case, choose $x_1 \in \overline{O(x)} \setminus O(x)$. Then "closed orbit Lemma" implies that $dim\, O(x_1) < dim\, O(x) = d$. Hence $x_1 \notin X_{\max}$; this implies that $\varphi(x_1) \in \varphi(X \setminus X_{\max})(= Y \setminus Y')$. Hence $\varphi(x_1) \notin Y'$. This implies that $\varphi(x) \notin Y'$ (since $x_1 \in \overline{O(x)}$), a contradiction (note that by hypothesis, $\varphi(x) \in Y'$).

Thus in both cases, we arrive at contradictions. Hence our assumption is wrong and the inclusion $\supseteq$ follows.

**Corollary 9.4.1.8** *Let notation be as in Lemma 9.4.1.7. Let $Y'$ be as in the proof of Lemma 9.4.1.7. Then $\varphi : X'_{\max} \to Y'$ is a geometric quotient.*

*Proof.* The facts that $Y'$ is open in $Y$, and $X'_{\max} = \varphi^{-1}(Y')$ imply that $\varphi : X'_{\max} \to Y'$ is a good quotient (cf. Proposition 9.4.0.2,(1)). Further, each orbit $O(x), x \in X'_{\max}$ being closed, we have that $Y'$ is the orbit space of $X'_{\max}$ for the action of $G$.

We next recall the relationship between closed orbits and proper morphisms. We shall first recollect the definition of proper morphisms, complete varieties and some basic properties of complete varieties.

**Definition 9.4.1.9** *A morphism $f : X \to Y$ of schemes is* proper *if it is separated (i.e., $\Delta : X \to X \times_Y X$ is a closed immersion), of finite type, and universally closed (by "universally closed" one means the following: $f$ is closed, and for any morphism $Y' \to Y$, the corresponding morphism $f' : X' \to Y'$ obtained by base extension is also closed).*

**Definition 9.4.1.10** *A variety $X$ is* complete *if the structure morphism $f : X \to$ Spec $K$ is proper (note that this is equivalent to $f$ being universally closed).*

**Proposition 9.4.1.11** *(Some basic properties of complete varieties)*

*(1) Image of a complete variety is complete.*
*(2) A complete subvariety is closed.*
*(3) A closed subvariety of a complete variety is complete.*
*(4) A complete quasi-projective variety is projective.*
*(5) Constants are the only regular functions on a connected complete variety.*
*(6) A connected complete affine variety is a point.*
*(7) Inverse image of a complete variety under a proper morphism is complete.*

For a proof, see Borel [7].

The following Proposition relates closed orbits and proper morphisms.

**Proposition 9.4.1.12** *Let $X$ be an algebraic variety together with an action by an (affine) algebraic group $G$. Let $x \in X$. Let $\sigma_x$ be the morphism $G \to X, g \mapsto g \cdot x$ The following are equivalent:*

*(1) $O(x)$ is closed and $\dim O(x) = \dim G$.*
*(2) $\sigma_x$ is a proper morphism.*

*Proof.* **The implication** $\Rightarrow$: Let $\sigma'_x$ be the morphism $G \to O(x), g \mapsto g \cdot x$. The hypothesis that $\dim O(x) = \dim G$ implies that fibers of $\sigma'_x$ are finite.

**Claim:** $\sigma'_x$ is a finite morphism (recall that a morphism $f : V \to W$ is *finite*, if for every open affine $U$ of $W$, $f^{-1}(U)$ is affine, and $K[f^{-1}(U)]$ is integral over $K[U]$).

To prove the claim, clearly it suffices to find a non-empty open set $U$ in $O(x)$ such that $\sigma'_x : \sigma'^{-1}_x(U) \to U$ is finite; for, we can then use the action of $G$ to cover $O(x)$ with such open sets. But now the existence of such a $U$ follows from the following

**Sublemma:** Let $f : V \to W$ be a morphism with finite fibers. Then there exists a non-empty open set $U$ in $W$ such that $f : f^{-1}(U) \to U$ is a finite morphism.
For a proof of the sublemma, see Newstead [96].
Claim now implies that $\sigma'_x$ is proper (since a finite morphism is proper). Further, the inclusion $i : O(x) \hookrightarrow X$ is a closed immersion (since $O(x)$ is closed), and hence is proper. Now $\sigma_x$ being $i \circ \sigma'_x$ (a composition of two proper morphisms) is proper.

**The implication** $\Leftarrow$: The hypothesis that $\sigma_x$ is proper implies that $\sigma_x(G) = O(x)$ is closed in $X$; further, $\sigma_x^{-1}(x)(= G_x)$ is complete (in view of Proposition 9.4.1.11, (7)). Thus $G_x$ is complete and affine (note that $G_x$ being a closed subgroup of $G$ is affine); hence $G_x$ is finite (in view of Proposition 9.4.1.11, (6)). This implies that $\dim O(x) = \dim G$.

## 9.5 Stable and semi-stable points

Projective quotients are not as straightforward as affine quotients. To motivate our definition, note for comparison that in the case of an action by a reductive group $G$ on an affine variety $X = Spec\ R \subset \mathbb{A}^n$, the categorical quotient of $X$ by $G$ exists, and is defined on all of $X$. However, if $X$ is a projective variety, then in general there may not exist any of the three types of quotients (defined so far) on all of $X$. The best we can do in general is find a quotient on some nice open subvariety of $X$ as described below.

Consider a projective variety $X = Proj\ R$, together with an action by an algebraic group $G$. Suppose $X$ has an open affine $G$-stable cover $X = \bigcup_{\alpha \in I} Spec\ R_\alpha$. Then $Spec\ R_\alpha^G$, $\alpha \in I$ could be glued to arrive at a quotient. Again, in general such a nice affine cover does not exist for a given $X$. Nevertheless we are now going to get a "good quotient" on an open $G$-stable subset of $X$, namely $X^{ss}$ := {the semistable points of $X$} which will admit a $G$-stable open affine cover.

### 9.5.1 Stable, semistable, and polystable points

These concepts are defined for $G$-*linear* actions: $X$ is an algebraic variety (affine or projective), i.e., $X \subseteq V$ or $X \subseteq \mathbb{P}(V)$, and $X$ is a closed subset of $V, \mathbb{P}(V)$ respectively, $V$ a $G$-module (i.e., we are given an algebraic group morphism $G \to Aut_K\ V$), and the action of $G$ on $X$ is induced from the action of $G$ on $V, \mathbb{P}(V)$ respectively.

For the rest of this section, we shall suppose that we have a linear action on $X$ by a reductive group $G$.

We first consider the affine case. Let $Z \subset K^n (= \mathbb{A}^n)$ be a closed subset and let $G$ act linearly on $Z$.

**Definition 9.5.1.1** *A point $z \in Z$ is* semistable *if* $0 \notin \overline{G \cdot z}$.

$$Z^{ss} := \{z \in Z | z \textit{ is semistable}\}.$$

**Definition 9.5.1.2** *A point $z \in Z$ is* polystable *if $G \cdot z$ is a closed orbit.*

$$Z^{ps} := \{z \in Z | z \textit{ is polystable}\}.$$

**Definition 9.5.1.3** *A point $z \in Z$ is* strongly stable *(or just* stable*) if*

*(i) $G \cdot z$ is a closed orbit.*
*(ii) $\dim (G \cdot z) = \dim G$.*

$$Z^s := \{z \in Z | z \textit{ is stable}\}.$$

**Proposition 9.5.1.4** *Let $Z \subset K^n (= \mathbb{A}^n)$ be a closed subset and let a reductive group $G$ act linearly on $Z$. Let $z \in Z$. We have,*

*(i) $z \in Z^{ss} \Leftrightarrow$ there exists a homogeneous $G$-invariant form $F$ of degree $> 0$ on $K^n$ such that $F(z) \neq 0$*
*(ii) $z \in Z^s \Leftrightarrow \sigma_z : G \to K^n, g \mapsto g \cdot z$ is proper.*

*Proof.* (i) Let $z \in Z^{ss}$. Then $0, \overline{O(z)}$ are two disjoint $G$-stable closed subsets of $K^n$. Hence by Proposition 9.2.0.1, there exists a $G$-invariant polynomial $F$ such that $F(0) = 0$, $F(\overline{O(z)}) = 1$. Now $F(0) = 0$ implies that the constant term in $F$ is zero. Hence, some homogeneous part of $F$ is non-zero at $z$ (note that any homogeneous part of $F$ is again $G$-invariant, in view of the fact that if $G$ acts on a graded $K$-algebra $R$, then $R^G$ is also graded).

Conversely, given a homogeneous $G$-invariant form $F$ of degree $> 0$ on $K^n$ such that $F(z) \neq 0$, clearly $0 \notin \overline{O(z)}$ (since $F$ is homogeneous of degree $> 0$, we have, $F(0) = 0$).

(ii) follows from the definition of stability and Proposition 9.4.1.12.

Next, we define the notions of semistability, polystability, and stability for actions on a projective variety $X$.

For the following definitions, we suppose

$$X = Proj(R) \hookrightarrow \mathbb{P}(V),$$

$V$ a $G$-module, $G$ reductive, and the $G$-action on $X$ is induced from that on $\mathbb{P}(V)$. We denote by $\widehat{X} (\hookrightarrow \mathbb{A}^n = K^n = V)$, *the cone Spec R* over $X$.

**Definition 9.5.1.5** *A point $x \in X$ is* semistable, polystable, or stable *if there exists a point $\hat{x}$ lying over $x$ such that $\hat{x}$ is semistable, polystable, or stable respectively. Denote*

$$X^{ss} := \{x \in X | x \text{ is semistable}\}. \tag{9.1}$$

$$X^{ps} := \{x \in X | x \text{ is polystable}\}. \tag{9.2}$$

$$X^{s} := \{x \in X | x \text{ is stable}\}. \tag{9.3}$$

Note that $X^s \subset X^{ps} \subset X^{ss}$.

### 9.5.2 Other characterizations of stability, semistability

We now describe few other characterizations of stability, semistability. We begin with the following Lemma:

**Lemma 9.5.2.1** *Let $X$ be an algebraic variety (affine or projective) together with a linear action by a reductive group $G$, i.e., $X \subseteq V$ or $X \subseteq \mathbb{P}(V)$, and $V$ a $G$-module (and $X$ is a closed $G$-stable subset of $V, \mathbb{P}(V)$ respectively). Let $f$ be a $G$-invariant form on $V$ of degree $> 0$. Then $X_f$ is $G$-stable.*

*Proof.* Let $x \in X_f$. Let $\hat{x} (\in V)$ lie over $x$, in case $X$ is projective. Denote $y = x$ or $\hat{x}$ according as $X$ is affine or projective. Then $f(y) \neq 0$. For $g \in G$, we have,

$$f(g \cdot y) = (g^{-1} f)(y) = f(y) \neq 0$$

(the second equality holds in view of $G$-invariance of $f$). This implies $g \cdot x \in X_f$, and the $G$-stability of $X_f$ follows.

Similar to Proposition 9.5.1.4,(i) we have the following:

**Proposition 9.5.2.2** *Let $X = Proj\ R \subset \mathbb{P}(V)$, $V = K^n$ be a closed subset and let a reductive group $G$ act linearly on $X$. Let $x \in X$. Then $x \in X^{ss} \Leftrightarrow$ there exists a homogeneous $G$-invariant form $F$ of degree $> 0$ on $V$ such that $F(x) \neq 0$.*

The result is immediate from Proposition 9.5.1.4,(i) and the definition of semistable points on $X$.

Next, we describe three other characterizations of stable points on projective varieties (for linear actions by reductive groups).

**Proposition 9.5.2.3** *(First characterization of stability) Let $X = Proj\ R \subset \mathbb{P}(V)$, $V = K^n$ be a closed subset and let a reductive group $G$ act linearly on $X$. Let $x \in X$. Then $x \in X^s \Leftrightarrow \sigma_{\hat{x}} : G \to K^n, g \mapsto g \cdot \hat{x}$ is proper, $\hat{x}$ being a point lying over $x$.*

The result is immediate from Proposition 9.5.1.4,(ii).

Before describing the other two characterizations for stability of points in $X = Proj\ R$, we consider the following:

**Lemma 9.5.2.4** *Let $X$ be a projective variety together with a linear action by a reductive group $G$, i.e., $X \subseteq \mathbb{P}(V)$, and $V$ a $G$-module (and $X$ is a closed subset of $\mathbb{P}(V)$). Let $x \in X$. Let $f$ be a $G$-invariant form on $V$ of degree $d > 0$ such that $f(x) \neq 0$. Let $\hat{x}$ be a point in $V$ lying over $x$. Let $\sigma_{\hat{x}}$, $(\sigma_x)_f$ be the morphisms*

$$\sigma_{\hat{x}} : G \to V, g \mapsto g \cdot \hat{x}, \ (\sigma_x)_f : G \to \mathbb{P}(V)_f, g \mapsto g \cdot x$$

*Then $(\sigma_x)_f$ is proper if and only if $\sigma_{\hat{x}}$ is.*

*Proof.* Let $f(\hat{x}) = \alpha(\neq 0)$, $Z_\alpha = \{y \in V \mid f(y) = \alpha\}$, and let $i_\alpha$ be the inclusion $Z_\alpha \hookrightarrow V$; note that $i_\alpha$ is a closed immersion. Let $(\sigma_{\hat{x}})_f$ be the morphism

$$(\sigma_{\hat{x}})_f : G \to Z_\alpha, g \mapsto g \cdot \hat{x}$$

Consider the commutative diagram

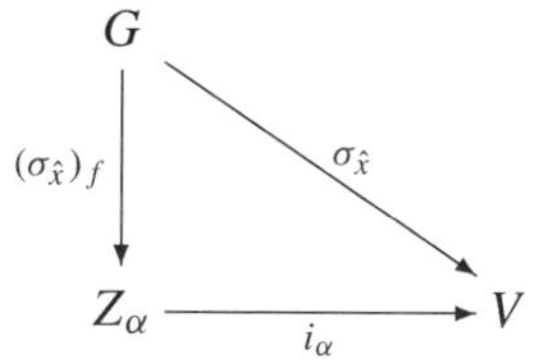


This implies a factorization

(∗) $$\sigma_{\hat{x}} = i_\alpha \circ (\sigma_{\hat{x}})_f$$

Let $\pi_f$ be the morphism

$$\pi_f : Z_\alpha \to \mathbb{P}(V)_f, z \mapsto [z]$$

We have, $\pi_f$ is a finite morphism (note that fibers of $\pi_f$ are the $d$-th roots of 1, $d$ being deg $f$). Consider the commutative diagram

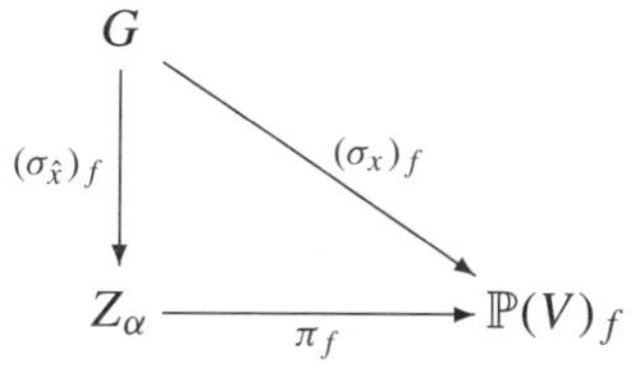


This implies a factorization

(∗∗) $$(\sigma_x)_f = \pi_f \circ (\sigma_{\hat{x}})_f$$

Consider the following (easy) facts:

- Composite of two proper morphisms is proper.
- Let $F$ be a proper morphism of algebraic varieties. Further, let $F = \theta \circ \delta$, with $\theta$ a separated morphism. Then $\delta$ is a proper morphism; in particular, if $\theta$ is proper, then $\delta$ is a proper morphism ($\theta \circ \delta$ being a proper morphism).
- A closed immersion is proper. (In particular, $i_\alpha$ is proper)
- A finite morphism is proper. (In particular, $\pi_f$ is proper.)

These facts together with (*), (**) imply the required result.

**Lemma 9.5.2.5** *(Second characterization of stability) Let $X, G$ be as in Lemma 9.5.2.4. Let $x \in X$. The following are equivalent:*

*(i) $x \in X^s$.*
*(ii) (a) There exists a $G$-invariant form $f$ on $V$ of degree $> 0$ such that $f(x) \neq 0$.*
*(b) $O(x)$ is closed in $X_f$.*
*(c) $dim\, O(x) = dim\, G$.*

***Note:*** *Condition (ii) is equivalent to the condition $x \in X^s_f$, for some $G$-invariant form $f$ on $V$ of degree $> 0$. Hence we have:*

***Reformulation:*** *$x \in X^s$ if and only if there exists a $G$-invariant form $f$ on $V$ of degree $> 0$ such that $x \in X^s_f$.*

*Proof.* Suppose that there exists a $G$-invariant form $f$ on $V$ of degree $> 0$, such that $x \in X_f$; note that $X_f$ is $G$-stable (cf. Lemma 9.5.2.1). Let $i_f$ be the closed immersion:

$$i_f : X_f :\hookrightarrow \mathbb{P}(V)_f.$$

Let $(\phi_x)_f$ be the map

$$(\phi_x)_f : G \to X_f, g \mapsto g \cdot x$$

Then $(\sigma_x)_f : G \to \mathbb{P}(V)_f, g \mapsto g \cdot x$, factors as

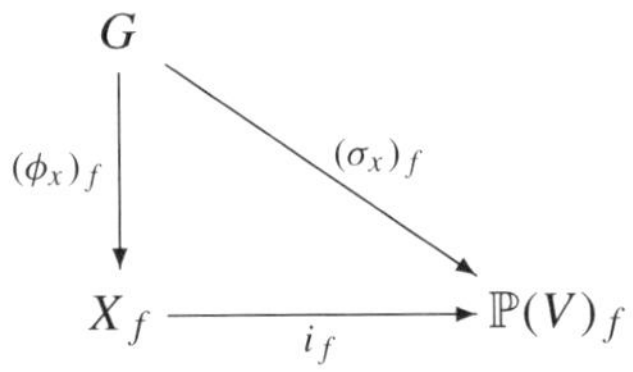


Thus we obtain

$$(\sigma_x)_f = i_f \circ (\phi_x)_f$$

Hence as in the proof of Lemma 9.5.2.4, we have

$$(*) \qquad (\sigma_x)_f \text{ is proper } \Leftrightarrow (\phi_x)_f \text{ is}$$

**(i) $\Rightarrow$ (ii):** The hypothesis $x \in X^s$ in particular implies that $x \in X^{ss}$, and assertion ii(a) follows from Proposition 9.5.2.2. Let $\hat{x}$ be a point in $V$ lying over $x$. Then the hypothesis that $x \in X^s$ implies that $\hat{x}$ is stable, and hence $\sigma_{\hat{x}} : G \to V, g \mapsto g \cdot \hat{x}$ is proper (cf. Lemma 9.4.1.12). This implies (in view of Lemma 9.5.2.4) that $(\sigma_x)_f$ is proper; hence $(\phi_x)_f$ is proper (cf. (*)). This together with Proposition 9.4.1.12 (applied to $X_f$) imply assertions (ii)(b),(c).

**(ii) $\Rightarrow$ (i):** The hypotheses (ii)(b),(c) imply that $(\phi_x)_f : G \to X_f$ is proper (in view of Proposition 9.4.1.12 applied to $X_f$), and hence $(\sigma_x)_f$ is proper (cf. (*)). This implies (in view of Lemma 9.5.2.4) that $\sigma_{\hat{x}} : G \to V$ is proper. It follows (in view of Proposition 9.4.1.12) that $\hat{x}$ (and hence $x$) is stable.

**Lemma 9.5.2.6** *(Third characterization of stability) Let $X, G$ be as in Lemma 9.5.2.4. Let $x \in X$. The following are equivalent:*

*(i)* $x \in X^s$.
*(ii)* *(a) There exists a G-invariant form* $\theta$ *on* $V$ *of degree* $d > 0$ *such that* $\theta(x) \neq 0$.
*(b) G-action on* $X_\theta$ *is closed.*
*(c)* $dim\, O(x) = dim\, G$

*Proof.* The implication $\Leftarrow$ follows from Lemma 9.5.2.5.

**The implication** $\Rightarrow$**:** The hypothesis that $x \in X^s$ implies that there exists a $G$-invariant form $f$ of degree $> 0$ such that $x \in X^s_f$ (by Lemma 9.5.2.5, rather its reformulation). Let $d = dim\, G$, $Z = \{y \in X_f \mid dim\, O(y) < d\}$. Then in view of Lemma 9.4.1.2,(2), $Z$ is a closed subset of $X_f$. Now, $O(x)$, $Z$ are two disjoint $G$-stable closed subsets of the affine variety $X_f = Spec\, R_{(f)}$ (here, $R$ is given by $X = Proj\, R$). Hence by (geometric reductivity of $G$), there exists a $F \in R^G_{(f)}$ such that $F(Z) = 0$, $F(O(x)) = 1$) (cf. Proposition 9.2.0.1). Writing $F = \frac{H}{f^r}$ (for suitable $H, r$), we have $H(\in R^G)$. Let $S = S(V^*)$, $R = S/I$; then $H^t \in S^G/(I \cap S^G)$ for a suitable $t$ (cf. Newstead [96], Lemma 3.4.1). Taking a lift $h$ for $H^t$ in $S^G$, we obtain $F^t = \frac{h}{f^s}$, for a suitable $s$. Let $\theta = fh$. We have $h(Z) = 0$ (since $F(Z) = 0$). Therefore $\theta(z) = 0$, for $z \in Z$. Hence

$$(*) \qquad Z \cap X_\theta = \emptyset$$

**Claim:** For any $y \in X_\theta$, $dim\, O(y) = dim\, G$.

Let $y \in X_\theta$. Then $f(y) \neq 0$ (since $\theta(y) \neq 0$). Hence $y \in X_f$, and therefore $X_\theta \subseteq X_f$. This together with (*) implies

$$X_\theta \subseteq X_f \setminus Z$$

Claim follows (by the definition of $Z$). Claim together with "closed orbit Lemma" implies that the $G$-action on $X_\theta$ is closed. Now $x \in X_\theta$ (since $f(x) \neq 0$, $F(x) \neq 0$, we have $h(x) \neq 0$, and therefore $\theta(x) \neq 0$), the $G$-action on $X_\theta$ is closed, and $dim\, O(x) = dim\, G$ (cf. Lemma 9.5.2.5,ii(c)). This completes the proof of the implication $\Rightarrow$, and hence of the Lemma.

Combining Proposition 9.5.2.3, Lemmas 9.5.2.5, 9.5.2.6, we obtain the following:

**Proposition 9.5.2.7** *Let X be a projective variety together with a linear action by a reductive group G, i.e.,* $X \subseteq \mathbb{P}(V)$*, and* $V (= K^n)$ *a G-module (and X is a closed subset of* $\mathbb{P}(V)$*). Let* $x \in X$*. The following are equivalent.*

*(i)* $x \in X^s$
*(ii)* $\sigma_{\hat{x}} : G \to K^n, g \mapsto g \cdot \hat{x}$ *is proper,* $\hat{x}$ *being a point lying over* $x$.
*(iii)* $dim\, O(x) = dim\, G$*, and there exists a G-invariant form f on V of degree* $> 0$ *such that* $x \in X_f$ *and* $O(x)$ *is closed in* $X_f$.
*(iv)* $dim\, O(x) = dim\, G$*, and there exists a G-invariant form* $\theta$ *on V of degree* $> 0$ *such that* $x \in X_\theta$ *and the G-action on* $X_\theta$ *is closed.*

As a consequence of Propositions 9.5.1.4, 9.5.2.2, we have the following:

**Proposition 9.5.2.8** *Let $X$ be an algebraic variety (affine or projective) together with a linear action by a reductive group $G$, i.e., $X \subseteq V$ or $X \subseteq \mathbb{P}(V)$, and $V$ a $G$-module (and $X$ is a closed subset of $V, \mathbb{P}(V)$ respectively). Then $X^{ss}$ is a $G$-stable open subset of $X$.*

*Proof.* In view of Propositions 9.5.1.4, 9.5.2.2, we have,

$$X^{ss} = \cup X_F$$

where the union is taken over all $G$-invariant forms of degree $> 0$ on $V$. Hence it follows that $X^{ss}$ is open. The $G$-stability of $X^{ss}$ follows from that of $X_F$ (cf. Lemma 9.5.2.1).

**Proposition 9.5.2.9** *Let $X$ be an algebraic variety (affine or projective) together with an action of a reductive group $G$. Then $X^s$ is a $G$-stable open subset of $X$.*

*Proof.* This requires a proof only when $X^s$ is non-empty.

If $X$ is affine, then the result follows from Lemma 9.4.1.7 (note that since $X^s$ is non-empty, we have $d = dim\, G$, $X'_{\max} = X^s$).

Let $X$ be projective, say, $X = Proj\, R \subseteq \mathbb{P}(V)$, and $V$ a $G$-module (and $X$ is a closed subset of $\mathbb{P}(V)$). Let

$$Z = \cup X_f$$

the union being taken over the set of all $G$-invariant homogeneous forms $f$ on $V$ of degree $> 0$ such that the action of $G$ on $X_f$ is closed. Then $Z$ is $G$-stable and open. Further, $X^s \subseteq Z$ (in view of Lemma 9.5.2.6); hence the maximum of the orbit dimensions on $Z$ equals $dim\, G$. In fact, we have $X^s = Z_{\max}$; for, if $x \in Z_{\max}$, then $dim\, O(x) = dim\, G$, $x \in X_f$ for some $G$-invariant homogeneous form $f$, and action of $G$ on $X_f$ is closed; hence $x \in X^s$ (cf. Proposition 9.5.2.7). Thus $X^s (= Z_{\max})$ is $G$-stable and open in $Z$ (cf. Lemma 9.4.1.3), and hence open in $X$.

## 9.6 Projective quotients

In the previous sections we have described several types of quotients. The class of categorical quotients includes the class of good quotients, which in turn includes the class of geometric quotients.

Let $X = Proj\, R \subset \mathbb{P}(V)$, where $X$ is closed in $\mathbb{P}^n$, and $V = K^n$.

**A good quotient for $X^{ss}$:** We will now see that

$$X^{ss} (= \{\text{all the semistable points of X}\})$$

admits a good quotient.

As seen in the proof of Proposition 9.5.2.8, we have

$$X^{ss} = \bigcup X_F$$

the union being taken over the set of all $G$-invariant homogeneous forms $f$ on $V$ of degree $> 0$. Now $X_F = Spec\, R_{(F)}$, where $R_{(F)}$ is the homogeneous localization of $R$ at $F$. Then $X_F /\!\!/ G$ exists, namely, $Spec\, R^G_{(F)}$, and in fact is a good quotient (recall that a categorical affine quotient is a good quotient). Hence by Proposition 9.4.0.2,(2), we have, $X^{ss}$ admits a good quotient; in fact, $X^{ss} /\!\!/ G$ is nothing but $Proj\, R^G$.

**A geometric quotient for** $X^s$**:** Let $Y = Proj\, R^G$, and $\varphi : X^{ss} \to Y$, the quotient map. Let

$$W = \bigcup Y_f$$

the union being taken over the set of all $G$-invariant homogeneous forms $f$ on $V$ of degree $> 0$ such that the action of $G$ on $X_f$ is closed. Let

$$Z = \varphi^{-1}(W) = \cup X_f$$

where the union is taken over the same set of $f$'s as above. As in the proof of Proposition 9.5.2.9, we have,

$$X^s = Z_{\max}$$

In particular, the maximum orbit dimension on $Z$ equals $dim\, G$. Denoting $Z'_{\max} := \{x \in Z_{\max} \mid O(x) \text{ is closed}\}$, we have that the inclusion $Z'_{\max} \subseteq Z_{\max}$ is in fact an equality; for, if $x \in Z_{\max}(= X^s)$, say $x \in X_f$, for some $G$-invariant homogeneous $f$, then $O(x)$ is closed in $X_f$ (by the definition of $Z$); hence, $O(x)$ is closed in $Z$ in view of the following easy observation:

*Suppose $Z = \cup U_\alpha$ is an open cover, and $T$ is a subset of $Z$ such that $U_\alpha \cap T$ is closed in $U_\alpha$ for all $\alpha$. Then $T$ is closed in $Z$.*

Thus we obtain

$$Z'_{\max} = Z_{\max}(= X^s)$$

Also, by the Claim in the proof of Lemma 9.4.1.7, we have,

$$(*) \qquad X^s(= Z'_{\max}) = \varphi^{-1}(W \setminus \varphi(Z \setminus Z_{\max})) = \varphi^{-1}(Y^s), \quad \texttt{say}$$

where, $Y^s = W \setminus \varphi(Z \setminus Z_{\max})$. Now $Y^s$ being open in $W$, we get (cf. Proposition 9.4.0.2, (1))

$$\varphi : X^s \to Y^s$$

is a good quotient. It is in fact a geometric quotient since the $G$-action on $X^s(= Z'_{\max})$ is closed, and hence $Y^s$ is the orbit space of $X^s$ (for the action of $G$).

**Remark 9.6.0.1.** Now $Y^s$ being open in $Y(= X^{ss} /\!\!/ G = Proj\, R^G)$, $X^{ss} /\!\!/ G$ may be regarded as a compactification of $X^s/G$.

From the discussion above, we obtain the following:

**Lemma 9.6.0.2** *Let $X, G$ be as in Lemma 9.5.2.4. Let $x \in X^{ss}$. The following are equivalent:*

*(i) $x \in X^s$.*

*(ii) $O(x)$ is closed in $X^{ss}$, and $dim\, O(x) = dim\, G$.*

*Proof.* **The implication** $(i) \Rightarrow (ii)$**:** Let $x \in X^s$. We have (cf. Proposition 9.5.2.7) $dim\, O(x) = dim\, G$.

Now $\overline{O(x)} \subseteq \varphi^{-1}(\varphi(x))$. Also, since $\varphi(x)$ is in $Y^s$, we have, $\varphi^{-1}\varphi(x) \subseteq \varphi^{-1}(Y^s) = X^s$ (cf. (*) above). Thus we obtain

$$(**) \qquad \overline{O(x)} \subseteq X^s$$

Hence

$$\overline{O(x)} \cap X^{ss} \subseteq X^s \cap X^{ss} = X^s$$

Thus, denoting $\overline{O(x)} \cap X^{ss}$ (the orbit closure in $X^{ss}$) by $C(x)$, we have, $C(x) \subseteq X^s$. This together with the "closed orbit Lemma" and the fact that $dim\, O(y) = dim\, G$, $y \in X^s$ implies that $C(x) = O(x)$, i.e., $O(x)$ is closed in $X^{ss}$.

**The implication** $(ii) \Rightarrow (i)$**:** The hypothesis that $O(x)$ is closed in $X^{ss}$ implies that for every $G$-invariant homogeneous form $f$ such that $f(x) \neq 0$, we have, $O(x)$ is closed in $X_f$ (since $X_f \subseteq X^{ss}$). This together with the hypothesis that $dim\, O(x) = dim\, G$ implies that $x \in X^s$ (cf. Proposition 9.5.2.7).

We now give some examples:

**Example 1:** Let $X = \mathbb{P}^{n-1}$. Consider the natural action of $G := SL_n(K)$ on $X$, we have

$$X^{ss} = \emptyset$$

(since constants are the only invariant polynomials).

**Example 2:** Let $G = GL_n(K)$, $X = \mathbb{M}_n(K)$, $G \circlearrowleft X$ by conjugation. Then

$$X^{ss} = \{\text{non-nilpotent matrices}\}$$

$$X^s = \{\text{non-nilpotent diagonalizable matrices}\}.$$

**Example 3:** Let $G = SL_d(K)$, $X = \mathbb{M}_{d\times n}(K)$, $G \circlearrowleft X$ by multiplication on the left. Then

$$X^s = X^{ss} = \{d \times n \text{ matrices of rank } d\}.$$

**Example 4:** Let $X = \{\text{quadrics in } \mathbb{P}^{n-1}\}$. Elements of $X$ are of the form $\sum a_{ij}x_ix_j$, where $i$ and $j$ range over $1, \ldots, n$ and $a_{ij} \in K$. We may identify each quadric with a symmetric $n \times n$ matrix. In other words, denoting $S := Sym\, \mathbb{M}_n$, we have a bijection $X \overset{bij}{\longleftrightarrow} \mathbb{P}(S)$,

$$\sum a_{ij}x_ix_j \longleftrightarrow A = (a_{ij}) \in S.$$

Thus we may identify $X$ with $\mathbb{P}(S)$.

Let $G = SL_n(K)$. Consider the action of $G$ on $X$ induced by the action of $G$ on $S$ by conjugation. The function $\Delta := det$ (i.e. the determinant) on $S$ is $G$-invariant. We have,

$$X^{ss} = \mathbb{P}(S)_\Delta (= \{\text{quadrics such that } |A| \neq 0\}), \;\; X^s = \emptyset$$

## 9.7 $L$-linear actions

More generally, One defines stability, semistability on a projective variety $X$ with respect to a $L$-*linear action* of $G$ on $X$, i.e., $L$ is a line bundle over $X$, together with a $G$-action such that under the natural projection $\pi : L \to X$, the following hold:

(i) $\pi(g \cdot y) = g \cdot \pi(y),\ y \in L,\ g \in G.$
(ii) For $x \in X$, $g \in G$, the natural map $L_x \to L_{g \cdot x}$ is linear.

Let $G$ be reductive, and $L$ an ample line bundle on $X$.

**Definition 9.7.0.1** *For a L-linear action of G on X, a point $x \in X$ is* semistable *if there exists a G-invariant section $f$ of $L^r$ (for some positive integer r) such that $f(x) \neq 0$, and $X_f$ is affine.*

**Definition 9.7.0.2** *For a L-linear action of G on X, a point $x \in X$ is* stable *if*

*(i) $\dim O(x) = \dim G$.*
*(ii) There exists a G-invariant section $f$ of $L^r$ (for some positive integer r) such that $f(x) \neq 0$, $X_f$ is affine, and the action of G on $X_f$ is closed.*

**Remark 9.7.0.3.** Note that Definition 9.5.1.5, (1),(3) coincide with Definitions 9.7.0.1, 9.7.0.2, with $L$ being the restriction to $X$ of the tautological line bundle on $\mathbb{P}(V)$.

Denote $X^{ss}(L)$, $X^s(L)$ the set of points given by Definitions 9.7.0.1, 9.7.0.2 respectively. We have results for $X^{ss}(L)$, $X^s(L)$ similar to the results for $X^{ss}$, $X^s$ as seen in §9.6:

**Theorem 9.7.0.4** *(i) There exists a good quotient $(Y_L, \varphi_L)$ of $X^{ss}(L)$ by G, and $Y_L$ is quasi-projective.*
*(ii) There exists an open subset $Y_L^s$ of $Y_L$, such that $\varphi^{-1}(Y_L^s) = X^s(L)$, and $(Y_L^s, \varphi)$ is a geometric quotient of $X^s(L)$ by G.*
*(iii) For $x_1, x_2 \in X^{ss}(L)$,*

$$\varphi(x_1) = \varphi(x_2) \Leftrightarrow \overline{O(x_1)} \cap \overline{O(x_2)} \cap X^{ss}(L) \neq \emptyset$$

*(iv) Let $x \in X^{ss}(L)$. The following are equivalent:*
*a) $x \in X^s(L)$.*
*b) $O(x)$ is closed in $X^{ss}(L)$, and $\dim O(x) = \dim G$.*

For a proof of Theorem 9.7.0.4, see Newstead [96], Theorem 3.21.

## 9.8 Hilbert-Mumford criterion

In this section, we recall *Hilbert-Mumford criterion.*

In practice it is a hard problem to find (calculate) $X^{ss}$ and $X^s$ for a general $X$ and $G$. The Hilbert-Mumford criterion gives a numerical criterion for the determination of $X^{ss}$ and $X^s$ in terms of one-parameter subgroups of $G$ given by algebraic group morphisms $\lambda : K^* \to G$. In the sequel, we shall abbreviate "one-parameter subgroup" as 1-PS.

Let $X = Proj R \subseteq \mathbb{P}(V)$ and $\widehat{X} \subset V(= K^n)$ be the cone over $X$; the actions of $G$ on $V, \mathbb{P}(V)$ induce (linear) actions on $\widehat{X}, X$ respectively. Fix $x \in X$. Given a 1-PS $\lambda : K^* \to G$, define $\varphi_\lambda : K^* \to K^n \backslash \{0\}$ by $t \overset{\varphi}{\longmapsto} \lambda(t) \cdot \hat{x}$.

**Question:** Does $\varphi_\lambda$ extend to all of $\mathbb{P}^1(= K \cup \{\infty\})$ ?

In general, a morphism $f : K^* \to Y$ may not extend to all of $\mathbb{P}^1$.

**Fact:** If $Y$ is projective, then $f$ does extend to all of $\mathbb{P}^1$, and the extension is unique. For a proof of the above Fact, see Borel [7].

In the present situation, with $f = \varphi_\lambda$, we have, $Y = K^n \backslash \{0\}$; hence, $\varphi_\lambda$ extends to all of $\mathbb{P}^1$ and the extension is unique.

Let $\varphi_0 = \varphi_\lambda(0)$, and $\varphi_\infty = \varphi_\lambda(\infty)$. One denotes $\varphi_0, \varphi_\infty$, also by $\lim_{t\to 0} \varphi_\lambda(t)$, $\lim_{t\to\infty} \varphi_\lambda(t)$ respectively.

The next question is how to compute $\varphi_0$ and $\varphi_\infty$. We have

$$\lambda : K^* \to G \to Aut\, V,$$

where $V = K^n$. Then $\lambda(K^*)$ is a family of commuting semisimple endomorphisms of $V$, and so the actions of $\lambda(K^*)$ on $V$ can be simultaneously diagonalized. We denote the corresponding basis by $\{e_i\}$. We have, $\lambda(t)e_i = t^{r_i} e_i$ for some $r_i \in \mathbb{Z}$. Let $\hat{x} = \sum c_i e_i$. Then

$$(*) \qquad \lambda(t) \cdot \hat{x} = \sum c_i t^{r_i} e_i$$

Define $\mu(\lambda, x) := -\min\{r_i\}$.

**Theorem 9.8.0.1 (Hilbert-Mumford criterion)** *We have*

*(i)* $x \in X^{ss}$ *if and only if* $\mu(\lambda, x) \geq 0$, *for every one-parameter subgroup* $\lambda$ *of* $G$.
*(ii)* $x \in X^s$ *if and only if* $\mu(\lambda, x) > 0$, *for every one-parameter subgroup* $\lambda$ *of* $G$.

**Sketch of proof of Hilbert-Mumford criterion:** we first note the following:

(i) $\mu(\lambda, x) > 0$ if and only if there exists a term $t^{r_i}$ with $r_i < 0$ on the right hand side of (*).
(ii) $\mu(\lambda, x) = 0$ if and only if there are no negative exponents (for $t$) on the right hand side of (*), and there is a term not involving $t$ on the right hand side of (*)
(iii) $\mu(\lambda, x) < 0$ if and only if all exponents (for $t$) on the right hand side of (*) are $> 0$.

We may reformulate these in terms of $\lim \lambda(t) \cdot \hat{x}$:

- $\mu(\lambda, x) > 0$ if and only if $\lim\limits_{t\to 0} \lambda(t) \cdot \hat{x}$ does not exist in $K^n$.
- $\mu(\lambda, x) \geq 0$ if and only if $\lim\limits_{t\to 0} \lambda(t) \cdot \hat{x}$ either does not exist in $K^n$, or the limit exists in $K^n$ and is non-zero.

We shall first see a proof of Hilbert-Mumford criterion for $G = K^*$:

**Theorem 9.8.0.2** *Let $G = K^*$. Then*

*(i) $x \in X^{ss}$ if and only if $\mu(\lambda, x) \geq 0$, for every one-parameter subgroup $\lambda$ of $G$.*
*(ii) $x \in X^s$ if and only if $\mu(\lambda, x) > 0$, for every one-parameter subgroup $\lambda$ of $G$.*

*Proof.* We first prove (ii). We have, by definition, $x \in X^s$ if and only if (i) $G \cdot \hat{x}$ is closed, and (ii) $dim\ G \cdot \hat{x} = dim\ G$. We have $\varphi_{\hat{x}} : K^* \to G \cdot \hat{x} \subset K^n$ defined by $t \mapsto t \cdot \hat{x}$. Since $dim\ K^* = 1$, $dim\ \varphi_{\hat{x}}(K^*)$ is either 0 or 1. If $dim\ \varphi_{\hat{x}}(K^*) = 0$, then $\hat{x}$ is $G$-fixed. Otherwise, $dim\ \varphi_{\hat{x}}(K^*) = 1$, in which case $x$ is stable if and only if $G \cdot \hat{x}$ is closed in $K^n$.

Now the closure of $G \cdot \hat{x}$ in $\mathbb{P}(V)$ is $G \cdot \hat{x} \cup \{\varphi_0, \varphi_\infty\}$. Hence for the closure of $G \cdot \hat{x}$ in $K^n$, we have four cases, depending on whether either or both of $\varphi_0$ and $\varphi_\infty$ belong to $K^n$ or not:

$$\overline{G \cdot \hat{x}} = \begin{cases} G \cdot \hat{x} & \text{if } \varphi_0, \varphi_\infty \notin K^n \\ G \cdot \hat{x} \cup \{\varphi_0, \varphi_\infty\} & \text{if } \varphi_0, \varphi_\infty \in K^n \\ G \cdot \hat{x} \cup \{\varphi_\epsilon\} & \text{if only } \varphi_\epsilon \in K^n,\ \epsilon = 0 \texttt{ or } \infty \end{cases}$$

Now $G \cdot \hat{x}$ is closed in $K^n$ if and only if neither $\varphi_0$ nor $\varphi_\infty$ is in $K^n$, which is to say that the limits $\lim\limits_{t \to 0} t \cdot \hat{x}$ and $\lim\limits_{t \to \infty} t \cdot \hat{x}$ do not exist in $K^n$. This holds if and only if there are both negative and positive exponents of $t$ in formula (*).

Now if we define $\lambda_1 : K^* \to K^*$ by $t \mapsto t$ and $\lambda_2 : K^* \to K^*$ by $t \mapsto t^{-1}$, then any one-parameter subgroup $\lambda : K^* \to K^*$ is given by $t \mapsto t^r$ for some $r \in \mathbb{Z}$. Hence, $\lambda = \lambda_i^r$ for some $i \in \{1, 2\}$ and $r \geq 0$.

Thus there are both negative and positive exponents of $t$ in formula (*) if and only if $\mu(x, \lambda_1) > 0$ and $\mu(x, \lambda_2) > 0$, which happens if and only if $\mu(x, \lambda) > 0$ for every one-parameter subgroup $\lambda$. This completes the proof of (ii).

The proof of (i) is similar. We have that $x \in X^{ss}$ if and only if $0 \notin \overline{G \cdot \hat{x}}$, which happens if and only if the limits $\lim\limits_{t \to 0} t \cdot \hat{x}$ and $\lim\limits_{t \to \infty} t \cdot \hat{x}$ are not zero. This happens if and only if $\mu(x, \lambda_1) \geq 0$ and $\mu(x, \lambda_2) \geq 0$, which happens if and only if $\mu(x, \lambda) \geq 0$ for every one-parameter subgroup $\lambda$. This completes the proof of (i).

**Proof of Hilbert-Mumford criterion:** Now we let $G$ be any reductive group. We first prove the implication ($\Rightarrow$) for (i) and (ii).

Let $\hat{x}$ be a point in $K^n$ lying over $x$.

(i) Let $x \in X^{ss}$. Then $0 \notin \overline{G \cdot \hat{x}}$. This implies that $0 \notin \overline{K^* \cdot \hat{x}}$ for every one-parameter subgroup $\lambda : K^* \to G$. Hence $x \in X^{ss}$ for the $K^*$ action on $X$, and this implies that $\mu(x, \lambda) \geq 0$ for every one-parameter subgroup $\lambda$ by Theorem 9.8.0.2, (i).

(ii) Let $x \in X^s$. Let $\lambda : K^* \to G$ be a (non-trivial one-parameter subgroup). Consider the commutative diagram

$$\begin{array}{ccc} K^* & \xrightarrow{\lambda} & G \\ {\scriptstyle\varphi_{\hat{x}}} \downarrow & & \downarrow {\scriptstyle\sigma_{\hat{x}}} \\ K^* \cdot \hat{x} & \hookrightarrow\ G \cdot \hat{x}\ \hookrightarrow & K^n \end{array}$$

The hypothesis that $x \in X^s$ implies that $dim\ G \cdot x = dim\ G$, and hence $G_{\hat{x}}$ is finite. This implies

$$(*) \qquad K^*_{\hat{x}} \text{ is finite}$$

Further $\lambda(K^*)$ is closed in $G$ (since the image of an algebraic group homomorphism is closed), and hence, $K^* \cdot \hat{x} (= \sigma_{\hat{x}}(\lambda(K^*)))$ is closed in $K^n$ (since $x \in X^s, \sigma_{\hat{x}}$ is proper (cf. Proposition 9.4.1.12), in particular closed). This together with (*) implies that $x$ is stable for the $K^*$ action, and the required result follows from this (in view of Theorem 9.8.0.2,(ii)).

**The implication** ($\Leftarrow$)**:** We are required to show the following:

$(i)$ `if` $x \notin X^s$, `then there exists a 1-PS`$\lambda$ `such that` $\mu(x, \lambda) \leq 0$,
$(ii)$ `if` $x \notin X^{ss}$, `then there exists a 1-PS`$\lambda$ `such that` $\mu(x, \lambda) < 0$.

In view of Theorem 9.8.0.2, it suffices to show that if $x$ is non-stable (resp. non-semistable), then there exists a 1-PS $\lambda$ such that $x$ is still non-stable (resp. non-semistable) for the induced action of $K^*$. Such a $\lambda$ is constructed using the "valuative criterion" for proper morphisms (see [37] for example), and Iwahori decomposition

$$G(L) = \bigcup_{\{\lambda_L\}} G(A)\lambda_L(L^*)G(A)$$

of $L$-valued points of (the reductive group) $G$, in terms of the $A$-valued points of $G$, where $A$ is a valuation ring and $L$ is the quotient field of $A$; the union is taken over all 1-PS's $\lambda$. Also, $\lambda_L$ denotes the 1-PS $L^* \to G(L)$ induced by $\lambda : K^* \to G(K)$. We skip the details (and refer the reader to [110]).

# 10

# Classical Invariant Theory

In this chapter, we describe classical invariant theory (cf. [115]) for the classical group actions. We give GIT-theoretic proofs of the results in [115]. By relating them to certain $G/P$'s (and their Schubert varieties), we describe "standard monomial bases" for these rings of invariants.

## 10.1 Preliminary lemmas

In this section, we first prove some Lemmas concerning quotients leading to the main Lemma (cf. Lemma 10.1.0.4), to be applied to the following situation:

Suppose, we have an action of a reductive group $G$ on an affine variety $X = Spec R$. Suppose that $S$ is a subalgebra of $R^G$. We give below (cf. Lemma 10.1.0.4) a set of sufficient conditions for the equality $S = R^G$. We start with recalling

**Theorem 10.1.0.1** *(**Zariski Main Theorem**, [86],III.9) Let $\varphi : X \to Y$ be a morphism such that*

*(i) $\varphi$ is surjective*
*(ii) fibers of $\varphi$ are finite*
*(iii) $\varphi$ is birational*
*(iv) $Y$ is normal*

*Then $\varphi$ is an isomorphism.*

Let $X$ = SpecR and a reductive group $G$ act linearly on $X$, i.e., we have a linear action of $G$ on an affine space $\mathbb{A}^r$ and we have a $G$-equivariant closed immersion $X \hookrightarrow \mathbb{A}^r$. Further, let $R$ be a graded $K$-algebra. Let $X^{ss}$ be the set of semi-stable points of $X$ (i.e., points $x$ such that $0 \notin \overline{G \cdot x}$). Let $X_1 = Proj\ R$, and $X_1^{ss}$, the set of semi-stable points of $X_1$ (i.e., points $y \in X_1$ such that if $\hat{x}$ is any point in $K^{n+1} \setminus 0$ lying over $y$, then $\hat{x}$ is in $X^{ss}$). Let $f_1, \cdots, f_N$ be homogeneous $G$-invariant elements in $R$. Let $S = K[f_1, \cdots, f_N]$. Then for the morphism $Spec\ R^G \to Spec\ S$, the hypothesis (2) in Theorem 10.1.0.1 may be concluded if $\{f_1, \cdots, f_N\}$ is base-point free on $X_1^{ss}$ as given by the following

**Lemma 10.1.0.2** *Suppose $f_1, \cdots, f_N$ are homogeneous $G$-invariant elements in $R$ such that for any $x \in X^{ss}$, $f_i(x) \neq 0$, for at least one $i$. Then Spec $R^G \to$ Spec $S$ has finite fibers.*

*Proof.* **Case 1:** Let $f_1, \cdots, f_N$ be of the same degree, say, $d$. Let $Y = Spec R^G$ ($= X /\!\!/ G$, the categorical quotient) and $\varphi : X \to Y$ be the canonical quotient map. Let $X_1 = Proj R$, and $X_1^{ss}$ be the set of semi-stable points of $X_1$. Let $Y_1 = Proj R^G$ ($= X_1^{ss} /\!\!/ G$), and $\varphi_1 : X_1^{ss} \to Y_1$ be the canonical quotient map. Consider $\psi : X \to \mathbb{A}^N, x \mapsto (f_1(x), \cdots, f_N(x))$. This induces a map $\rho : Y \to \mathbb{A}^N$ (since $f_1, \cdots, f_N$ are $G$-invariant). The commutative diagram

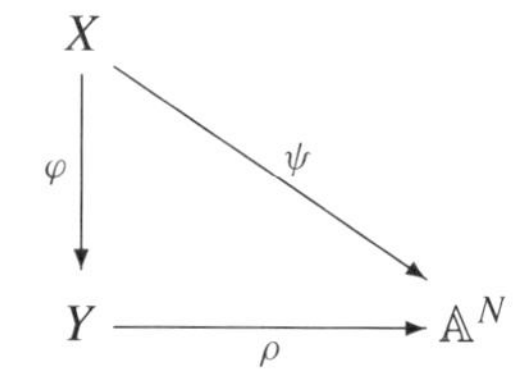


induces the commutative diagram

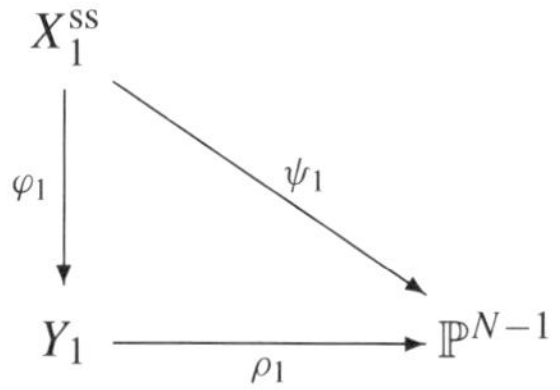


Note that $\psi_1 : X_1^{ss} \to \mathbb{P}^{N-1}$ is defined in view of the hypothesis that for any $x \in X^{ss}$, $f_i(x) \neq 0$, for at least one $i$. Note also that $f_1, \ldots, f_N$ are sections of the ample line bundle $\mathcal{O}_{X_1}(d)$ as well as the basic fact from GIT that this line bundle descends to an ample line bundle on $Y_1$, which we denote by $\mathcal{O}_{Y_1}(d)$.
**Claim 1:** $\rho_1$ is a finite morphism.
**Proof of Claim 1:** Since $f_1, \cdots, f_N$ are $G$-invariant, we get that $f_i \in H^0(Y_1, \mathcal{O}_{Y_1}(d))$. Hence we obtain that

$$\rho_1^*(\mathcal{O}_{\mathbb{P}^{N-1}}(1)) = \mathcal{O}_{Y_1}(d)$$

Thus, $\rho_1^*(\mathcal{O}_{\mathbb{P}^{N-1}}(1))$ is ample, and hence $\rho_1$ is finite (over any fiber $(\rho_1)_z, z \in \mathbb{P}^{N-1}$, $\rho_1^*(\mathcal{O}_{\mathbb{P}^{N-1}}(1)) \mid_{(\rho_1)_z}$ is both ample and trivial, and hence $\dim(\rho_1)_z$ is zero), and Claim 1 follows.
**Claim 2:** $\rho$ is a finite morphism.
**Proof of Claim 2:** Let $S' = R^G$. Let $S'^{(d)} = \oplus_n S'_{nd}$. We have $\mathbb{P}^{N-1} = $ ProjK[$\mathrm{x}_1, \ldots, \mathrm{x}_\mathrm{N}$]. Since $\rho_1$ is finite, $\mathcal{O}_{Y_1}$ is a coherent $\mathcal{O}_{\mathbb{P}^{N-1}}$-module.

We see that

$$H^0(\mathbb{P}^{N-1}, \mathcal{O}_{Y_1} \otimes \mathcal{O}_{\mathbb{P}^{N-1}}(n)) \simeq H^0(Y_1, \rho_1^*(\mathcal{O}_{\mathbb{P}^{N-1}}(n))$$

since the direct image of $\rho_1^*(\mathcal{O}_{\mathbb{P}^{N-1}}(n))$ by $\rho_1$ is $\mathcal{O}_{Y_1} \otimes \mathcal{O}_{\mathbb{P}^{N-1}}(n)$ and $\rho_1$ is a finite morphism. On the other hand we have

$$\rho_1^*(\mathcal{O}_{\mathbb{P}^{N-1}}(n)) \simeq \mathcal{O}_{Y_1}(nd).$$

Thus we have

$$H^0(\mathbb{P}^{N-1}, \mathcal{O}_{Y_1} \otimes \mathcal{O}_{\mathbb{P}^{N-1}}(n)) \simeq H^0(Y_1, \mathcal{O}_{Y_1}(nd)) \simeq S'_{nd}.$$

Thus the graded $K[x_1, \ldots, x_N]$-module associated to the coherent sheaf $\mathcal{O}_{Y_1}$ on $\mathbb{P}^{N-1}$ is $S'^{(d)}$ and by the basic theorems of Serre, $S'^{(d)}$ is of finite type over $K[x_1, \ldots, x_N]$. Now a $d$-th power of any homogeneous element of $S'$ is in $S'^{(d)}$ and thus $S'$ is integral over $K[x_1, \ldots, x_N]$, which proves that $\rho$ is finite. Claim 2 and hence the required result follows from this.

**Case 2:** Let $f_1, \cdots, f_N$ be homogeneous possibly of different degrees, say, $\deg f_i = d_i$. Let $d = l.c.m.\{d_i\}, e_i = \frac{d}{d_i}$. Set $g_i = f_i^{e_i}, 1 \leq i \leq N$. Then $\{g_1, \cdots, g_N\}$ is again base-point free on $(Proj R)^{ss}$. Hence by Case 1, we have that $R^G$ is a finite $K[g_1, \cdots, g_N]$-module, and hence a finite $K[f_1, \cdots, f_N]$-module (note that $K[g_1, \cdots, g_N] \hookrightarrow K[f_1, \cdots, f_N] \hookrightarrow R^G$).

In the Lemma below, we describe a set of sufficient conditions for (3) of Theorem 10.1.0.1, namely, birationality.

**Lemma 10.1.0.3** *Suppose $F : X \to Y$ is a surjective morphism of (irreducible) algebraic varieties, and $U$ is an open subset of $X$ such that*

*(i) $F\mid_U : U \to Y$ is an immersion*
*(ii) $\dim U = \dim Y$.*

*Then $F$ is birational.*

*Proof.* Hypothesis (1) implies that $F(U)$ is locally closed in $Y$. This fact together with Hypothesis (2) implies that $F(U)$ is open in $Y$, and the result follows.

We now return to the situation of a linear action of a reductive group $G$ on an affine variety $X = Spec R$ with $R$ a graded $K$-algebra. Let $f_1, \cdots, f_N$ be homogeneous $G$-invariant elements in $R$. Let $S = K[f_1, \cdots, f_N]$. Combining Theorem 10.1.0.1 with Lemmas 10.1.0.2, 10.1.0.3, we arrive at the following Lemma which gives a set of sufficient conditions for the equality $S = R^G$. Before stating the lemma, let us observe the following. Suppose that $U$ is a non-empty $G$-stable open subset in $X$. Since $\varphi : X \longrightarrow Spec\, R^G$ is surjective, $\varphi(U)$ contains a non-empty open subset. Hence by shrinking $U$, if necessary, we can suppose that $\varphi(U)$ is open. We suppose that this is the case and denote $\varphi(U)$ by $U /\!\!/ G$.

**Lemma 10.1.0.4** *Let notation be as above. Let $\psi : X \to \mathbb{A}^N$ be the map, $x \mapsto (f_1(x), \cdots, f_N(x))$. Denote $D = Spec S$. Then $D$ is the categorical quotient of $X$ by $G$ and $\psi : X \to D$ is the canonical quotient map, if the following conditions are satisfied:*

*(i) For $x \in X^{ss}$, $\psi(x) \neq (0)$.*

*(ii) There is a $G$-stable open subset $U$ of $X$ such that $\psi\mid_{U/\!\!/G} : U /\!\!/ G \to D$ is an immersion.*

*(iii)* $\dim D = \dim U /\!\!/ G$.

*(iv) $D$ is normal.*

**Remark 10.1.0.5.** Suppose that $U$ is a (non-empty) $G$-stable open subset of $X$, $G$ operates freely with $U/G$ as quotient, and $\psi$ induces an immersion of $U/G$ in $A^N$. Then (ii) is satisfied:

This assertion is immediately seen, for we have

$$U/G \longrightarrow U/\!\!/G \longrightarrow \mathbb{A}^N$$

and the fact that $U/G \longrightarrow A^N$ is an immersion implies that $U/\!\!/G \longrightarrow \mathbb{A}^N$ immersion.

In the following sections, using Lemma 10.1.0.4, we give a GIT-theoretic proof of the First and Second Fundamental Theorems of Classical invariant theory (cf. Weyl [115]) for the classical group actions.

## 10.2 $SL_d(K)$-action

In this section, we give a GIT-theoretic proof of the results for $SL_d(K)$-actions appearing in classical invariant theory (cf. Weyl [115]).

Let $X = M_{n,d}$, the space of $n \times d$ matrices with entries in $K$; further, let $n > d$. Let $R = K[x_{ij}, 1 \le i \le n, 1 \le j \le d]$ be the coordinate ring of $X$.

### 10.2.1 The functions $f_\tau$

Let $A \in X$. Consider the $d \times d$-submatrices of $A$; these are indexed by

$$I_{d,n} := \{\underline{i} = (i_1 < \cdots < i_d)\}$$

namely, for $\underline{i} \in I_{d,n}$, let $A_{\underline{i}}$ denote the $d \times d$-submatrix of $A$ with row indices given by $\{i_1, \ldots, i_d\}$. Then $\{A_{\underline{i}}, \underline{i} \in I_{d,n}\}$ give all the $d \times d$-submatrices of $A$. For $\tau \in I_{d,n}$, let $f_\tau$ be the regular function on $X$:

$$f_\tau : X \to K,\ A \mapsto |A_\tau|\ (\text{determinant of } A_\tau)$$

**$SL_d(K)$-action:** Let $G = SL_d(K)$. Consider the action of $G$ on $X$ by right multiplication. For $\tau \in I_{d,n}, A \in X, g \in G$, we have, $f_\tau(A \cdot g) = |A_\tau||g| = |A_\tau|$. Thus

$$(*) \qquad f_\tau \in R^G,\ \tau \in I_{d,n}$$

**Proposition 10.2.1.1** *Let $X, G$ be as above. Let $X^\circ$ be the set of matrices in $X$ of rank d (the maximum possible rank). Then $X^s = X^{ss} = X^\circ$*

*Proof.* We first

**Claim 1:** We have

$$(i) \qquad X^{ss} = X^\circ$$

**The inclusion $X^{ss} \subseteq X^\circ$:** Let $A \notin X^\circ$. We shall show that $A \notin X^{ss}$. Let $r = rank(A)(< d)$. Then there exists a $g \in GL_d(K)$ such that $Ag$ is in reduced row echelon form:

$$Ag = \begin{pmatrix} 1 & 0 & \cdots & 0 & 0 & \cdots & 0 \\ 0 & 1 & \cdots & \cdots & 0 & \cdots & 0 \\ \vdots & \vdots & \vdots & \vdots & 0 & \vdots & \vdots \\ 0 & 0 & \cdots & 1 & 0 & \cdots & 0 \\ 0 & 0 & \cdots & 0 & 0 & \cdots & 0 \\ \vdots & \vdots & \vdots & \vdots & \vdots & \vdots & \vdots \\ 0 & 0 & \cdots & 0 & 0 & \cdots & 0 \end{pmatrix}$$

with precisely $r$ non-zero columns (which appear as the first $r$ columns). Let $c = |g|$; pick $\alpha$, an $n$-th root of $c$. Then $h := \frac{1}{\alpha} g$ is in $G$. Then $Ag = A(\alpha h)$; replacing $A$ by $Ah(\in O(A))$, the $SL_d(K)$-orbit of $A$), we may suppose that

$$A = \frac{1}{\alpha} \begin{pmatrix} 1 & 0 & \cdots & 0 & 0 & \cdots & 0 \\ 0 & 1 & \cdots & \cdots & 0 & \cdots & 0 \\ \vdots & \vdots & \vdots & \vdots & 0 & \vdots & \vdots \\ 0 & 0 & \cdots & 1 & 0 & \cdots & 0 \\ 0 & 0 & \cdots & 0 & 0 & \cdots & 0 \\ \vdots & \vdots & \vdots & \vdots & \vdots & \vdots & \vdots \\ 0 & 0 & \cdots & 0 & 0 & \cdots & 0 \end{pmatrix}$$

Choose integers $\{a_1, \cdots, a_d\}$ such that

$$(**) \qquad \sum a_i = 0, \quad \text{and} \quad a_i > 0, 1 \le i \le r$$

Consider the 1-PS $\Gamma = \{D_t \in G, t \neq 0\}$, where $D_t$ is the diagonal matrix $diag(t^{a_1}, \cdots, t^{a_d})$. Then

$$A \cdot D_t = \frac{1}{\alpha} \begin{pmatrix} t^{a_1} & 0 & \cdots & 0 & 0 & \cdots & 0 \\ 0 & t^{a_2} & \cdots & \cdots & 0 & \cdots & 0 \\ \vdots & \vdots & \vdots & \vdots & 0 & \vdots & \vdots \\ 0 & 0 & \cdots & t^{a_r} & 0 & \cdots & 0 \\ 0 & 0 & \cdots & 0 & 0 & \cdots & 0 \\ \vdots & \vdots & \vdots & \vdots & \vdots & \vdots & \vdots \\ 0 & 0 & \cdots & 0 & 0 & \cdots & 0 \end{pmatrix}$$

Hence we obtain (in view of (**))

$$\lim_{t \to 0} A \cdot D_t = 0(\in X).$$

Thus the origin 0 is in the closure of $O(A)$, and consequently $A$ is not semi-stable, and the inclusion $X^{ss} \subseteq X^\circ$ follows.

**The inclusion $X^\circ \subseteq X^{ss}$:**

Let $A \in X^\circ$. We have that for some $\tau \in I_{d,n}$, $f_\tau(A) \neq 0$. This together with the $G$-invariance of $f_\tau$ (cf. (*) above) implies that $A \in X^{ss}$.

This completes the proof of Claim 1.

**Claim 2:** Let $A \in X^\circ$. Then the stabilizer of $A$ equals $e = Id_{d\times d}$.

Let $\tau \in I_{d,n}$ be such that $|A_\tau| \neq 0$. For $g \in G$, we have, $(A \cdot g)_\tau = A_\tau g$. Hence if $A \cdot g = A$, then we have,

$$A_\tau = (A \cdot g)_\tau = A_\tau g$$

This together with the invertibility of $A_\tau$ implies that stabilizer of $A$ is just $\{e\}$, and Claim 2 follows.

We have (by Claim 2) that $O(A)$ (the $G$-orbit through $A$) is isomorphic to $G$; in particular, we have,

$$(ii) \qquad dim\, O(x) = dim\, G, \; x \in X^\circ$$

**Claim 3:** Let $x \in X^\circ$. Then

$$(iii) \qquad O(x) = \overline{O(x)},$$

i.e, $O(x)$ is closed.

If possible, assume $O(x) \subsetneq \overline{O(x)}$. Pick a point $y \in \overline{O(x)} \setminus O(x)$. Then (ii) together with "closed orbit Lemma" (cf. Chapter 9, Lemma 9.4.0.3) implies that $dim\, O(y) < dim\, G$; hence $y \notin X^\circ$ (again by (ii)). This together with Claim 1 (cf. (i) above) implies that $y$ is not semi-stable, and therefore $0 \in \overline{O(y)} (\subseteq \overline{O(x)})$, a contradiction (note that in view of (i), $x$ is semi-stable).

Now (ii) and (iii) imply $X^s = X^\circ$. This together with (i) implies

$$X^s = X^{ss} = X^\circ$$

### 10.2.2 The First and Second Fundamental Theorems of Classical Invariant Theory (cf. Weyl [115]) for the $SL_d(K)$-action:

Let $N = \binom{n}{d} (= \# I_{d,n})$, and let $\psi$ be the map

$$\psi : X \to \mathbb{A}^N, \; A \mapsto (|A_\tau|, \tau \in I_{d,n})$$

**Theorem 10.2.2.1** *Let $X, G$ be as above. The categorical quotient $X/\!\!/G$ is isomorphic to $\widehat{G}_{d,n}$, where, $\widehat{G}_{d,n}$ denotes the cone over the Grassmannian $G_{d,n}$.*

*Proof.* Let $\psi : X \to \mathbb{A}^N$ be as above. We shall verify the hypotheses of Lemma 10.1.0.4 for $X$ and $\psi$.

First observe that $D :=$ `image of` $\psi = \widehat{G}_{d,n}$ (since, $K[G_{d,n}]$ is generated by $|A_\tau|$ (cf. Chapter 4)). Hence (iv) of Lemma 10.1.0.4 is satisfied. As above let $X^\circ$ be the set of $n \times d$ matrices of rank $d$. Then we have (cf. Proposition 10.2.1.1) that if $A \in X^{ss}$, then $A \in X^\circ$; hence, $\psi(A) \neq 0$. Thus (i) of Lemma 10.1.0.4 is satisfied.

Next we show that (ii) of Lemma 10.1.0.4 holds. Denote $G' = GL_d(K)$. Let

$$U = \{A \in X \mid A_{id} \in G'\}$$

where $id$ is the $d$-tuple $(1, 2, \cdots, d)$, i.e., $A_{id}$ is invertible. Clearly, $U$ is a $G$-stable open subset of $X$ (note that $U = \{A \in X \mid |A_{id}| \neq 0\}$).
**Claim:** $G$ operates freely on $U$, $U \to U\, mod\, G$ is a $G$-principal fiber space, and $\psi$ induces an immersion $U/G \to \mathbb{A}^N$.
**Proof of Claim:** We have a $G$-equivariant isomorphism

$$(\dagger) \qquad U \cong G' \times M_{s,d}$$

where $s = n - d$. (Here, $M_{s,d}$ denotes the space of $s \times d$ matrices with entries in $K$.) From this it is clear that and $G$ operates freely on $U$. Further, we see that $U\, mod\, G$ may be identified with the fiber space with base $G'\, mod\, G$, and fiber $M_{s,d}$, associated to the principal fiber space $G' \to G'/G$. It remains to show that $\psi$ induces an immersion $U/G \to \mathbb{A}^N$, i.e., to show that the map $\psi : U/G \to \mathbb{A}^N$ and its differential $d\psi$ are both injective. We first prove that $\psi : U/G \to \mathbb{A}^N$ is injective. Let $x, x'$ in $U/G$ be such that $\psi(x) = \psi(x')$. Let $\eta, \eta' \in U$ be lifts for $x, x'$ respectively. Using the identification ($\dagger$) above, we may write

$$\eta = (A, M),\ A \in G', M \in M_{s,d},$$
$$\eta' = (A', M'),\ A' \in G', M' \in M_{s,d}$$

The fact that $\psi(x) = \psi(x')$ implies $|A| = |A'|$. Hence $g := A^{-1}A'$ is in $G(= SL_d(K))$. Thus we obtain $A' = A \cdot g$. Hence on $U/G$, we may suppose that

$$x = (A, M),\ x' = (A, M')$$

Let $|A| = c(\neq 0)$; pick $\alpha$, an $n$th root of $c$. Then $A = \alpha g$, where $g := \frac{1}{\alpha}A$ is in $G$. Hence we may suppose (on $U/G$)

$$x = (\alpha Id_{d\times d}, M),\ x' = (\alpha Id_{d\times d}, M')$$

Thus, we have lifts $B, B'$ for $x, x'$ respectively, where the $n \times d$ matrices $B, B'$ are given by $\alpha Id_{d\times d}$ sitting on the top of the $s \times d$ matrices $M, M'$ respectively:

$$B = \begin{pmatrix} \alpha Id_{d\times d} \\ M \end{pmatrix},\ B' = \begin{pmatrix} \alpha Id_{d\times d} \\ M' \end{pmatrix}$$

Let $i > d$ and $1 \leq j \leq d$; let $\tau$ be the $d$-tuple $(1, 2, \cdots, j-1, j+1, \cdots, d, j)$. We have, $|B_\tau| = |B'_\tau|$ (since $\psi(x) = \psi(x')$); but

$$|B_\tau| = \varepsilon\alpha^{d-1}b_{ij},\ |B'_\tau| = \varepsilon\alpha^{d-1}b'_{ij},\ \varepsilon = \pm 1$$

Hence we obtain

$$b_{ij} = b'_{ij},\ i > d, 1 \leq j \leq d$$

This implies $M = M'$ and hence $x = x'$ which proves the injectivity of $\psi : U/G \to \mathbb{A}^N$.

To prove that the differential $d\psi$ is injective, we merely note that the above argument remains valid for the points over $K[\epsilon]$, the algebra of dual numbers ($= K \oplus K\epsilon$, the $K$-algebra with one generator $\epsilon$, and one relation $\epsilon^2 = 0$), i.e., it remains valid if we replace $K$ by $K[\epsilon]$, or in fact by any $K$-algebra.
Thus condition (ii) of Lemma 10.1.0.4 holds.

It remains to show $dim\, U /\!\!/ G = dim\, \widehat{G}_{d,n}$. We have

$$dim\, U /\!\!/ G = dim\, U - dim\, G = nd - (d^2 - 1) = d(n-d) + 1 = dim\, \widehat{G}_{d,n}$$

Thus all of the conditions of Lemma 10.1.0.4 hold for $X$, $\psi$, and hence Lemma 10.1.0.4 implies that $X /\!\!/ G = \widehat{G}_{d,n}$.

Using the identification $X /\!\!/ G = \widehat{G}_{d,n}$ and the results from Chapter 4, we have the following Theorem:

**Theorem 10.2.2.2** *(i)* **First Fundamental Theorem**
*The ring of invariants $R^G$ is generated by $\{f_\tau, \tau \in I_{d,n}\}$.*
*(ii)* **Second Fundamental Theorem**
*The ideal of relations among the generators in (i) is generated by the Plücker quadratic relations.*

Defining standard monomials in $f_\tau$'s as in Chapter 4, we have (cf. Chapter 4) the following:

**Theorem 10.2.2.3 A standard monomial basis for the ring of invariants:** *The ring of invariants $R^G$ has a basis consisting of standard monomials in the regular functions $f_\tau$, $\tau \in I_{d,n}$.*

## 10.3 $GL_n(K)$-action:

In this section, we give a GIT-theoretic proof of the results for $GL_n(K)$-actions appearing in classical invariant theory (cf. DeConcini-Procesi [22], Weyl [115]).

Let $V = K^n$, $X = \underbrace{V \oplus \cdots \oplus V}_{m \text{ copies}} \oplus \underbrace{V^* \oplus \cdots \oplus V^*}_{q \text{ copies}}$.

**The $GL(V)$-action on $X$:** Writing $\underline{u} = (u_1, u_2, ..., u_m)$ with $u_i \in V$ and $\underline{\xi} = (\xi_1, \xi_2, ..., \xi_q)$ with $\xi_i \in V^*$, we shall denote the elements of $X$ by $(\underline{u}, \underline{\xi})$. The (natural) action of $GL(V)$ on $V$ induces an action of $GL(V)$ on $V^*$, namely, for $\xi \in V^*$, $g \in GL(V)$, denoting $g \cdot \xi$ by $\xi^g$, we have

$$\xi^g(v) = \xi(g^{-1}v),\ v \in V$$

The diagonal action of $GL(V)$ on $X$ is given by

$$g \cdot (\underline{u}, \underline{\xi}) = (g\underline{u}, \underline{\xi}^g) = (gu_1, gu_2, ..., gu_m, \xi_1^g, \xi_2^g, ..., \xi_q^g),\ g \in G,\ (\underline{u}, \underline{\xi}) \in X$$

The induced action on $K[X]$ is given by

$$(g \cdot f)(\underline{u}, \underline{\xi}) = f(g^{-1}(\underline{u}, \underline{\xi})), \ f \in K[X], \ g \in GL(V)$$

Consider the functions $\varphi_{ij} : X \longrightarrow K$ defined by $\varphi_{ij}((\underline{u}, \underline{\xi})) = \xi_j(u_i)$, $1 \le i \le m$, $1 \le j \le q$. Each $\varphi_{ij}$ is a $GL(V)$-invariant: For $g \in GL(V)$, we have,

$$\begin{aligned}(g \cdot \varphi_{ij})((\underline{u}, \underline{\xi})) &= \varphi_{ij}(g^{-1}(\underline{u}, \underline{\xi}))\\ &= \varphi_{ij}((g^{-1}u, \xi^{g^{-1}})\\ &= \xi_j^{g^{-1}}(g^{-1}u_i)\\ &= \xi_j(u_i)\\ &= \varphi_{ij}((\underline{u}, \underline{\xi}))\end{aligned}$$

It is convenient to have a description of the above action in terms of coordinates. So with respect to a fixed basis, we write the elements of $V$ as row vectors and those of $V^*$ as column vectors. Thus denoting by $M_{a,b}$ the space of $a \times b$ matrices with entries in $K$, $X$ can be identified with the affine space $M_{m,n} \times M_{n,q}$. The action of $GL_n(K)$ $(= GL(V))$ on $X$ is then given by

$$A \cdot (U, W) = (UA, A^{-1}W), \ A \in GL_n(K), \ U \in M_{m,n}, \ W \in M_{n,q}$$

And the action of $GL_n(K)$ on $K[X]$ is given by

$$(A \cdot f)(U, W) = f\left(A^{-1}(U, W)\right) = f\left(UA^{-1}, AW\right), \ f \in K[X]$$

Writing $U = (u_{ij})$ and $W = (\xi_{kl})$ we denote the coordinate functions on $X$, by $u_{ij}$ and $\xi_{kl}$. Further, if $u_i$ denotes the $i$-th row of $U$ and $\xi_j$ the $j$-th column of $W$, the invariants $\varphi_{ij}$ described above are nothing but the entries $\langle u_i, \xi_j \rangle$ $(= \xi_j(u_i))$ of the product $UW$.

In the sequel, we shall denote $\varphi_{ij}(\underline{u}, \underline{\xi})$ also by $\langle u_i, \xi_j \rangle$.

**The Function** $p(A, B)$: For $A \in I(r, m)$, $B \in I(r, q)$, $1 \le r \le n$, let $p(A, B)$ be the regular function on $X$: $p(A, B)((\underline{u}, \underline{\xi}))$ is the determinant of the $r \times r$-matrix $(\langle u_i, \xi_j \rangle)_{i \in A, \, j \in B}$. Let $S$ be the subalgebra of $R^G$ generated by $\{p(A, B)\}$. We shall now show (using Lemma 10.1.0.4) that $S$ is in fact equal to $R^G$.

### 10.3.1 The First and Second Fundamental Theorems of Classical Invariant Theory (cf. [115]) for the action of $GL_n(K)$:

**Theorem 10.3.1.1** *Let $G = GL_n(K)$. Let $X$ be as above. The morphism $\psi : X \to M_{m,q}$, $(\underline{u}, \underline{\xi}) \mapsto \left(\varphi_{ij}(\underline{u}, \underline{\xi})\right)$ $(= (\langle u_i, \xi_j \rangle))$ maps $X$ into the determinantal variety $D_{n+1}(M_{m,q})$, and the induced homomorphism $\psi^* : K[D_{n+1}(M_{m,q})] \to K[X]$ between the coordinate rings induces an isomorphism $\psi^* : K[D_{n+1}(M_{m,q})] \to K[X]^G$, i.e. the determinantal variety $D_{n+1}(M_{m,q})$ is the categorical quotient of $X$ by $G$.*

*Proof.* Clearly, $\psi(X) \subseteq D_{n+1}(M_{m,q})$ (since, $\psi(X) = Spec\, S$, and clearly $Spec\, S \subseteq D_{n+1}(M_{m,q})$ (since any $n+1$ vectors in $V$ are linearly independent)). We shall prove the result using Lemma 10.1.0.4. To be very precise, we shall first check the conditions (i)-(iii) of Lemma 10.1.0.4 for $\psi : X \to M_{m,q}$, deduce that the inclusion $Spec\, S \subseteq D_{n+1}(M_{m,q})$ is in fact an equality, and hence conclude the normality of $Spec\, S$ (condition (iv) of Lemma 10.1.0.4).

(i) Let $x = (\underline{u}, \underline{\xi}) = (u_1, \ldots, u_m, \xi_1, \ldots, \xi_q) \in X^{ss}$. Let $W_{\underline{u}}$ be the subspace of $V$ spanned by $x_i$'s and $W_{\underline{\xi}}$ the subspace of $V^*$ spanned by $\xi_j$'s. Assume if possible that $\psi((\underline{u}, \underline{\xi})) = 0$, i.e. $\langle u_i, \xi_j \rangle = 0$ for all $i, j$.

**Case (a)**: $W_{\underline{\xi}} = 0$, i.e., $\xi_j = 0$ for all $j$.

Consider the one parameter subgroup $\Gamma = \{g_t, t \neq 0\}$ of $GL(V)$, where $g_t = tI_n$, $I_n$ being the $n \times n$ identity matrix. Then $g_t \cdot x = g_t \cdot (\underline{u}, \underline{0}) = (t\underline{u}, \underline{0})$, so that $g_t \cdot x \to (0)$ as $t \to 0$. Thus the origin 0 is in the closure of $G \cdot x$, and consequently $x$ is not semi-stable, which is a contradiction.

**Case (b)**: $W_{\underline{\xi}} \neq 0$.

Since the case $W_{\underline{u}} = 0$ is similar to Case (a), we may assume that $W_{\underline{u}} \neq 0$. Also the fact that $W_{\underline{\xi}} \neq 0$ together with the assumption that $\langle x_i, \xi_j \rangle = 0$ for all $i, j$ implies that $\dim W_{\underline{u}} < n$. Let $r = dim\, W_{\underline{u}}$ so that we have $0 < r < n$. Hence, we can choose a basis $\{e_1, \ldots, e_n\}$ of $V$ such that $W_{\underline{u}} = \langle e_1, \ldots, e_r \rangle$, $r < n$, and $W_{\underline{\xi}} \subset \langle e^*_{r+1}, \ldots, e^*_n \rangle$, where $\{e^*_1, \ldots, e^*_n\}$ is the dual basis in $V^*$. Consider the one parameter subgroup $\Gamma = \{g_t, t \neq 0\}$ of $GL(V)$, where

$$g_t = \begin{pmatrix} tI_r & 0 \\ 0 & t^{-1}I_{n-r} \end{pmatrix}.$$

We have $g_t \cdot (\underline{u}, \underline{\xi}) = (t\underline{u}, t\underline{\xi}) \to 0$ as $t \to 0$. Thus, by the same reasoning as in Case (a), the point $(\underline{u}, \underline{\xi})$ is not semi-stable, which leads to a contradiction. Hence we obtain $\psi((\underline{u}, \underline{\xi})) \neq 0$.

Thus (i) of Lemma 10.1.0.4 holds.

(ii) Let

$$U = \{(\underline{u}, \underline{\xi}) \in X \mid \{u_1, \ldots, u_n\}, \{\xi_1, \ldots, \xi_n\} \text{ are linearly independent}\}$$

Clearly, $U$ is a $G$-stable open subset of $X$.

**Claim:** $G$ operates freely on $U$, $U \to U\, mod\, G$ is a $G$-principal fiber space, and $\psi$ induces an immersion $U/G \to M_{m,q}$.

**Proof of Claim:** We have a $G$-equivariant identification

$$(\dagger) \qquad U \cong G \times G \times \underbrace{V \times \cdots \times V}_{(m-n) \text{ copies}} \times \underbrace{V^* \times \cdots \times V^*}_{(q-n) \text{ copies}}$$

from which it is clear that and $G$ operates freely on $U$. Further, we see that $U\, mod\, G$ may be identified with the fiber space with base $G \times G\, mod\, G$ ($G$ acting on $G \times G$ as $g \cdot (g_1, g_2) = (g_1 g, g^{-1} g_2)$, $g, g_1, g_2 \in G$), and fiber $\underbrace{V \times \cdots \times V}_{(m-n) \text{ copies}} \times \underbrace{V^* \times \cdots \times V^*}_{(q-n) \text{ copies}}$ associated to the principal fiber space $G \times G \to (G \times G)\,/\,G$. It remains to show that

$\psi$ induces an immersion $U/G \to \mathbb{A}^N$, i.e., to show that the map $\psi : U/G \to M_{m,q}$ and its differential $d\psi$ are both injective. We first prove that $\psi : U/G \to M_{m,q}$ is injective. Let $x, x'$ in $U/G$ be such that $\psi(x) = \psi(x')$. Let $\eta, \eta' \in U$ be lifts for $x, x'$ respectively. Using the identification (†) above, we may write

$$\eta = (A, u_{n+1}, \cdots, u_m, B, \xi_{n+1}, \cdots, \xi_q),\ A, B \in G$$
$$\eta' = (A', u'_{n+1}, \cdots, u'_m, B', \xi'_{n+1}, \cdots, \xi'_q),\ A', B' \in G$$

(here, $u_i$, $1 \le i \le n$ are given by the rows of $A$, while $\xi_i$, $1 \le i \le n$ are given by the columns of $B$; similar remarks on $u'_i, \xi'_i$). The hypothesis that $\psi(x) = \psi(x')$ implies in particular that

$$\langle u_i, \xi_j \rangle = \langle u'_i, \xi'_j \rangle,\ 1 \le i, j \le n$$

which may be written as

$$AB = A'B'$$

This implies that $A' = A \cdot g$, where $g = BB'^{-1}$. Hence on $U/G$, we may suppose that

$$x = (u_1, \cdots, u_n, u_{n+1}, \cdots, u_m, \xi_1, \cdots, \xi_q)$$
$$x' = (u_1, \cdots, u_n, u'_{n+1}, \cdots, u'_m, \xi'_1, \cdots, \xi'_q)$$

where $\{u_1, \ldots, u_n\}$ is linearly independent.

For a given $j$, we have,

$$\langle u_i, \xi_j \rangle = \langle u_i, \xi'_j \rangle,\ 1 \le i \le n,\ \text{implies}\ \xi_j = \xi'_j$$

(since $\{u_1, \ldots, u_n\}$ is linearly independent). Thus we obtain

(*) $$\xi_j = \xi'_j,\ \text{for all}\ j$$

On the other hand, we have (by definition of $U$) that $\{\xi_1, \ldots, \xi_n\}$ is linearly independent. Hence fixing an $i$, $n+1 \le i \le m$, we get

$$\langle u_i, \xi_j \rangle = \langle u'_i, \xi_j \rangle (= \langle u'_i, \xi'_j \rangle),\ 1 \le j \le n,\ \text{implies}\ u_i = u'_i$$

Thus we obtain

(**) $$u_i = u'_i,\ \text{for all}\ i$$

The injectivity of $\psi : U/G \to M_{m,q}$ follows from (*) and (**).

To prove that the differential $d\psi$ is injective, we merely note that the above argument remains valid for the points over $K[\epsilon]$, the algebra of dual numbers ($= K \oplus K\epsilon$, the $K$-algebra with one generator $\epsilon$, and one relation $\epsilon^2 = 0$), i.e., it remains valid if we replace $K$ by $K[\epsilon]$, or in fact by any $K$-algebra.
Thus (ii) of Lemma 10.1.0.4 holds.

(iii) We have

$$\dim U/G = \dim U - \dim G = (m+q)n - n^2 = \dim D_{n+1}(M_{m,q}).$$

The immersion $U/G \hookrightarrow Spec\, S(\subseteq D_{n+1}(M_{m,q}))$ together with the fact above that $\dim U/G = \dim D_{n+1}(M_{m,q})$ implies that $Spec\, S$ in fact equals $\dim D_{n+1}(M_{m,q})$. Thus (iii) of Lemma 10.1.0.4 holds.

(iv) The normality of $Spec\, S(= D_{n+1}(M_{m,q}))$ follows from Chapter 5, Corollary 5.2.2.8.

Combining the above Theorem with Chapter 5, Theorem 5.2.4.2, we obtain the following

**Corollary 10.3.1.2** *Let $X$ and $G$ be as above. Let $\varphi_{ij}$ denote the regular function $(\underline{u}, \underline{\xi}) \mapsto \langle u_i, \xi_j \rangle$ on $X$, $1 \le i \le m$, $1 \le j \le q$, and let $f$ denote the $m \times q$ matrix $(\varphi_{ij})$. The ring of invariants $K[X]^G$ has a basis consisting of standard monomials in the regular functions $p_{\lambda,\mu}(f)$ with $\#\lambda \le n$, where $\#\lambda = t$ is the number of elements in the sequence $\lambda = (\lambda_1, \dots, \lambda_t)$ and $p_{\lambda,,\mu}(f)$ is the $t$-minor with row indices $\lambda_1, \dots, \lambda_t$ and column indices $\mu_1, \dots, \mu_t$.*

As a consequence of the above Theorem and the Corollary, we obtain the first and Second Fundamental Theorems of classical invariant theory (cf. [115]). Let notation be as above.

**Theorem 10.3.1.3** *(i)* **First Fundamental Theorem**
*The ring of invariants $K[X]^{GL(V)}$ is generated by $\varphi_{ij}$, $1 \le i \le m, 1 \le j \le q$.*
*(ii)* **Second Fundamental Theorem**
*The ideal of relations among the generators in (1) is generated by the $(n+1)$-minors of the $m \times q$-matrix $(\varphi_{ij})$.*

Further we have

**Theorem 10.3.1.4 A standard monomial basis for the ring of invariants:** *The ring of invariants $K[X]^{GL(V)}$ has a basis consisting of standard monomials in the regular functions $p_(A, B)$, $A \in I(r, m)$, $B \in I(r, q)$, $r \le n$.*

## 10.4 $O_n(K)$-action

In this section, we give a GIT-theoretic proof of the results for $O_n(K)$-actions appearing in classical invariant theory (cf. [22], [115]).

Let $V = K^n$, together with a non-degenerate, symmetric bilinear form $\langle , \rangle$. Let $G = O(V)$ (the orthogonal group consisting of linear automorphisms of $V$ preserving $\langle , \rangle$). We shall take the matrix of the form $\langle , \rangle$ to be $J_n :=$ $\texttt{anti-diagonal}(1, \cdots, 1)$. Then a matrix $A \in GL(V)$ is in $O(V)$ if and only if

$$ {}^tAJ_nA = J_n, \text{ i.e, } J_n^{-1}({}^tA^{-1})J_n = A, \text{ i.e, } J_n({}^tA^{-1})J_n = A \text{ (note that } J_n^{-1} = J_n) $$

**Remark 10.4.0.1.** In particular, note that a diagonal matrix $A = \texttt{diagonal}(t_1, \cdots, t_n)$ is in $O(V)$ if and only if $t_{n+1-i} = t_i^{-1}$.

Let $X$ denote the affine space $V^{\oplus m} = V \oplus \cdots \oplus V$, where $m > n$. For $\underline{u} = (u_1, u_2, ..., u_m) \in X$, writing $u_i = (u_{i1}, \cdots, u_{in})$ (with respect to the standard basis for $K^n$), we shall identify $\underline{u}$ with the $m \times n$ matrix $U := (u_{ij})_{m\times n}$. Thus, we identify $X$ with $M_{m,n}(K)$, the space of $m \times n$ matrices (with entries in $K$). Consider the diagonal action of $O(V)$ on $X$. The induced action on $K[X]$ is given by

$$(A \cdot f)(\underline{u}) = f(A \cdot \underline{u}) = f(UA),\ \underline{i} \in X,\ f \in K[X],\ A \in O_n(K)(= O(V))$$

**The Functions** $\varphi_{ij}$**:** Consider the functions $\varphi_{ij} : X \to K$ defined by $\varphi_{ij}((\underline{u})) = \langle u_i, u_j \rangle$, $1 \le i, j \le m$. Each $\varphi_{ij}$ is clearly in $K[X]^{O(V)}$.

Let $S$ be the subalgebra of $K[X]^{O(V)}$ generated by $\{\varphi_{ij}\}$. We shall now show (using Lemma 10.1.0.4) that $S$ is in fact equal to $K[X]^{O(V)}$.

**Theorem 10.4.0.2** *The morphism $\psi : X \to Sym\, M_m$, $\underline{u} \mapsto (\langle u_i, u_j \rangle)$ is $O(V)$-invariant. Further, it maps $X$ onto $D_{n+1}(Sym\, M_m)$ and identifies the categorical quotient $X /\!\!/ O(V)$ with $D_{n+1}(Sym\, M_m)$ (here, $D_{n+1}(Sym\, M_m)$ is as in Chapter 6, namely the determinantal variety contained in $Sym\, M_m$, the space of symmetric $m \times m$ matrices, consisting of matrices of rank $\le n$).*

*Proof.* Clearly, $\psi(X) \subseteq D_{n+1}(Sym\, M_m)$ (since, $\psi(X) = Spec\, S$, and clearly $Spec\, S \subseteq D_{n+1}(Sym\, M_m)$ (since any $n+1$ vectors in $V$ are linearly independent)). We shall prove the result using Lemma 10.1.0.4. To be very precise, we shall first check the conditions (i)-(iii) of Lemma 10.1.0.4 for $\psi : X \to Sym\, M_m$, deduce that the inclusion $Spec\, S \subseteq D_{n+1}(Sym\, M_m)$ is in fact an equality, and hence conclude the normality of $Spec\, S$ (condition (iv) of Lemma 10.1.0.4).

(i) Let $x = \underline{u} = (u_1, \ldots, u_m) \in X^{ss}$. Let $W_x$ be the subspace of $V$ spanned by $u_i$'s. Let $r = dim\, W_x$. Then $r > 0$ (since $x \in X^{ss}$). Assume if possible that $\psi(x) = 0$, i.e. $\langle u_i, u_j \rangle = 0$ for all $i, j$. This implies in particular that $W_x$ is totally isotropic; hence, $r \le [\frac{n}{2}]$, integral part of $\frac{n}{2}$. Hence we can choose a basis $\{e_1, \cdots, e_n\}$ of $V$ such that $W_x =$ the $K$-span of $\{e_1, \cdots, e_r\}$. Writing each vector $u_i$ as a row vector (with respect to this basis), we may represent $\underline{u}$ by the $m \times n$ matrix $\mathcal{U}$ given by

$$\mathcal{U} := \begin{pmatrix} u_{11} & u_{12} & \ldots & u_{1r} & 0 \ldots & 0 \\ u_{21} & u_{22} & \ldots & u_{2r} & 0 \ldots & 0 \\ \vdots & \vdots & & \vdots & \vdots \quad \vdots & \vdots \\ u_{m1} & u_{m2} & \ldots & u_{mr} & 0 \ldots & 0 \end{pmatrix}$$

Choose integers $a_1, \cdots, a_r, a_{r+1}, \cdots, a_n$, so that $a_i > 0, i \le [\frac{n}{2}]$, and $a_{n+1-i} = -a_i, i \le [\frac{n}{2}]$ (if $n$ is odd, say, $n = 2\ell + 1$, then, we take $a_{\ell+1}$ to be 0).

Let $g_t$ be the diagonal matrix $g_t = diag(t^{a_1}, \cdots t^{a_r}, t^{a_{r+1}}, \cdots, t^{a_n})$ (note that $g_t \in O(V)$ (cf. Remark 10.4.0.1)). Consider the one parametric subgroup $\{g_t, t \in K^*\}$. We have, $g_t x = g_t \cdot \mathcal{U} = \mathcal{U} g_t = \mathcal{U}_t$, where

$$\mathcal{U}_t = \begin{pmatrix} t^{a_1}u_{11} & t^{a_2}u_{12} & \ldots & t^{a_r}u_{1r} & 0 \ldots & 0 \\ t^{a_1}u_{21} & t^{a_2}u_{22} & \ldots & t^{a_r}u_{2r} & 0 \ldots & 0 \\ \vdots & \vdots & & \vdots & \vdots \quad \vdots & \vdots \\ t^{a_1}u_{m1} & t^{a_2}u_{m2} & \ldots & t^{a_r}u_{mr} & 0 \ldots & 0 \end{pmatrix}$$

Hence $g_t x \to 0$ as $t \to 0$ (note that $r \leq [\frac{n}{2}]$, and hence $a_i > 0, i \leq r$), and this implies that $0 \in \overline{G \cdot x}$ ($G$ being $O_n(K)$) which is a contradiction to the hypothesis that $x$ is semi-stable. Therefore our assumption that $\psi(x) = 0$ is wrong and (i) of Lemma 10.1.0.4 is satisfied.

(ii) Let

$$U = \{\underline{u} \in X \mid \{u_1, \ldots, u_n\}\text{are linearly independent}\}$$

Clearly, $U$ is a $G$-stable open subset of $X$.
**Claim:** $G$ operates freely on $U$, $U \to U\, mod\, G$ is a $G$-principal fiber space, and $\psi$ induces an immersion $U/G \to Sym\, M_m$.
**Proof of Claim:** Let $H = GL_n(K)$. We have a $G(= O_n(K))$-equivariant identification

$$(*) \qquad U \cong H \times \underbrace{V \times \cdots \times V}_{(m-n)\text{ copies}} = H \times F, \text{ say}$$

where $F = \underbrace{V \times \cdots \times V}_{(m-n)\text{ copies}}$. From this it is clear that $G$ operates freely on $U$. Further, we see that $U\, mod\, G$ may be identified with the fiber space with base $H\, mod\, G$, and fiber $\underbrace{V \times \cdots \times V}_{(m-n)\text{ copies}}$ associated to the principal fiber space $H \to H\,/\,G$. It remains to show that $\psi$ induces an immersion $U/G \to \mathbb{A}^N$, i.e., to show that the map $\psi : U/G \to \mathbb{A}^N$ and its differential $d\psi$ are both injective. We first prove the injectivity of $\psi : U/G \to \mathbb{A}^N$. Let $x, x'$ in $U/G$ be such that $\psi(x) = \psi(x')$. Let $\eta, \eta' \in U$ be lifts for $x, x'$ respectively. Using the identification (*) above, we may write

$$\eta = (A, u_{n+1}, \cdots, u_m),\ A \in H$$
$$\eta' = (A', u'_{n+1}, \cdots, u'_m),\ A' \in H$$

(here, $u_i$, $1 \leq i \leq n$, are given by the rows of $A$, while $u'_i$, $1 \leq i \leq n$, are given by the rows of $A'$. The hypothesis that $\psi(x) = \psi(x')$ implies in particular that

$$\langle u_i, u_j\rangle = \langle u'_i, u'_j\rangle,\ 1 \leq i, j \leq n$$

which may be written as

$$A J_n\, {}^t A = A' J_n\, {}^t A'$$

where $J_n$ is the matrix of the form $\langle , \rangle$ (note that since we are writing a vector $v \in V$ as a row vector, $\langle v_i, v_j\rangle = v_i J_n\, {}^t v_j$). Hence we obtain

$$(A'^{-1} A) J_n\, {}^t(A'^{-1} A) = J_n,\ i.e.,\ A'^{-1} A \in G$$

This implies that

$$(**) \qquad A = A' \cdot g, \text{ for some g} \in \text{G}.$$

Hence on $U/G$, we may suppose that

$$x = (u_1, \cdots, u_n, u_{n+1}, \cdots, u_m)$$
$$x' = (u_1, \cdots, u_n, u'_{n+1}, \cdots, u'_m)$$

where $\{u_1, \cdots, u_n\}$ is linearly independent.

For a given $j, n+1 \leq j \leq m$, we have,

$$\langle u_i, u_j\rangle = \langle u_i, u'_j\rangle, 1 \leq i \leq n, \text{ implies}, u_j = u'_j$$

(since, $\{u_1, \cdots, u_n\}$ is linearly independent and the form $\langle , \rangle$ is non-degenerate). Thus we obtain

(†) $$u_j = u'_j, \text{ for all } j.$$

The injectivity of $\psi : U/G \to \mathbb{A}^N$ follows from (†).

To prove that the differential $\mathrm{d}\psi$ is injective, we merely note that the above argument remains valid for the points over $K[\epsilon]$, the algebra of dual numbers ($= K \oplus K\epsilon$, the $K$-algebra with one generator $\epsilon$, and one relation $\epsilon^2 = 0$), i.e., it remains valid if we replace $K$ by $K[\epsilon]$, or in fact by any $K$-algebra.
Thus (ii) of Lemma 10.1.0.4 holds.

(iii) We have

$$\dim U/G = \dim U - \dim G = mn - \frac{1}{2}n(n-1) = \dim D_{n+1}(Sym\ M_m)$$

(cf. Chapter 6, Corollary 6.2.5.3 with $t = n+1$). The immersion $U/G \hookrightarrow Spec\ S (\subseteq D_{n+1}(Sym\ M_m))$ together with the fact above that $\dim U/G = \dim D_{n+1}(Sym\ M_m)$ implies that $Spec\ S$ in fact equals $D_{n+1}(Sym\ M_m)$.
Thus (iii) of Lemma 10.1.0.4 holds.

(iv) The normality of $Spec\ S (= D_{n+1}(Sym\ M_m))$ follows from Chapter 6, Corollary 6.2.5.3.

Using the identification $X /\!\!/ O(V) \cong D_{n+1}(Sym\ M_m)$, we get

**Theorem 10.4.0.3** *(i)* **First Fundamental Theorem**
*The ring of invariants $K[X]^{O(V)}$ is generated by $\varphi_{ij} = \langle u_i, u_j\rangle,\ 1 \leq i, j \leq m$.*
*(ii)* **Second Fundamental Theorem**
*The ideal of relations among the generators in (i) is generated by the $(n+1)$-minors of the symmetric $m \times m$-matrix $(\varphi_{ij})$.*

Further, using the results of Chapter 6 (especially §6.2.9 of Chapter 6), with notations as in loc.cit, we obtain

**Theorem 10.4.0.4** *(A Standard Monomial Basis for $K[X]^{O(V)}$) The ring of invariants $K[X]^{O(V)}$ has a basis consisting of standard monomials in the regular functions $p(A, B),\ A, B \in I(r, m), A \geq B,\ r \leq n$.*

## 10.5 $Sp_{2\ell}(K)$-action

In this section, we give a GIT-theoretic proof of the results for $SP_{2\ell}(K)$-actions appearing in classical invariant theory (cf. [22], [115]).

Let $V = K^n, n = 2\ell$, together with a non-degenerate, skew-symmetric bilinear form $\langle , \rangle$. Let $G = Sp(V)$ (the symplectic group consisting of linear automorphisms of $V$ preserving $\langle , \rangle$) on $X$. We shall take the matrix of the form $\langle , \rangle$ to be $J_n :=$ `anti-diagonal`$(\underbrace{1, \cdots, 1}_{l \text{ times}}, \underbrace{-1, \cdots, -1}_{l \text{ times}})$. Then a matrix $A \in GL(V)$ is in $Sp(V)$ if and only if

$$ {}^tAJ_nA = J_n, \ i.e, \ J_n^{-1}({}^tA^{-1})J_n = A, \ i.e, \ J_n({}^tA^{-1})J_n = -A \text{ (note that } J_n^{-1} = -J_n) $$

Let $X$ denote the affine space $V^{\oplus m} = V \oplus \cdots \oplus V$, where $m > n$. For $\underline{u} = (u_1, u_2, ..., u_m) \in X$, writing $u_i = (u_{i1}, \cdots, u_{in})$ (with respect to the standard basis for $K^n$), we shall identify $\underline{u}$ with the $m \times n$ matrix $U := (u_{ij})_{m \times n}$. Thus, we identify $X$ with $M_{m,n}$, the space of $m \times n$ matrices (with entries in $K$). Consider the diagonal action of $Sp(V)$ on $X$. The induced action on $K[X]$ is given by

$$ (A \cdot f)(\underline{u}) = f(A \cdot \underline{u}) = f(UA), \ \underline{i} \in X, \ f \in K[X], \ A \in Sp_n(K)(= Sp(V)) $$

**The Functions $\varphi_{ij}$:** Consider the functions $\varphi_{ij} : X \to K$ defined by $\varphi_{ij}((\underline{u})) = \langle u_i, u_j \rangle$, $1 \le i, j \le m$. Each $\varphi_{ij}$ is clearly in $K[X]^{Sp(V)}$.

**Theorem 10.5.0.1** *Let $Sk\,M_m$ denote the space of skew symmetric matrices with entries in $K$. The morphism $\psi : X \to Sk\,M_m$, $\underline{u} \mapsto (\langle u_i, u_j \rangle)$ is $Sp(V)$-invariant. Further, it maps $X$ onto $D_{n+1}(Sk\,M_m)$ and identifies the categorical quotient $X /\!\!/ Sp(V)$ with $D_{n+1}(Sk\,M_m)$ (here, $D_{n+1}(Sk\,M_m)$ is the determinantal variety inside $Sk\,M_m$, the space of skew symmetric $m \times m$ matrices, consisting of matrices of rank $\le n$).*

The proof is exactly same as that of Theorem 10.4.0.2 (with appropriate modifications).

Using the identification $X /\!\!/ Sp(V) \cong D_{n+1}(Sk\,M_m)$, and the results of Chapter 7 (especially, §7.2.6) of Chapter 7, we obtain, with notations as in loc.cit

**Theorem 10.5.0.2** *(i)* **First Fundamental Theorem**

*The ring of invariants $K[X]^{Sp(V)}$ is generated by $\varphi_{ij} = \langle u_i, u_j \rangle$, $1 \le i, j \le m$.*

*(ii)* **Second Fundamental Theorem**

*The ideal of relations among the generators in (i) is generated by the $(n+1)$-minors of the skew symmetric $m \times m$-matrix $(\varphi_{ij})$.*

*(iii)* ***A Standard Monomial Basis for*** $K[X]^{Sp(V)}$

*The ring of invariants $K[X]^{Sp(V)}$ has a basis consisting of standard monomials $q(A_1) \cdots q(A_r)$, $A_1 \le \cdots \le A_r$, $r \in \mathbb{N}$, $A_1 \ge (1, \cdots, t-1)$, $A_i \in \mathcal{A}_{n,2n}$ in the Pfaffians (notations as in Chapter 7, §7.2.6).*

# 11

# $SL_n(K)$-action

In this chapter, we consider the $SL_n(K)$-action on $X = \underbrace{V \oplus \cdots \oplus V}_{m \text{ copies}} \oplus \underbrace{V^* \oplus \cdots \oplus V^*}_{q \text{ copies}}$, $V = K^n$. We present a proof (cf. [72]) of the Cohen-Macaulayness of $K[X]^{SL_n(K)}$. While *Spec* $K[X]^{GL_n(K)}$ (being a certain determinantal variety (cf. Chapter 10)) gets identified with the opposite cell in a certain Schubert variety, *Spec* $K[X]^{SL_n(K)}$ does not have a direct relationship to Schubert varieties. Nevertheless, certain "Schubert variety connections" are used in a crucial way (cf. §11.3). As mentioned in the Introduction, in recent times, among the several techniques of proving the Cohen-Macaulayness of algebraic varieties, two techniques have proven to be quite effective, namely, Frobenius-splitting technique and deformation technique.

Frobenius-splitting technique is used in [82], for example, for proving the (arithmetic) Cohen-Macaulayness of Schubert varieties. Frobenius-splitting technique is also used in [80, 81], [84] for proving the Cohen-Macaulayness of certain varieties. The deformation technique consists in constructing a flat family over $\mathbb{A}^1$, with the given variety as the generic fiber (corresponding to $t \in K$ invertible). If the special fiber (corresponding to $t = 0$) is Cohen-Macaulay, then one may conclude the Cohen-Macaulayness of the given variety. Hodge algebras (cf. [20]) are typical examples where the deformation technique affords itself very well. Deformation technique is also used in [12, 17, 21, 33, 46]. The philosophy behind these works is that if there is a "standard monomial basis" for the co-ordinate ring of the given variety, then the deformation technique will work well in general (using the "straightening relations"). It is this philosophy that is adopted in [72] in proving the Cohen-Macaulayness of $K[X]^{SL_n(K)}$. To make this more precise, the proof of the Cohen-Macaulayness of $K[X]^{SL_n(K)}$ is accomplished in the following steps:

- One first constructs a $K$-subalgebra $S$ of $K[X]^{SL_n(K)}$ by prescribing a set of algebra generators $\{f_\alpha, \alpha \in H\}$, $H$ being a finite partially ordered set and $f_\alpha \in K[X]^{SL_n(K)}$.
- One then constructs a "standard monomial" basis for $S$ by

(i) defining "standard monomials" in the $f_\alpha$'s

(ii) writing down the straightening relation for a non-standard (degree 2) monomial $f_\alpha f_\beta$

(iii) proving linear independence of standard monomials (by relating the generators of $S$ to certain determinantal varieties)

(iv) proving the generation of $S$ (as a vector space) by standard monomials (using (ii)). In fact, to prove the generation for $S$, one first proves generation for a "graded version" $R(D)$ of $S$, where $D$ is a distributive lattice obtained by adjoining $\mathbf{1}, \mathbf{0}$ (the largest and the smallest elements of $D$) to $H$. One then deduces the generation for $S$. In fact, one constructs a "standard monomial" basis for $R(D)$. While the generation by standard monomials for $S$ is deduced from the generation by standard monomials for $R(D)$, the linear independence of standard monomials in $R(D)$ is deduced from the linear independence of standard monomials in $S$ (cf. (iii) above).

- One proves the normality and Cohen-Macaulayness of $R(D)$ by showing that Spec $R(D)$ flatly degenerates to the toric variety associated to the distributive lattice $D$.
- One deduces the normality and Cohen-Macaulayness of $S$ from the normality and Cohen-Macaulayness of $R(D)$ (cf. Theorem 11.4.4.4).
- Using the normality of $S$ and a crucial Lemma concerning GIT (cf. Chapter 10, Lemma 10.1.0.4 which gives a set of sufficient conditions for a normal sub algebra of $K[X]^{SL_n(K)}$ to equal $K[X]^{SL_n(K)}$), one shows that $S$ is in fact $K[X]^{SL_n(K)}$, and hence concludes that $K[X]^{SL_n(K)}$ is Cohen-Macaulay.

As a consequence, one obtains the following:

- **First Fundamental Theorem for $SL_n(K)$-invariants**, i.e., describing algebra generators for $K[X]^{SL_n(K)}$.
- **Second Fundamental Theorem for $SL_n(K)$-invariants**, i.e., describing generators for the ideal of relations among these algebra generators for $K[X]^{SL_n(K)}$.
- **A Standard Monomial Basis for $K[X]^{SL_n(K)}$**

As a by-product one recovers Theorem 3.3 of DeConcini-Procesi [22] (which describes a set of algebra generators for $K[X]^{SL_n(K)}$).

## 11.1 Quadratic relations

In this section, we want to derive certain quadratic relations on a determinantal variety. As in Chapter 5, let $D_t$ be the determinantal variety in $M_{r,d}(K)$ ($\cong \mathbb{A}^{rd}$), defined by the vanishing of $t$-minors. As in Chapter 5, we have an identification of $D_t$ with the opposite cell $Y(\phi)$ in $G_{d,n}$, $n = r + d$, where $\phi$ is the $d$-tuple $([t, d], [n+2-t, n])$. Given $s \le min\{r, d\}$, and a pair of $s$-tuples, $A, B$, we shall denote by $p(A, B)$, the regular function on $D_t$ which is just the $s$-minor (of the $r \times d$ generic matrix $(x_{ij})$) with row and column indices given by $A, B$ respectively. Recall that a monomial $p(A_1, B_1) \cdots p(A_m, B_m)$ is standard if $A_1 \succeq B_1 \succeq \cdots \succeq A_m \succeq B_m$, the partial order $\succeq$ being as in Chapter 5, §5.2.4. Recall the following

**Theorem 11.1.0.1** *Standard monomials in $p(A, B)$'s with $\# A \le t-1$ form a basis for $K[D_t]$, the algebra of regular functions on $D_t$.*

Also, using the bijection $\theta$ (cf. Chapter 5, §5.2.3) between the minors of $(x_{ij})$ and Plücker co-ordinates (via the identification of $M_{r,d}(K)$ with the opposite cell in $G_{d,n}$, $n = r + d$), we have (in view of Chapter 4, Proposition 4.3.3.3), the following

**Proposition 11.1.0.2** *Let $p(A_1, A_2)$, $p(B_1, B_2)$ (in $K[D_t]$) be not comparable. Let*

$$p(A_1, A_2)p(B_1, B_2) = \sum a_i p(C_{i1}, C_{i2})p(D_{i1}, D_{i2}), a_i \in K^* \qquad (*)$$

*be the straightening relation in $K[D_t]$. Then for every $i$, $C_{i1}, C_{i2}, D_{i1}, D_{i2}$ have cardinalities $\leq t - 1$; further,*

*(i) $C_{i1} \geq$ both $A_1$ and $B_1$; $D_{i1} \leq$ both $A_1$ and $B_1$.*
*(ii) $C_{i2} \geq$ both $A_2$ and $B_2$; $D_{i2} \leq$ both $A_2$ and $B_2$.*
*(iii) The term $p((A_1, A_2) \vee (B_1, B_2))p((A_1, A_2) \wedge (B_1, B_2))$ occurs in (*) with coefficient 1.*

### 11.1.1 The partially ordered set $H_{r,d}$

Let

$$H_{r,d} = \bigcup_{0 \leq s \leq \min\{r,d\}} I(s, r) \times I(s, d)$$

where our convention is that $(\emptyset, \emptyset)$ is the element of $H_{r,d}$ corresponding to $s = 0$. We define a partial order $\succeq$ on $H_{r,d}$ as follows:

- We declare $(\emptyset, \emptyset)$ as the largest element of $H_{r,d}$.
- For $(A, B)$, $(A', B')$ in $H_{r,d}$, say, $A = \{a_1, \cdots, a_s\}$, $B = \{b_1, \cdots, b_s\}$, $A' = \{a'_1, \cdots, a'_{s'}\}$, $B' = \{b'_1, \cdots, b'_{s'}\}$ for some $s, s' \geq 1$, we define $(A, B) \succeq (A', B')$ if $s \leq s'$, $a_j \geq a'_j$, $b_j \geq b'_j$, $1 \leq j \leq s$.

Note that via the bijection $\theta$ (cf. Chapter 5, §5.2.3), join and meet of two non-comparable elements $(A_1, A_2)$, $(B_1, B_2)$ of $H_{r,d}$ exist, and in fact are given by $(A_1, A_2) \vee (B_1, B_2) = (A_1 \vee B_1, A_2 \vee B_2)$, $(A_1, A_2) \wedge (B_1, B_2) = (A_1 \wedge B_1, A_2 \wedge B_2)$.

**Remark 11.1.1.1.** On the R.H.S. of (*), $C_{i1}, C_{i2}$ could both be the empty set (in which case $p(C_{i1}, C_{i2})$ is understood as 1). For example, with $X$ being a $2 \times 2$ matrix of indeterminates, we have

$$p_{1,2}p_{2,1} = p_{2,2}p_{1,1} - p_{\emptyset,\emptyset}p_{12,12}$$

**Remark 11.1.1.2.** In the sequel, while writing a straightening relation as in Proposition 11.1.0.2, if for some $i$, $C_{i1}$, $C_{i2}$ are both the empty set, we keep the corresponding $p(C_{i1}, C_{i2})$ on the right hand side of the straightening relation (even though its value is 1) in order to have homogeneity in the relation.

Taking $t = d + 1$ (in which case $D_t = Z = M_{r,d}(K)$) in Theorem 11.1.0.1 and Proposition 11.1.0.2, we obtain

**Theorem 11.1.1.3** *(i) Standard monomials in $p(A, B)$'s form a basis for $K[Z]$ ($\cong K[x_{ij}, 1 \leq i \leq r, 1 \leq j \leq d]$).*
*(ii) Relations similar to those in Proposition 11.1.0.2 hold on $Z$.*

**Remark 11.1.1.4.** Note that Theorem 11.1.1.3,(1) recovers the result of Doubilet-Rota-Stein (cf. [26], Theorem 2):

**Remark 11.1.1.5.** Theorem 11.1.0.1 is also proved in DeConcini-Procesi [22] (Theorem 1.2 in [22]). But we had taken the Schubert variety theoretic approach to deduce Theorem 11.1.0.1 from the results on Schubert varieties as discussed in chapter 4 in order to derive the straightening relations as given by Proposition 11.1.0.2(which are crucial for the discussion in §11.3).

## 11.2 The $K$-algebra $S$

Let $V = K^n$, $X = \underbrace{V \oplus \cdots \oplus V}_{m \text{ copies}} \oplus \underbrace{V^* \oplus \cdots \oplus V^*}_{q \text{ copies}}$.

**The $GL(V)$-action on $X$:** Writing $\underline{u} = (u_1, u_2, ..., u_m)$ with $u_i \in V$ and $\underline{\xi} = (\xi_1, \xi_2, ..., \xi_q)$ with $\xi_i \in V^*$, we shall denote the elements of $X$ by $(\underline{u}, \underline{\xi})$. The (natural) action of $GL(V)$ on $V$ induces an action of $GL(V)$ on $V^*$, namely, for $\xi \in V^*$, $g \in GL(V)$, denoting $g \cdot \xi$ by $\xi^g$, we have

$$\xi^g(v) = \xi(g^{-1}v),\ v \in V$$

The diagonal action of $GL(V)$ on $X$ is given by

$$g \cdot (\underline{u}, \underline{\xi}) = (g\underline{u}, \underline{\xi}^g) = (gu_1, gu_2, ..., gu_m, \xi_1^g, \xi_2^g, ..., \xi_q^g),\ g \in G,\ (\underline{u}, \underline{\xi}) \in X$$

The induced action on $K[X]$ is given by

$$(g \cdot f)(\underline{u}, \underline{\xi}) = f(g^{-1}(\underline{u}, \underline{\xi})),\ f \in K[X],\ g \in GL(V)$$

Consider the functions $\varphi_{ij} : X \longrightarrow K$ defined by $\varphi_{ij}((\underline{u}, \underline{\xi})) = \xi_j(u_i)$, $1 \le i \le m$, $1 \le j \le q$. Each $\varphi_{ij}$ is a $GL(V)$-invariant: For $g \in GL(V)$, we have,

$$\begin{aligned}(g \cdot \varphi_{ij})((\underline{u}, \underline{\xi})) &= \varphi_{ij}(g^{-1}(\underline{u}, \underline{\xi})) \\ &= \varphi_{ij}((g^{-1}u, \xi^{g^{-1}}) \\ &= \xi_j^{g^{-1}}(g^{-1}u_i) \\ &= \xi_j(u_i) \\ &= \varphi_{ij}((\underline{u}, \underline{\xi}))\end{aligned}$$

It is convenient to have a description of the above action in terms of coordinates. So with respect to a fixed basis, we write the elements of $V$ as row vectors and those of $V^*$ as column vectors. Thus denoting by $M_{a,b}$ the space of $a \times b$ matrices with entries in $K$, $X$ can be identified with the affine space $M_{m,n} \times M_{n,q}$. The action of $GL_n(K) (= GL(V))$ on $X$ is then given by

$$A \cdot (U, W) = (UA, A^{-1}W),\ A \in GL_n(K),\ U \in M_{m,n},\ W \in M_{n,q}$$

As seen in chapter 10, we have an action of $GL_n(K)$ on $K[X]$ given by

$$(A \cdot f)(U, W) = f\left(A^{-1}(U, W)\right) = f\left(UA^{-1}, AW\right), \; f \in K[X]$$

Writing $U = \left(u_{ij}\right)$ and $W = (\xi_{kl})$ we denote the coordinate functions on $X$, by $u_{ij}$ and $\xi_{kl}$. Further, if $u_i$ denotes the $i$-th row of $U$ and $\xi_j$ the $j$-th column of $W$, the invariants $\varphi_{ij}$ described above are nothing but the entries $\langle u_i, \xi_j\rangle$ $(= \xi_j(u_i))$ of the product $UW$.

In the sequel, we shall denote $\varphi_{ij}(\underline{u}, \underline{\xi})$ also by $\langle u_i, \xi_j\rangle$.

### 11.2.1 The $SL_n(K)$-action

Let $X$ be as above. We shall denote $K[X]$ by $R$ so that $R = K[u_{ij}, \xi_{kl}\; 1 \leq i \leq m,\; 1 \leq j, k \leq n, 1 \leq l \leq q]$. We consider the action of $SL_n(K)$ on $X$ induced from the $GL_n(K)$-action.

**The Function** $p(A, B)$**:** For $A \in I(r, m)$, $B \in I(r, q)$, $1 \leq r \leq n$, let $p(A, B)$ be the regular function on $X$: $p(A, B)((\underline{u}, \underline{\xi}))$ is the determinant of the $r \times r$-matrix $(\langle u_i, \xi_j\rangle)_{i\in A,\, j\in B}$.

**The Functions** $u(I), \xi(J)$**:** As above, let $U = \left(u_{ij}\right)_{1\leq i\leq m,\, 1\leq j\leq n}$ and $W = (\xi_{kl})_{1\leq k\leq n,\, 1\leq l\leq q}$. For $I \in I(n, m)$, $J \in I(n, q)$, let $u(I), \xi(J))$ denote the following regular functions on $X$:

$u(I)((\underline{u}, \underline{\xi}))$ = the $n$-minor of $U$ with row indices given by $I$.

$\xi(J)((\underline{u}, \underline{\xi}))$ = the $n$-minor of $W$ with column indices given by $J$.

Note that for the diagonal action of $SL_n(K)$ $(= SL(V))$ on $X$, we have, $u(I), \xi(J)$ are in $R^{SL_n(K)}$.

**The** $K$**-algebra** $S$**:** Let $S$ be the $K$-subalgebra of $R$ generated by $\{u(I), \xi(J), p(A, B), I \in I(n, m), J \in I(n, q), A \in I(r, m), B \in I(r, q), 1 \leq r \leq n\}$. We shall denote the set $I(n, m)$ indexing the $u(I)$'s by $H_u$ and the set $I(n, q)$ indexing the $\xi(J)$'s by $H_\xi$. Also, we shall denote $H_p := \bigcup_{1\leq r\leq n} (I(r, m) \times I(r, q))$, and set

$$\begin{aligned} H &= H_u \dot\cup H_\xi \cup H_p \\ &= I(n, m) \dot\cup I(n, q) \cup \bigcup_{1\leq r\leq n} (I(r, m) \times I(r, q)), \end{aligned}$$

where $\dot\cup$ denotes a disjoint union. (If $m = q$, then $H_u$, $H_\xi$ are to be considered as two disjoint copies of $I(n, m)$.) Then the algebra generators $\{u(I), \xi(J), p(A, B), I \in I(n, m), J \in I(n, q), A \in I(r, m), B \in I(r, q), 1 \leq r \leq n\}$ of $S$ are indexed by the set $H$. Clearly $S \subseteq R^{SL(V)}$.

**Remark 11.2.1.1.** The $K$-algebra $S$ could have been simply defined as the $K$-subalgebra of $R^G$ generated by $\{\langle u_i, \xi_j\rangle\}$ (i.e., by $\{p(A, B), \#A = \#B = 1\}$) and $\{u(I), \xi(J)\}$. But we have a purpose in defining it as above, namely, the standard monomials (in $S$) will be built out of the $p(A, B)$'s with $\#A \leq n$, the $u(I)$'s and $\xi(J)$'s.

**Our goal is to show that $S$ equals $R^{SL(V)}$.**

**A Partial Order on $H$:** Define a partial order on $H$ as follows:

(i) The partial order on $H_p$ is as in §11.1.1 (note that $H_p \subset H_{m,q}$)
(ii) The partial order on $H_u$ and $H_\xi$ are as in Chapter 4, §4.1.1.
(iii) Any element of $H_u$ and any element of $H_\xi$ are not comparable.
(iv) No element of $H_u$, $H_\xi$ is greater than any element of $H_p$.
(v) For $I \in H_u$ and $(A, B) \in H_p$, we define $I \leq (A, B)$ if $I \leq A$ (the partial order being as in Chapter 4, §4.1.1). Similarly, for $J \in H_\xi$ and $(A, B) \in H_p$, we define $J \leq (A, B)$ if $J \leq B$.

**Lemma 11.2.1.2** *$H$ is a ranked poset of rank $d := (m+q)n - n^2$, i.e., all maximal chains in $H$ have the same cardinality $= (m+q)n - n^2 + 1$.*

*Proof.* Clearly, $H$ is a ranked poset (since it is composed of ranked posets). To compute the rank of $H$, we consider the maximal chain consisting of $\tau_1, \cdots, \tau_N$, where the first $q$ of them are given by $(m, q), (m, q-1), \cdots, (m, 1)$ (of $H_p$),
the next $(m-1)$ of them are given by $(m-1, 1), (m-2, 1), \cdots, (1, 1)$ (of $H_p$).
(thus contributing $m+q-1$ to the cardinality of the chain).
This is now followed by the $q-1$ elements of $H_p$:
$((1, m), (1, q)), ((1, m), (1, q-1)), \cdots, ((1, m), (1, 2))$,
followed by the $m-2$ elements of $H_p$:
$((1, m-1), (1, 2)), ((1, m-2), (1, 2)), \cdots, ((1, 2), (1, 2))$
(thus contributing $m+q-3$ to the cardinality of the chain).
Thus proceeding, finally, we end up with $((1, 2, \cdots, n), (1, 2, \cdots, n))$ (in $H_p$). This is now followed by either $(1, 2, \cdots, n)$ of $H_u$ or $(1, 2, \cdots, n)$ of $H_\xi$.
The number of elements in the above chain equals
$[m+q-1+(m+q-3)+\cdots+m+q-(2n-1)]+1 = (m+q)n - n^2 + 1$.

## 11.3 Standard monomials in the $K$-algebra $S$

**Definition 11.3.0.1** *A monomial $F$ in the $p(A, B)$'s, $u(I)$'s, and $\xi(J)$'s, is said to be* standard *if $F$ satisfies the following conditions:*

*(i) If $F$ involves $u(I)$, for some $I$ (resp. $\xi(J)$ for some $J$), then $F$ does not involve $\xi(J')$ for any $J'$ (resp. $u(I')$, for any $I'$).*
*(ii) If $F = p(A_1, B_1) \cdots p(A_r, B_r) u(I_1) \cdots u(I_s)$ (resp. $p(A_1, B_1) \cdots p(A_r, B_r) \xi(J_1) \cdots \xi(J_t)$), where $r, s, t$ are integers $\geq 0$, then*

$$A_1 \geq \cdots \geq A_r \geq I_1 \geq \cdots \geq I_s \text{ ( resp. } B_1 \geq \cdots \geq B_r \geq J_1 \geq \cdots \geq J_t)$$

### 11.3.1 Quadratic relations

In this subsection, we describe certain straightening relations to be used while proving the linear independence of standard monomials and generation (of $S$ as a $K$-vector space) by standard monomials.

**Theorem 11.3.1.1** *(i) Let $I \in H_u$, $J \in H_\xi$. We have*

$$u(I)\xi(J) = p(I, J)$$

*(ii) Let $I, I' \in H_u$ be not comparable. We have,*

$$u(I)u(I') = \sum_r b_r u(I_r)u(I'_r),\ b_r \in K^*$$

*where for all $r$, $I_r \geq$ both $I, I'$, and $I'_r \leq$ both $I, I'$.*

*(iii) Let $J, J' \in H_\xi$ be not comparable. We have,*

$$\xi(J)\xi(J') = \sum_s c_s \xi(J_s)\xi(J'_s),\ c_s \in K^*$$

*where for all $s$, $J_s \geq$ both $J, J'$, and $J'_s \leq$ both $J, J'$.*

*(iv) Let $(A_1, A_2), (B_1, B_2) \in H_p$ be not comparable. Then we have*

$$p(A_1, A_2)p(B_1, B_2) = \sum_i a_i p(C_{i1}, C_{i2})p(D_{i1}, D_{i2}), a_i \in K^*,$$

*where $(C_{i1}, C_{i2}), (D_{i1}, D_{i2})$ belong to $H_p$; further, for every $i$, we have*

*a) $C_{i1} \geq$ both $A_1$ and $B_1$; $D_{i1} \leq$ both $A_1$ and $B_1$.*

*b) $C_{i2} \geq$ both $A_2$ and $B_2$; $D_{i2} \leq$ both $A_2$ and $B_2$.*

*(v) Let $I \in H_u$, $(A, B) \in H_p$ be such that $A \not\geq I$. We have,*

$$p(A, B)u(I) = \sum_t d_t p(A_t, B_t)u(I_t),\ d_t \in K^*$$

*where for every $t$, we have, $A_t \geq$ (resp. $I_t \leq$) both $A$ and $I$, and $B_t \geq B$.*

*(vi) Let $J \in H_\xi$, $(A, B) \in H_p$ be such that $B \not\geq J$. We have,*

$$p(A, B)\xi(J) = \sum_l e_l p(A_l, B_l)\xi(J_l),\ e_l \in K^*$$

*where for every $l$, we have, $A_l \geq A$, and $B_l \geq$ (resp. $J_l \leq$) both $B$ and $J$.*

*Proof.* In the course of the proof, we will be repeatedly using the fact that the subalgebra generated by $\{p(A, B),\ A \in I(r, m),\ B \in I(r, q)\ 1 \leq r \leq n\}$ being $R^{GL(V)}$ ($\cong K[D_{n+1}(M_{m,q})]$ (cf. Chapter 10, §10.3)), the results given in Proposition 11.1.0.2 apply to this subalgebra.

(i) is clear from the definitions of $u(I)$, $\xi(J)$ and $p(I, J)$.

(ii). We shall denote a minor of $U(=(u_{ij})_{1\le i\le m,\,1\le j\le n})$ with rows and columns given by $I, J$ (where $I, J \in I(r,m)$ for some $r \le n$) by $\Delta(I, J)$. Observe that if $\#I = n$, then $J = (1, 2, \cdots, n)$ necessarily (since $U$ has size $m \times n$). Thus for $I \in H_u$, we have that $u(I) = \Delta(I, I_n), u(I') = \Delta(I', I_n)$ (as minors of $U$), where $I_n = (1, 2, \cdots, n)$, we have, in view of Proposition 11.1.0.2, (ii),

$$u(I)u(I') = \Delta(I, I_n)\Delta(I', I_n) = \sum_i b_i \Delta(C_{i1}, C_{i2})\Delta(D_{i1}, D_{i2}), a_i \in K^*,$$

where we have for every $i$, $C_{i1} \ge$ both $I$ and $I'$; $D_{i1} \le$ both $I$ and $I'$; $C_{i2} \ge I_n$; $D_{i2} \le I_n$ which forces $\#D_{i2} = n$ (in view of the partial order (cf. §11.1.1); note that $D_{i2}$ being the column indices of a minor of the $m \times n$ matrix $U$, we have that $\#D_{i2} \le n$). Hence we obtain that $D_{i2} = I_n$, for all $i$. In particular, we obtain that $\#D_{i1}(=\#D_{i2}) = n$. This in turn implies (by consideration of the degrees in $u_{ij}$'s of the terms in the above sum) that $\#C_{i1} = \#C_{i2} = n$. Hence $C_{i2} = I_n$ (again note that $C_{i2}$ gives the column indices of the $n$-minor $\Delta(C_{i1}, C_{i2})$ of the $m \times n$ matrix $U$). Thus the above relation becomes

$$u(I)u(I') = \sum_i b_i u(C_{i1})u(D_{i1}),$$

with $C_{i1} \ge$ both $I$ and $I'$; $D_{i1} \le$ both $I$ and $I'$. This proves (ii).
Proof of (iii) is similar to that of (ii).
(iv) is a direct consequence of Proposition 11.1.0.2.
(v). If $\#A = n = \#B$, then $p(A, B)u(I) = u(A)u(I)\xi(B)$. By (ii),

$$u(A)u(I) = \sum_i d_i u(C_i)u(D_i),\ d_i \in K^*$$

where $C_i \ge$ both $A, I$, and $D_i \le$ both $A, I$. Hence

$$p(A, B)u(I) = \sum_i d_i u(C_i)u(D_i)\xi(B) = \sum_i d_i p(C_i, B)u(D_i)$$

where $C_i \ge$ both $A, I$, and $D_i \le$ both $A, I$, and the result follows.

Let then $\#A < n$. By (i), we have $u(I)\xi(I_n) = p(I, I_n)$. Hence, $p(A, B)u(I)\xi(I_n) = p(A, B)p(I, I_n)$. The hypothesis that $A \not\ge I$ implies that $p(A, B)p(I, I_n)$ is not standard (note that the facts that $\#A < n, \#I = n$ implies that $I \not\ge A$). Hence (iv) implies that

$$p(A, B)p(I, I_n) = \sum a_i p(C_{i1}, C_{i2})p(D_{i1}, D_{i2}), a_i \in K^*,$$

where $(C_{i1}, C_{i2}), (D_{i1}, D_{i2})$ belong to $H_p$; further, for every $i$, $C_{i1} \ge$ both $A$ and $I$; $D_{i1} \le$ both $A$ and $I$; $C_{i2} \ge$ both $B$ and $I_n$; $D_{i2} \le$ both $B$ and $I_n$ which forces $D_{i2} = I_n$ (note that in view of Theorem 10.3.1.4, all minors in the above relation have size $\le n$); and hence $\#D_{i1} = n$, for all $i$. Hence $p(D_{i1}, D_{i2}) = u(D_{i1})\xi(I_n)$, for all $i$. Hence cancelling $\xi(I_n)$, we obtain

$$p(A,B)u(I)=\sum a_i p(C_{i1},C_{i2})u(D_{i1}),$$

where $C_{i1}\geq$ both $A$ and $I$, $D_{i1}\leq$ both $A$ and $I$, and $C_{i2}\geq B$. This proves (v). Proof of (vi) is similar to that of (v).

### 11.3.2 Linear independence of standard monomials

In this subsection, we prove the linear independence of standard monomials.

**Lemma 11.3.2.1** *Let* $(A,B)\in H_p$, $I\in H_u$, $J\in H_\xi$.

*(i) The set of standard monomials in the* $p(A,B)$*'s is linearly independent.*
*(ii) The set of standard monomials in the* $u(I)$*'s is linearly independent.*
*(iii) The set of standard monomials in the* $\xi(J)$*'s is linearly independent.*

*Proof.* (i) follows from Theorem 11.1.0.1.
(ii), (iii) follow from Theorem 11.1.1.3,(1) applied to $K[u_{ij}, 1\leq i\leq m, 1\leq j\leq n]$, $K[\xi_{kl}, 1\leq k\leq n, 1\leq l\leq q]$ respectively.

**Proposition 11.3.2.2** *Standard monomials are linearly independent.*

*Proof.* For a monomial $M$, by *u-degree* (resp. *$\xi$-degree*) of $M$, we shall mean the degree of $M$ in the variables $u_{ij}$'s (resp. $\xi_{kl}$'s ). We have

$$\begin{aligned}&u\texttt{-degree of }p(A_1,B_1)\cdots p(A_r,B_r)\\&\quad=\xi\texttt{-degree of }p(A_1,B_1)\cdots p(A_r,B_r)=\sum_i \#A_i\\&u\texttt{-degree of }u(I_1)\cdots u(I_s)=ns,\ \xi\texttt{-degree of }u(I_1)\cdots u(I_s)=0\\&\xi\texttt{-degree of }\xi(J_1)\cdots\xi(J_t)=nt,\ u\texttt{-degree of }\xi(J_1)\cdots\xi(J_t)=0\end{aligned}$$

By considering the $u$-degree and the $\xi$-degree, and using Lemma 11.3.2.1 we see that $\{p(A_1,B_1)\cdots p(A_r,B_r), u(I_1)\cdots u(I_s), \xi(J_1)\cdots\xi(J_t), r,s,t\geq 0\}$ is linearly independent.

Let

$$(*)\qquad F:=R+S=0$$

be a relation among standard monomials, where $R=\sum a_iM_i$, $S=\sum b_iN_i$ such that each $M_i$ (resp. $N_i$) is a standard monomial of the form $p(A_1,B_1)\cdots p(A_{r_i},B_{r_i})$ (resp. $p(A_1,B_1)\cdots p(A_{q_i},B_{q_i})u(I_1)\cdots u(I_{s_i})\xi(J_1)\cdots\xi(J_{t_i})$, $q_i\geq 0$, and at least one of $\{s_i,t_i\}>0$ ). Let $g$ be in $GL_n(K)$ such that $(det\,g)^{s_i+t_i}\neq 1$, for all $i$. Then using the facts that $g\cdot p(A,B)=p(A,B)$, $g\cdot u(I)=(det\,g)u(I)$, $g\cdot\xi(J)=(det\,g)\xi(J)$, we have, $F-gF=\sum b_i(1-(det\,g)^{s_i+t_i})N_i=0$. Hence if we show that the $N_i$'s are linearly independent, then (in view of Lemma 11.3.2.1,(1)), we would obtain that (*) is the trivial relation. Thus we may suppose that

$$(**)\qquad F=\sum b_iN_i=0,$$

where each $N_i$ is a standard monomial of the form

$$p(A_1, B_1) \cdots p(A_r, B_r) u(I_1) \cdots u(I_s) \xi(J_1) \cdots \xi(J_t)$$

where $r \geq 0$ and at least one of $\{s, t\} > 0$; in fact, $N_i$'s being standard, in any $N_i$, precisely one of $\{s_i, t_i\}$ is non-zero.

We first multiply (**) by $u(I_n)^N$ ($I_n$ being $(1, 2, \cdots, n)$), for a sufficiently large $N$ ($N$ could be taken to be any integer greater than all of the $t$'s, appearing in the $\xi(J_1) \cdots \xi(J_t)$'s); we then replace a $\xi(J)u(I_n)$ by $p(I_n, J)$ (cf. Theorem 11.3.1.1, (1)). Then in the resulting sum, any monomial will involve only the $p(A, B)$'s and the $u(I)$'s. Thus we may suppose that (**) is of the form

$$(* * *) \qquad G := \sum c_i G_i = 0$$

where each $G_i$ is of the form $p(A_1, B_1) \cdots p(A_r, B_r) u(I_1) \cdots u(I_s)$. Note that for each standard monomial $M = p(A_1, B_1) \cdots p(A_r, B_r) u(I_1) \cdots u(I_s)$ (resp. $p(A_1, B_1) \cdots p(A_r, B_r) \xi(J_1) \cdots \xi(J_t)$) appearing in (**), $M u(I_n)^N$ is again standard. Again, considering $G - gG$, $g \in GL_n(K)$, with $\det g \neq$ a root of unity, as above, we may suppose that in each monomial $p(A_1, B_1) \cdots p(A_r, B_r) u(I_1) \cdots u(I_s)$ appearing in (***), $s > 0$. Further, in view of Lemma 11.3.2.1,(2), we may suppose that for at least one monomial $r > 0$. Now considering the $\xi$-degree of the monomials, we may suppose (in view of Lemma 11.3.2.1,(2)) that in each monomial $p(A_1, B_1) \cdots p(A_r, B_r) u(I_1) \cdots u(I_s)$ appearing in (***), $r > 0$.

Thus, for each monomial $p(A_1, B_1) \cdots p(A_r, B_r) u(I_1) \cdots u(I_s)$ appearing in (***), we have, $r, s > 0$. Now the $\xi$-degree (as well as the $u$-degree) being the same for all of the monomials in (***), for any two monomials $G_i, G_{i'}$, say

$$\begin{aligned} G_i &= p(A_1, B_1) \cdots p(A_r, B_r) u(I_1) \cdots u(I_s), \\ G_{i'} &= p(A'_1, B'_1) \cdots p(A'_{r'}, B'_{r'}) u(I'_1) \cdots u(I'_{s'}) \end{aligned}$$

we have $\sum\limits_{1 \leq i \leq r} \# A_i = \sum\limits_{1 \leq i \leq r'} \# A'_i$. This together with the fact that the $u$-degree is the same for all of the terms $G_k$'s in (***) implies that $s = s'$. Thus we obtain that in all of the monomials $p(A_1, B_1) \cdots p(A_r, B_r) u(I_1) \cdots u(I_s)$ in (***), the integer $s$ is the same (and $s > 0$). Now we multiply (***) through out by $\xi(I_n)^s$ (where $I_n = (1, 2, \cdots, n)$) to arrive at a linear sum

$$\sum d_i H_i = 0$$

where each $H_i$ is a standard monomial in the $p(A, B)$'s (note that $H_i = p(A_1, B_1) \cdots p(A_r, B_r) p(I_1, I_n) \cdots p(I_s, I_n)$ is standard). Now the required result follows from the linear independence of $p(A, B)$'s (cf. Lemma 11.3.2.1,(1)).

### 11.3.3 The algebra $S(D)$

To prove the generation of $S$ (as a $K$-vector space) by standard monomials, we define a $K$-algebra $S(D)$, construct a standard monomial basis for $S(D)$ and deduce the results for $S$ (in fact, it will turn out that $S(D) \cong S$). We first define the $K$-algebra $R(D)$ as follows:

Let

$$D = H \cup \{\mathbf{1}\} \cup \{\mathbf{0}\}$$

$H$ being as in the beginning of §11.2.1. Extend the partial order on $H$ to $D$ by declaring $\{\mathbf{1}\}$ (resp. $\{\mathbf{0}\}$) as the largest (resp. smallest) element. Let $P(D)$ be the polynomial algebra

$$P(D) := K[X(A, B), Y(I), Z(J), X(\mathbf{1}), X(\mathbf{0}), (A, B) \in H_p, I \in H_u, J \in H_\xi]$$

Let $\mathfrak{a}(D)$ be the homogeneous ideal in the polynomial algebra $P(D)$ generated by the six relations of Theorem 11.3.1.1 ($X(A, B), Y(I), Z(J)$ replacing $p(A, B), u(I)$, $\xi(J)$ respectively), with relations (1) and (4) homogenized as follows: (1) is homogenized as

$$(*) \qquad X(I)Y(J) = X(I, J)X(\mathbf{0})$$

while (4) is homogenized as

$$X(A_1, A_2)X(B_1, B_2) = \sum a_i X(C_{i1}, C_{i2})X(D_{i1}, D_{i2})$$

where $X(C_{i1}, C_{i2})$ is to be understood as $X(\mathbf{1})$ if both $C_{i1}, C_{i2}$ equal the empty set (cf. Remark 11.1.1.2). Let

$$R(D) = P(D)/\mathfrak{a}(D)$$

We shall denote the classes of $X(A, B), Y(I), Z(J), X(\mathbf{1}), X(\mathbf{0})$ in $R(D)$ by $x(A, B), y(I), z(J), x(\mathbf{1}), x(\mathbf{0})$ respectively.

**The Algebra $M(D)$:** Set $M(D) = R(D)_{(x(\mathbf{0}))}$, the homogeneous localization of $R(D)$ at $x(\mathbf{0})$. We shall denote $\frac{x(\mathbf{1})}{x(\mathbf{0})}, \frac{x(A,B)}{x(\mathbf{0})}, \frac{y(I)}{x(\mathbf{0})}, \frac{z(J)}{x(\mathbf{0})}$ (in $M(D)$) by $q(\mathbf{1}), r(A, B)$, $s(I), t(J)$ respectively.

**A Grading for $M(D)$:** We give a grading for $M(D)$ by assigning degree one to $s(I), t(J)$, and degree 2 to $q(\mathbf{1}), r(A, B)$, where as above $I \in H_u$, $J \in H_\xi$, $(A, B) \in H_p$.

**The Algebra $S(D)$:** Set $S(D) = M(D)_{(q(\mathbf{1}))}$, the homogeneous localization of $M(D)$ at $q(\mathbf{1})$. We shall denote $\frac{r(A,B)}{q(\mathbf{1})}, \frac{s(I)}{q(\mathbf{1})}, \frac{t(J)}{q(\mathbf{1})}$ (in $S(D)$) by $c(A, B), d(I), e(J)$ respectively.

Let $\varphi_D : S(D) \to S$ be the map, $\varphi_D(c(A, B)) = p(A, B), \varphi_D(d(I)) = u(I), \varphi_D(e(J)) = \xi(J)$. Consider the canonical maps

$$\theta_D : R(D) \to M(D), \delta_D : M(D) \to S(D)$$

Denote $\gamma_D : R(D) \to S$ as the composite $\gamma_D = \varphi_D \circ \delta_D \circ \theta_D$.

### 11.3.4 A standard monomial basis for $R(D)$

We define a monomial in $x(A, B), y(I), z(J), x(\mathbf{1}), x(\mathbf{0})$ (in $R(D)$) to be standard in exactly the same way as in Definition 11.3.0.1 (we declare $x(\mathbf{1})$ (resp. $x(\mathbf{0})$) as the largest (resp. smallest)).

**Proposition 11.3.4.1** *The standard monomials in the $x(A, B)$'s, $y(I)$'s, $z(J)$'s, $x(\mathbf{1})$'s, $x(\mathbf{0})$'s are linearly independent.*

*Proof.* The result follows by considering $\gamma_D : R(D) \to S$, and using the linear independence of standard monomials in $S$ (cf. Proposition 11.3.2.2).

**Generation of $R(D)$ by Standard Monomials:** We shall now show that any non-standard monomial $F$ in $R(D)$ is a linear sum of standard monomials. Observe that if $M$ is a standard monomial, then $x(\mathbf{1})^l M$ (resp. $Mx(\mathbf{0})^l$) is again standard; hence we may suppose $F$ to be:

$$F = x(A_1, B_1)\cdots x(A_r, B_r)y(I_1)\cdots y(I_s)z(J_1)\cdots z(J_t)$$

Using the relations $y(I)z(J) = x(I, J)x(\mathbf{0})$, we may suppose that $F = x(A_1, B_1)\cdots x(A_r, B_r)y(I_1)\cdots y(I_s)$ or $F = x(A_1, B_1)\cdots x(A_r, B_r)z(J_1)\cdots z(J_t)$, say, $F = x(A_1, B_1)\cdots x(A_r, B_r)y(I_1)\cdots y(I_s)$.

Fix an integer $N$ sufficiently large. To each element $A \in \cup_{r=1}^{n} I(r, m)$, we associate an $(n+1)$-tuple as follows: Let $A \in I(r, m)$, for some $r$, say, $A = (a_1, \cdots, a_r)$. To $A$, we associate the $n + 1$-tuple

$$\overline{A} := (a_1, \cdots, a_r, m, m, \cdots, m, 1)$$

Similarly, for $B \in \cup_{r=1}^{n} I(r, q)$, say, $B = (b_1, \cdots, b_r)$, we associate the $n + 1$-tuple

$$\overline{B} := (b_1, \cdots, b_r, q, q, \cdots, q, 1)$$

To $F$, we associate the integer $n_F$ (and call it the *weight of* $F$) which has the entries of $\overline{A_1}, \overline{B_1}, \overline{A_2}, \overline{B_2}, \cdots, \overline{A_r}, \overline{B_r}, \overline{I_1}, \cdots, \overline{I_s}$ as digits (in the $N$-ary presentation). The hypothesis that $F$ is non-standard implies that either $x(A_i, B_i)x(A_{i+1}, B_{i+1})$ is non-standard for some $i \leq r-1$, or, $x(A_r, B_r)y(I_1)$ is non-standard or $y(I_j)y(I_{j+1})$ is non-standard for some $j \leq s-1$. Straightening these using Theorem 11.3.1.1, we obtain that $F = \sum a_i F_i$ where $n_{F_i} > n_F, \forall i$, and the result follows by decreasing induction on $n_F$ (note that while straightening a degree 2 relation using Theorem 11.3.1.1, (4), if $x(\mathbf{1})$ occurs in a monomial $G$, then the digits in $n_G$ corresponding to $x(\mathbf{1})$ are taken to be $(\underbrace{m, m\cdots, m}_{n+1 \text{ times}}, \underbrace{q, q\cdots, q}_{n+1 \text{ times}})$. Also note that the largest $F$ of degree $r$ in $x(A, B)$'s and degree $s$ in the $y(I)$'s is $x(\{m\}, \{q\})^r u(I_0)^s$ (where $I_0$ is the $n$-tuple $(m+1-n, m+2-n, \cdots, m)$) which is clearly standard (the starting point of the decreasing induction).

Hence we obtain

**Proposition 11.3.4.2** *Standard monomials in $x(A, B), y(I), z(J), x(\mathbf{1}), x(\mathbf{0})$ generate $R(D)$ as a $K$-vector space.*

Combining Propositions 11.3.4.1, 11.3.4.2, we obtain

**Theorem 11.3.4.3** *Standard monomials in* $x(A, B), y(I), z(J), x(\mathbf{1}), x(\mathbf{0})$ *give a basis for the* $K$*-vector space* $R(D)$.

### 11.3.5 Standard monomial bases for $M(D)$, $S(D)$

Standard monomials in
$r(A, B), s(I), t(J), q(\mathbf{1}))$ in $M(D)$ (resp. $c(A, B), d(I), e(J))$ in $S(D)$) are defined in exactly the same way as in Definition 11.3.0.1.

**Proposition 11.3.5.1** *Standard monomials in* $r(A, B), s(I), t(J), q(\mathbf{1})$ *give a basis for the* $K$*-vector space* $M(D)$.

*Proof.* The linear independence of standard monomials follows as in the proof of Prop 11.3.4.1 by considering $\varphi_D \circ \delta_D : M(D) \to S$, and using the linear independence of standard monomials in $S$ (cf. Proposition 11.3.2.2).
To see the generation of $M(D)$ by standard monomials, consider a non-standard monomial $F$ in $M(D)$. Since $q(\mathbf{1})^l$ is the largest monomial of a given degree $l$, we may suppose $F$ to be:

$$F = r(A_1, B_1) \cdots r(A_i, B_i) s(I_1) \cdots s(I_k) t(J_1) \cdots t(J_l)$$

In view of Theorem 11.3.1.1, (1), we may suppose that
$F = r(A_1, B_1) \cdots r(A_i, B_i) s(I_1) \cdots s(I_k)$ or $r(A_1, B_1) \cdots r(A_i, B_i) t(J_1) \cdots t(J_l)$, say, $F = r(A_1, B_1) \cdots r(A_i, B_i) s(I_1) \cdots s(I_k)$. Then $F = \theta_D(H)$, where
$H = x(A_1, B_1) \cdots x(A_i, B_i) y(I_1) \cdots y(I_k)$. The required result follows from Proposition 11.3.4.2.

**Proposition 11.3.5.2** *Standard monomials in* $c(A, B), d(I), e(J)$ *give a basis for the* $K$*-vector space* $S(D)$.

The proof is completely analogous to that of Proposition 11.3.5.1 (in view of the fact that $S(D) = M(D)_{(q(\mathbf{1}))}$).

**Theorem 11.3.5.3** *Standard monomials in* $p(A, B), u(I), \xi(J)$ *form a basis for the* $K$*-vector space* $S$.

*Proof.* We already have established the linear independence of standard monomials (cf. Proposition 11.3.2.2). The generation by standard monomials follows by considering the surjective map $\varphi_D : S(D) \to S$ and using the generation of $S(D)$ by standard monomials (cf. Proposition 11.3.5.2).

**Theorem 11.3.5.4** *The map* $\varphi_D : S(D) \to S$ *is an isomorphism of* $K$*-algebras.*

*Proof.* Under $\varphi_D$, the standard monomials in $S(D)$ are mapped bijectively onto the standard monomials in $S$. The result follows from Proposition 11.3.5.2 and Theorem 11.3.5.3.

**Theorem 11.3.5.5** ***A Presentation for*** *$S$:*

*(i) The $K$-algebra $S$ is generated by $\{p(A, B), u(I), \xi(J), (A, B) \in H_p, I \in H_u, J \in H_\xi\}$.*
*(ii) The ideal of relations among the generators $\{p(A, B), u(I), \xi(J), (A, B) \in H_p, I \in H_u, J \in H_\xi\}$ is generated by the six type of relations as given by Theorem 11.3.1.1.*

*Proof.* The result follows from Theorem 11.3.5.4.

## 11.4 Normality and Cohen-Macaulayness of the $K$-algebra $S$

In this section, we present a proof (cf. [72]) of the normality and Cohen-Macaulayness of *Spec* $S$ by relating it to a toric variety. From §11.2, we have

- $\{u(I), \xi(J), p(A, B), I \in H_u, J \in H_\xi, (A, B) \in H_p\}$ generates $S$ as a $K$-algebra.
- Standard monomials in $\{u(I), \xi(J), p(A, B), I \in H_u, J \in H_\xi, (A, B) \in H_p\}$ form a $K$-basis for $S$.
- Considering $S$ as a quotient of the polynomial algebra

$$K[X(A, B), Y(I), Z(J), (A, B) \in H_p, I \in H_u, J \in H_\xi]$$

the ideal $\mathfrak{a}$ of relations is generated by the six kinds of quadratic relations as given in Theorem 11.3.1.1.

### 11.4.1 The algebra associated to a distributive lattice

**Definition 11.4.1.1** *A* lattice *is a partially ordered set $(\mathcal{L}, \leq)$ such that, for every pair of elements $x, y \in \mathcal{L}$, there exist elements $x \vee y$, $x \wedge y$, called the* join, *respectively the* meet *of $x$ and $y$, satisfying:*

$$x \vee y \geq x,\ x \vee y \geq y,\ \textit{and if } z \geq x \textit{ and } z \geq y,\ \textit{then } z \geq x \vee y,$$
$$x \wedge y \leq x,\ x \wedge y \leq y,\ \textit{and if } z \leq x \textit{ and } z \leq y,\ \textit{then } z \leq x \wedge y.$$

**Definition 11.4.1.2** *A lattice is called* distributive *if the following identities hold:*

$$x \wedge (y \vee z) = (x \wedge y) \vee (x \wedge z)$$
$$x \vee (y \wedge z) = (x \vee y) \wedge (x \vee z)$$

**Definition 11.4.1.3** *Given a finite lattice $\mathcal{L}$, the* ideal associated to $\mathcal{L}$, *denoted by $I(\mathcal{L})$, is the ideal of the polynomial algebra $K[\mathcal{L}](= K[x_\alpha, \alpha \in \mathcal{L}])$ generated by the set of binomials*

$$\mathcal{G}_{\mathcal{L}} = \{xy - (x \wedge y)(x \vee y) \mid x, y \in \mathcal{L} \textit{ non-comparable}\}.$$

*Set $A(\mathcal{L}) = K[\mathcal{L}]/I(\mathcal{L})$,* the algebra associated to $\mathcal{L}$.

**The chain lattice $\mathcal{C}(n_1, \ldots, n_d)$:** Given an integer $n \geq 1$, let $\mathcal{C}(n)$ denote the chain $\{1 < \cdots < n\}$, and for $n_1, \ldots, n_d > 1$, let $\mathcal{C}(n_1, \ldots, n_d)$ denote the chain product lattice $\mathcal{C}(n_1) \times \cdots \times \mathcal{C}(n_d)$ consisting of all $d$-tuples $(i_1, \ldots, i_d)$, with $1 \leq i_1 \leq n_1, \ldots, 1 \leq i_d \leq n_d$. For $(i_1, \ldots, i_d), (j_1, \ldots, j_d)$ in $\mathcal{C}(n_1, \ldots, n_d)$, we define

$$(i_1, \ldots, i_d) \leq (j_1, \ldots, j_d) \iff i_1 \leq j_1, \ldots, i_d \leq j_d.$$

We have

$$(i_1, \ldots, i_d) \vee (j_1, \ldots, j_d) = (\max\{i_1, j_1\}, \ldots, \max\{i_d, j_d\})$$
$$(i_1, \ldots, i_d) \wedge (j_1, \ldots, j_d) = (\min\{i_1, j_1\}, \ldots, \min\{i_d, j_d\}).$$

Clearly, $\mathcal{C}(n_1, \ldots, n_d)$ is a finite distributive lattice.

### 11.4.2 Flat degenerations of certain $K$-algebras

Let $\mathcal{L}$ be a finite lattice, and $R(\mathcal{L})$ a $K$-algebra with generators $\{p_\alpha \mid \alpha \in \mathcal{L}\}$.

**Definition 11.4.2.1** *A monomial $p_{\alpha_1} \cdots p_{\alpha_r}$ is said to be standard if $\alpha_1 \geq \cdots \geq \alpha_r$.*

Suppose that the standard monomials form a $K$-basis for $R(\mathcal{L})$. Given any non-standard monomial $F$, the expression

$$F = \sum c_i F_i, \qquad c_i \in K^*$$

for $F$ as a sum of standard monomials will be referred to as a *straightening relation*. Consider the surjective map

$$\pi : K[\mathcal{L}] \to R(\mathcal{L}), \qquad x_\alpha \mapsto p_\alpha.$$

Let us denote $\ker \pi$ by $I$.

For $\alpha, \beta \in H$ with $\alpha > \beta$, we set

$$]\beta, \alpha[= \{\gamma \in \mathcal{L} \mid \alpha > \gamma > \beta\}.$$

Recall the following theorem (cf. Gonciulea-Lakshmibai [33], Theorem 5.2)

**Theorem 11.4.2.2** *Let $\mathcal{L}$, $R(\mathcal{L})$, $I$ be as above. Suppose that there exists a lattice embedding $\mathcal{L} \hookrightarrow \mathcal{C}$, where $\mathcal{C} = \mathcal{C}(n_1, \ldots, n_d)$ for some $n_1, \ldots, n_d \geq 1$, such that the entries of the $d$-tuple $(\theta_1, \ldots, \theta_d)$ representing an element $\theta$ of $\mathcal{L}$ form a non-decreasing sequence, i.e. $\theta_1 \leq \cdots \leq \theta_d$. Suppose that $I$ is generated as an ideal by elements of the form $x_\tau x_\varphi - \sum c_{\alpha\beta} x_\alpha x_\beta$(where $\tau, \varphi$ are non-comparable, and $\alpha \geq \beta$). Further suppose that in the straightening relation*

$$p_\tau p_\phi = \sum c_{\alpha\beta} p_\alpha p_\beta, \qquad (*)$$

*the following hold:*

*(a) $p_{\tau\vee\phi} p_{\tau\wedge\phi}$ occurs on the right-hand side of (*) with coefficient* 1.

*(b)* $\tau, \phi \in ]\beta, \alpha[$, *for every pair* $(\alpha, \beta)$ *appearing on the right-hand side of* (*).

*(c) Under the embedding* $\mathcal{L} \hookrightarrow \mathcal{C}$, *we have* $\tau \dot{\cup} \phi = \alpha \dot{\cup} \beta$, *for every* $(\alpha, \beta)$ *on the right-hand side of* (*).

*Then there exists a flat family over Spec* $K[t]$ *whose special fiber* ($t = 0$) *is Spec* $A(\mathcal{L})$ *and general fiber (*$t$ *invertible) is Spec* $R(\mathcal{L})$.

**Corollary 11.4.2.3** *Spec* $R(\mathcal{L})$ *flatly degenerates to a (normal) toric variety. In particular, Spec* $R(\mathcal{L})$ *is normal and Cohen-Macaulay.*

*Proof.* We have (cf. [39]) that $A(\mathcal{L})$ is a normal domain. Hence we obtain that $I(\mathcal{L})$ is a binomial prime ideal. On the other hand, we have (cf. [28]) that a binomial prime ideal is a toric ideal (in the sense of [114]). It follows that *Spec* $A(\mathcal{L})$ is a (normal) toric variety and we obtain the first assertion. The first assertion together with Theorem 11.4.2.2 and the fact that a toric variety is normal and Cohen-Macaulay implies that *Spec* $R(\mathcal{L})$ is normal and Cohen-Macaulay.

### 11.4.3 The distributive lattice $D$

Consider the partially ordered set

$$D = H \cup \{\mathbf{1}\} \cup \{\mathbf{0}\}$$

defined in §11.3.3. We equip $D$ with the structure of a distributive lattice by embedding it inside the chain lattice $\mathcal{C}(\underline{m}, \underline{q}) := \mathcal{C}(\underbrace{m, m \cdots, m}_{n+1 \text{ times}}, \underbrace{q, q \cdots, q}_{n+1 \text{ times}})$, as follows:

To each element of $D$, we associate a $2n+2$-tuple:

For $A = (a_1, \cdots, a_r) \in I(r, m)$, $B = (b_1, \cdots, b_r) \in I(r, q)$, let $\overline{A}, \overline{B}$ denote the $n+1$-tuples:

$$\overline{A} := (a_1, \cdots, a_r, m, m, \cdots, m, 1), \overline{B} := (b_1, \cdots, b_r, q, q, \cdots, q, 1)$$

(i) Let $(A, B) \in H_p$, say, $A \in I(r, m)$, $B \in I(r, q)$, for some $r$, $1 \leq r \leq n$. We let $\overline{(A, B)}$ be the $(2n+2)$-tuple: $\overline{(A, B)} = (\overline{A}, \overline{B})$.

(ii) Let $I \in H_u$, say, $I = (i_1, \cdots, i_n)$ $(\in I(n, m))$ . We let $\tilde{I}$ be the $(2n+2)$-tuple: $\tilde{I} = (i_1, \cdots, i_n, 1, \underbrace{1, \cdots, 1}_{n+1 \text{ times}})$

(iii) Let $\xi \in H_\xi$, say, $J = (j_1, \cdots, j_n)$ $(\in I(n, m))$), we let $\tilde{J}$ be the $(2n+2)$-tuple: $\tilde{J} = (\underbrace{1, \cdots, 1}_{n+1 \text{ times}}, j_1, \cdots, j_n, 1)$.

(iv) Corresponding to $\mathbf{1}, \mathbf{0}$, we let $\tilde{\mathbf{1}}, \tilde{\mathbf{0}}$ be the $(2n+2)$-tuples:

$$\tilde{\mathbf{1}} = (\underbrace{m, m \cdots, m}_{n+1 \text{ times}}, \underbrace{q, q \cdots, q}_{n+1 \text{ times}}), \ \tilde{\mathbf{0}} = (\underbrace{1, \cdots, 1}_{2n+2 \text{ times}})$$

This induces a canonical embedding of $D$ inside the chain lattice $\mathcal{C}(\underbrace{m, m \cdots, m}_{n+1 \text{ times}}, \underbrace{q, q \cdots, q}_{n+1 \text{ times}})$.

**Lemma 11.4.3.1** *Let $\tau_1, \tau_2 \in \mathcal{C}(\underline{m}, \underline{q})$. Suppose $\tau_1, \tau_2 \in D$. Then $\tau_1 \vee \tau_2, \tau_1 \wedge \tau_2$ are also in $D$. Thus $D$ acquires the structure of a distributive lattice.*

*Proof.* Clearly the Lemma requires a proof only when $\tau_1, \tau_2$ are non-comparable. We consider the following cases. For two $s$-tuples $E = \{e_1, \cdots, e_s\}$, $F = \{f_1, \cdots, f_s\}$, we shall denote

$$E \vee F := (\max\{e_1, f_1\}, \ldots, \max\{e_s, f_s\})$$
$$E \wedge F := (\min\{e_1, f_1\}, \ldots, \min\{e_s, f_s\}).$$

**Case 1:** $\tau_1, \tau_2 \in H_p$, say $\tau_1 = (\overline{A_1}, \overline{B_1})$, $\tau_2 = (\overline{A_2}, \overline{B_2})$. We have

$$\tau_1 \vee \tau_2 = (\overline{A_1} \vee \overline{A_2}, \overline{B_1} \vee \overline{B_2}), \ \ \tau_1 \wedge \tau_2 = (\overline{A_1} \wedge \overline{A_2}, \overline{B_1} \wedge \overline{B_2})$$

Clearly $\tau_1 \vee \tau_2, \tau_1 \wedge \tau_2$ are in $H_p$, and hence in $D$.
**Case 2:** $\tau_1 \in H_p, \tau_2 \in H_u$, say $\tau_1 = (\overline{A}, \overline{B})$, $\tau_2 = \tilde{I}$ (for some $I \in H_u$). Let $\overline{I}$ be the $n+1$-tuple $(I, 1)$ (entries of $I$ followed by 1). We have

$$\tau_1 \vee \tau_2 = (\overline{A} \vee \overline{I}, \overline{B}), \ \ \tau_1 \wedge \tau_2 = (\overline{A} \wedge \overline{I}, \underbrace{(1, \cdots, 1)}_{n+1 \text{ times}})$$

Clearly $\tau_1 \vee \tau_2 \in H_p$, $\tau_1 \wedge \tau_2 \in H_u$.
**Case 3:** $\tau_1 \in H_p, \tau_2 \in H_\xi$, say $\tau_1 = (\overline{A}, \overline{B})$, $\tau_2 = \tilde{J}$ (for some $J \in H_\xi$). Let $\overline{J}$ be the $n+1$-tuple $(J, 1)$ (entries of $I$ followed by 1). We have

$$\tau_1 \vee \tau_2 = (\overline{A}, \overline{B} \vee \overline{J}), \ \ \tau_1 \wedge \tau_2 = (\underbrace{1, \cdots, 1}_{n+1 \text{ times}}, \overline{B} \wedge \overline{J})$$

Clearly $\tau_1 \vee \tau_2 \in H_p$, $\tau_1 \wedge \tau_2 \in H_\xi$.
**Case 4:** $\tau_1, \tau_2 \in H_u$, say $\tau_1 = \tilde{I_1}$, $\tau_2 = \tilde{I_2}$ (for some $I_1, I_2 \in H_u$). We have

$$\tau_1 \vee \tau_2 = \widetilde{I_1 \vee I_2}, \ \ \tau_1 \wedge \tau_2 = \widetilde{I_1 \wedge I_2}$$

Clearly $\tau_1 \vee \tau_2, \tau_1 \wedge \tau_2$ are in $H_u$.
**Case 5:** $\tau_1, \tau_2 \in H_\xi$.
This case is similar to Case 4.
**Case 6:** $\tau_1 \in H_u, \tau_2 \in H_\xi$, say $\tau_1 = \tilde{I}$, $\tau_2 = \tilde{J}$ (for some $I, J$ in $H_u, H_\xi$ respectively). We have

$$\tau_1 \vee \tau_2 = (\overline{I}, \overline{J}), \ \ \tau_1 \wedge \tau_2 = \tilde{0}$$

Clearly $\tau_1 \vee \tau_2 \in H_p$, $\tau_1 \wedge \tau_2 \in D$.

**Lemma 11.4.3.2** *We have rank $(D) = (m+q)n - n^2 + 2$ $(= d + 2$, where $d = (m+q)n - n^2)$. In particular, dim $A(D) = d + 3$*

This is immediate from Lemma 11.2.1.2.

### 11.4.4 Flat degeneration of Spec $R(D)$ to the toric variety Spec $A(D)$

In this subsection, we show that Spec $R(D)$ flatly degenerates to the toric variety Spec $A(D)$, where $A(D)$ is the algebra associated as in 11.4.1 to the distributive lattice $D$, by showing that $R(D)$ satisfies the hypotheses of Theorem 11.4.2.2. We first prove some preparatory Lemmas.

**Lemma 11.4.4.1** *Let $\tau, \phi$ be two non-comparable elements of $H$. Then in the straightening relation for $p_\tau p_\phi$ as given by Theorem 11.3.1.1, $p_{\tau\vee\phi}p_{\tau\wedge\phi}$ occurs with coefficient 1 (here for an element $\varphi$ of $H$, $p_\varphi$ stands for $p(A,B)$, $u(I)$ or $\xi(J)$ according as $\varphi=(A,B)\in H_p$, $I\in H_u$ or $J\in H_\xi$).*

*Proof.* The assertion is clear if the relation is of the type (1) of Theorem 11.3.1.1. If the relation is of the type (4) of Theorem 11.3.1.1, then the result follows from Proposition 11.1.0.2,(3).

Similarly, if the relation is of the type (2) (resp. (3)) of Theorem 11.3.1.1, then the result follows as above (in view of Proposition 11.1.0.2,(3))
Let then the relation be of the type (5) or (6) of Theorem 11.3.1.1, say of type (5) (the proof is similar if it is of type (6)):

$$(*) \qquad p(A,B)u(I)=\sum_t c_t p(A_t,B_t)u(I_t)$$

where $I\in I(n,m)$, $(A,B)\in H_p$, and $A\not\geq I$. As in the proof of Theorem 11.3.1.1, (5), we multiply through out by $\xi(I_n)$ to arrive at

$$(**) \qquad p(A,B)p(I,I_n)=\sum a_i p(C_{i1},C_{i2})p(D_{i1},D_{i2}),\ a_i\in K^*$$

where $(C_{i1},C_{i2})$, $(D_{i1},D_{i2})$ belong to $H_p$. As above, using Proposition 11.1.0.2,(3), we obtain that $p((A,B)\vee(I,I_n))p((A,B)\wedge(I,I_n))$ occurs in (**) with coefficient 1. We have (in view of Lemma 11.4.3.1, rather its proof),

$$p((A,B)\vee(I,I_n))p((A,B)\wedge(I,I_n))=p(\overline{A}\vee\overline{I},\overline{B})p(\overline{A}\wedge\overline{I},\overline{I_n})=p(\overline{A}\vee\overline{I},\overline{B})u(\overline{A}\wedge\overline{I})\xi(\overline{I_n})$$

Also from the proof of Theorem 11.3.1.1, (5), we have, for every $i$, $D_{i2}=I_n$ (in (**)). Hence writing $p(D_{i1},D_{i2})=u(D_{i1})\xi(I_n)$, cancelling out $\xi(I_n)$ (note that L.H.S. of (**)= $p(A,B)u(I)\xi(I_n)$), we obtain that $p(\overline{A}\vee\overline{I},\overline{B})u(\overline{A}\wedge\overline{I})$ occurs in (*) with coefficient 1 (note that by Case 2 in the proof of Lemma 11.4.3.1, we have $(A,B)\vee I=(\overline{A}\vee\overline{I},\overline{B}),\ (A,B)\wedge I=(\overline{A}\wedge\overline{I},\underbrace{(1,\cdots,1)}_{n+1 \text{ times}})$).

Thus the result follows if the relation is of the type (5) (or (6)) of Theorem 11.3.1.1.

**Lemma 11.4.4.2** *Let $\tau, \phi$ be two non-comparable elements of $D$. Then for every $(\alpha,\beta)$ on the right-hand side of the straightening relation (in $R(D)$, as given by Theorem 11.3.1.1), we have*

*(i)* $\tau,\phi\in]\beta,\alpha[$,

*(ii)* $\tau \dot{\cup} \phi = \alpha \dot{\cup} \beta$
*(here,* $\dot{\cup}$ *denotes a disjoint union).*

*Proof.* The assertions follow from Theorem 11.3.1.1 (and the identification of $D$ as a sublattice of $\mathcal{C}(\underline{m}, \underline{q})$).

**Theorem 11.4.4.3** *There exists a flat family over $\mathbb{A}^1$, with Spec $R(D)$ as the generic fiber and Spec $A(D)$ as the special fiber. In particular, $R(D)$ is a normal Cohen-Macaulay ring of dimension $d+3$ (where $d = (m+q)n - n^2$).*

*Proof.* In view of Theorem 11.4.2.2, and Corollary 11.4.2.3, it suffices to show that (a)- (c) of Theorem 11.4.2.2 hold for $R(D)$.

(a) follows from Lemma 11.4.4.1; (b) and (c) follow from Lemma 11.4.4.2.

Clearly $R(D)$ has dim $d+3$ (since dim $A(D) = d+3$ (cf. Lemma 11.4.3.2)).

**Theorem 11.4.4.4** *The $K$-algebra $S$ is normal, Cohen-Macaulay of dimension $(m+q)n - n^2 + 1$.*

*Proof.* The algebra $M(D)(= R(D)_{(x(\mathbf{0}))})$ being a homogeneous localization of the normal, Cohen-Macaulay ring $R(D)$, is a normal, Cohen-Macaulay ring of dim $d+2$.

Considering $M(D)$ as a graded ring (cf. §11.3.3), we have $S(D) = M(D)_{(x(\mathbf{1}))}$. Hence $S(D)$ being a homogeneous localization of the normal, Cohen-Macaulay ring $M(D)$, is a normal, Cohen-Macaulay ring of dimension $d+1$. This together with Theorem 11.3.5.4 implies that $S$ is a normal, Cohen-Macaulay ring of dimension $d+1$ (note that $d = (m+q)n - n^2$).

## 11.5 The ring of invariants $K[X]^{SL_n(K)}$

We preserve the notation of §11.2. In this section, we shall show that the inclusion $S \subseteq R^{SL_n(K)}$ is in fact an equality, i.e., $S = R^{SL_n(K)}$.

We now apply Chapter 10, Lemma 10.1.0.4 to our situation. Let $G = SL_n(K)$. Consider

$$X = \underbrace{V \oplus \cdots \oplus V}_{m \text{ copies}} \oplus \underbrace{V^* \oplus \cdots \oplus V^*}_{q \text{ copies}} = Spec\, R,\ \mathbb{A}^N = M_{m,q}(K) \times K^{\binom{m}{n}} \times K^{\binom{q}{n}}$$

Let $\{\langle u_i, \xi_j \rangle), 1 \le i \le m, 1 \le j \le q, u(I), \xi(J),\ I \in H_u, J \in H_\xi\}$ be denoted by $\{f_1, \cdots, f_N\}$ (note that $f_1, \cdots, f_N$ are $G$-invariant elements in $R$). Let $x = (\underline{u}, \underline{\xi}) \in X$. Let $\psi : X \to \mathbb{A}^N$ be the map, $\psi(x) = (f_1(x), \cdots, f_N(x))$. Clearly $\psi(\overline{X}) = Spec\, S$. Let us denote $Y = Spec\, S$.

**Proposition 11.5.0.1** *With $X, \mathbb{A}^N, \psi, Y$ as above, the hypotheses of Chapter 10, Lemma 10.1.0.4 are satisfied.*

*Proof.* (i) Let $x \in X^{ss}$. We need to show that $\psi(x) \neq 0$. If possible, let us assume that $\psi(x) = 0$. Let $x = (\underline{u}, \underline{\xi})$. Let $W_u$ (resp. $W_\xi$) be the span of $\{u_1, \cdots, u_m\}$ (resp.$\{\xi_1, \cdots, \xi_q\}$). Further, let dim $W_u = r$, dim $W_\xi = s$. The assumption that $\psi(x) = 0$ implies in particular that $u(I)(x) = 0, \forall I \in I(n, m), \xi(J)(x) = 0, \forall J \in I(n, q)$. Hence, $W_u$ (resp. $W_\xi$) is not equal to $V$ (resp.$V^*$). Therefore, we get $r < n, s < n$. Also at least one of $\{r, s\}$ is non-zero; otherwise, $r = 0 = s$ would imply $u_i = 0, \forall i, \xi_j = 0, \forall j$, i.e., $x = 0$ which is not possible, since $x \in X^{ss}$. Let us suppose that $r \neq 0$. (The proof is similar if $s \neq 0$.) The assumption that $\psi(x) = 0$ implies in particular that $\langle u_i, \xi_j \rangle = 0$, for all $i, j$; hence, $W_\xi \subseteq W_u^\perp$. Therefore, $s \leq n - r$. Hence we can choose a basis $\{e_1, \cdots, e_n\}$ of $V$ such that $W_u =$ the $K$-span of $\{e_1, \cdots, e_r\}$, and $W_\xi \subseteq$ the $K$-span of $\{e_{r+1}^*, \cdots, e_n^*\}$. Writing each vector $u_i$ as a row vector (with respect to this basis), we may represent the $u$'s by the $m \times n$ matrix $\mathcal{U}$ given by

$$\mathcal{U} := \begin{pmatrix} u_{11} & u_{12} & \ldots & u_{1r} & 0 & \ldots & 0 \\ u_{21} & u_{22} & \ldots & u_{2r} & 0 & \ldots & 0 \\ \vdots & \vdots & & \vdots & \vdots & \vdots & \vdots \\ u_{m1} & u_{m2} & \ldots & u_{mr} & 0 & \ldots & 0 \end{pmatrix}$$

Similarly, writing each vector $\xi_j$ as a column vector (with respect to the above basis), we may represent $\xi$'s by the $n \times q$ matrix $\Lambda$ given by

$$\Lambda := \begin{pmatrix} 0 & 0 & \ldots & 0 \\ \vdots & \vdots & & \vdots \\ 0 & 0 & \ldots & 0 \\ \xi_{r+1\,1} & \xi_{r+1\,2} & \ldots & \xi_{r+1\,q} \\ \vdots & \vdots & & \vdots \\ \xi_{n\,1} & \xi_{n\,2} & \ldots & \xi_{n\,q} \end{pmatrix}$$

Choose integers $a_1, \cdots, a_r, b_{r+1}, \cdots, b_n$, all of them $> 0$ so that $\sum a_i = \sum b_j$. Let $g_t$ be the diagonal matrix in $G(= SL_n(K))$, $g_t = diag(t^{a_1}, \cdots t^{a_r}, t^{-b_{r+1}}, \cdots, t^{-b_n})$. We have, $g_t x = g \cdot (\mathcal{U}, \Lambda) = (\mathcal{U} g_t, g_t^{-1} \Lambda) = (\mathcal{U}_t, \Lambda_t)$, where

$$\mathcal{U}_t = \begin{pmatrix} t^{a_1} u_{11} & t^{a_2} u_{12} & \ldots & t^{a_r} u_{1r} & 0 & \ldots & 0 \\ t^{a_1} u_{21} & t^{a_2} u_{22} & \ldots & t^{a_r} u_{2r} & 0 & \ldots & 0 \\ \vdots & \vdots & & \vdots & \vdots & \vdots & \vdots \\ t^{a_1} u_{m1} & t^{a_2} u_{m2} & \ldots & t^{a_r} u_{mr} & 0 & \ldots & 0 \end{pmatrix}$$

and

$$\Lambda_t = \begin{pmatrix} 0 & 0 & \ldots & 0 \\ \vdots & \vdots & & \vdots \\ 0 & 0 & \ldots & 0 \\ t^{b_{r+1}} \xi_{r+1\,1} & t^{b_{r+1}} \xi_{r+1\,2} & \ldots & t^{b_{r+1}} \xi_{r+1\,q} \\ \vdots & \vdots & & \vdots \\ t^{b_n} \xi_{n\,1} & t^{b_n} \xi_{n\,2} & \ldots & t^{b_n} \xi_{n\,q} \end{pmatrix}$$

Hence $g_t x \to 0$ as $t \to 0$, and this implies that $0 \in \overline{G \cdot x}$ which is a contradiction to the hypothesis that $x$ is semi-stable. Therefore our assumption that $\psi(x) = 0$ is wrong and (i) of Lemma 10.1.0.4, Chapter 10 is satisfied.
(ii) Let

$$U = \{(\underline{u}, \underline{\xi}) \in X \mid \{u_1, \ldots, u_n\}, \{\xi_1, \ldots, \xi_n\} \text{ are linearly independent}\}$$

Clearly, $U$ is a $G$-stable open subset of $X$.
**Claim:** $G$ operates freely on $U$, $U \to U\, mod\, G$ is a $G$-principal fiber space, and $F$ induces an immersion $U/G \to \mathbb{A}^N$.
**Proof of Claim:** Let $H = GL_n(K)$. We have a $G$-equivariant identification

$$(*) \qquad U \cong H \times H \times \underbrace{V \times \cdots \times V}_{(m-n)\text{ copies}} \times \underbrace{V^* \times \cdots \times V^*}_{(q-n)\text{ copies}} = E \times F, \text{ say}$$

where $E = H \times H$, $F = \underbrace{V \times \cdots \times V}_{(m-n)\text{ copies}} \times \underbrace{V^* \times \cdots \times V^*}_{(q-n)\text{ copies}}$. From this it is clear that $G$ operates freely on $U$. Further, we see that $U\, mod\, G$ may be identified with the fiber space with base $(H \times H)\, mod\, G$ ($G$ acting on $H \times H$ as $g \cdot (h_1, h_2) = (h_1 g, g^{-1} h_2)$, $g \in G, h_1, h_2 \in H$), and fiber $\underbrace{V \times \cdots \times V}_{(m-n)\text{ copies}} \times \underbrace{V^* \times \cdots \times V^*}_{(q-n)\text{ copies}}$ associated to the principal fiber space $H \times H \to (H \times H)\, /\, G$. It remains to show that $\psi$ induces an immersion $U/G \to \mathbb{A}^N$, i.e., to show that the map $\psi : U/G \to \mathbb{A}^N$ and its differential $d\psi$ are both injective. We first prove the injectivity of $\psi : U/G \to \mathbb{A}^N$. Let $x, x'$ in $U/G$ be such that $\psi(x) = \psi(x')$. Let $\eta, \eta' \in U$ be lifts for $x, x'$ respectively. Using the identification (*) above, we may write

$$\eta = (A, u_{n+1}, \cdots, u_m, B, \xi_{n+1}, \cdots, \xi_q), \ A, B \in H$$
$$\eta' = (A', u'_{n+1}, \cdots, u'_m, B', \xi'_{n+1}, \cdots, \xi'_{q'}), \ A', B' \in H$$

(here, $u_i$, $1 \le i \le n$ are given by the rows of $A$, while $\xi_i$, $1 \le i \le n$ are given by the columns of $B$; similar remarks on $u'_i, \xi'_i$). The hypothesis that $\psi(x) = \psi(x')$ implies in particular that

$$\langle u_i, \xi_j \rangle = \langle u'_i, \xi'_j \rangle, 1 \le i, j \le n$$

which may be written as $AB = A'B'$. This implies that

$$(**) \qquad A' = A \cdot g,$$

where $g = BB'^{-1} (\in H)$. Further, the hypothesis that $u(I)(x) = u(I)(x')$, $\forall I$, implies in particular that $u(I_n)(x) = u(I_n)(x')$ (where $I_n = (1, 2, \cdots, n)$). Hence we obtain

$$(***) \qquad det A = det A'$$

Now (**) and (***) imply that $g$ in fact belongs to $G (= SL_n(K))$. Hence on $U/G$, we may suppose that

$$x = (u_1, \cdots, u_n, u_{n+1}, \cdots, u_m, \xi_1, \cdots, \xi_q)$$
$$x' = (u_1, \cdots, u_n, u'_{n+1}, \cdots, u'_m, \xi'_1, \cdots, \xi'_q)$$

where $\{u_1, \cdots, u_n\}$ is linearly independent.

For a given $j$, we have,

$$\langle u_i, \xi_j \rangle = \langle u_i, \xi'_j \rangle, \ 1 \le i \le n, \text{ implies, } \xi_j = \xi'_j$$

(since, $\{u_1, \cdots, u_n\}$ is linearly independent). Thus we obtain

$$(\dagger) \qquad \xi_j = \xi'_j, \text{ for all } j$$

On the other hand, we have (by definition of $U$) that $\{\xi_1, \ldots, \xi_n\}$ is linearly independent. Hence fixing an $i$, $n+1 \le i \le m$, we get

$$\langle u_i, \xi_j \rangle = \langle u'_i, \xi_j \rangle (= \langle u'_i, \xi'_j \rangle), \ 1 \le j \le n \text{ implies, } u_i = u'_i.$$

Thus we obtain

$$(\dagger\dagger) \qquad u_i = u'_i, \text{ for all } i$$

The injectivity of $\psi : U/G \to \mathbb{A}^N$ follows from (†),(††).

To prove that the differential $d\psi$ is injective, we merely note that the above argument remains valid for the points over $K[\epsilon]$, the algebra of dual numbers ($= K \oplus K\epsilon$, the $K$-algebra with one generator $\epsilon$, and one relation $\epsilon^2 = 0$), i.e., it remains valid if we replace $K$ by $K[\epsilon]$, or in fact by any $K$-algebra.
Thus condition (ii) of Lemma 10.1.0.4, Chapter 10 holds.

(iii) The above Claim implies in particular that $\dim U/G = \dim U - \dim G = (m+q)n - (n^2 - 1) = \dim Spec\, S$ (cf. Theorem 11.4.4.4).
Thus condition (iii) of Lemma 10.1.0.4, Chapter 10 holds.

The condition (iv) of Lemma 10.1.0.4, Chapter 10 follows from Theorem 11.4.4.4.

**Theorem 11.5.0.2** *Let $V = K^n$, $X = \underbrace{V \oplus \cdots \oplus V}_{m \text{ copies}} \times \underbrace{V^* \oplus \cdots \oplus V^*}_{q \text{ copies}}$, where $m, q > n$. Then for the diagonal action of $G := SL_n(K)$, we have*

*(i)* ***First Fundamental Theorem for $SL_n(K)$-invariants:*** *$K[X]^G$ is generated by $\{p(A, B), u(I), \xi(J), (A, B) \in H_p, I \in H_u, J \in H_\xi\}$.*
*(ii)* ***Second Fundamental Theorem for $SL_n(K)$-invariants:*** *The ideal of relations among the generators $\{p(A, B), u(I), \xi(J), (A, B) \in H_p, I \in H_u, J \in H_\xi\}$ is generated by the six type of relations as given by Theorem 11.3.1.1.*
*(iii)* ***A Standard Monomial Basis for $SL_n(K)$-invariants:*** *Standard monomials in $\{p(A, B), u(I), \xi(J), (A, B) \in H_p, I \in H_u, J \in H_\xi\}$ form a $K$-basis for $K[X]^G$.*
*(iv) $K[X]^G$ is Cohen-Macaulay.*

*Proof.* Chapter 10, Lemma 10.1.0.4 implies (in view of Proposition 11.5.0.1) that $Spec\, S$ is the categorical quotient of $X$ by $G$ and $\psi : X \to Spec\, S$ is the canonical quotient map. Assertions (i) and (ii) follow from Theorem 11.3.5.5. Assertion (iii) follows from Theorem 11.3.5.3. Assertion (iv) follows from Theorem 11.4.4.4.

# 12

# $SO_n(K)$-action

In this chapter, we consider the $SO_n(K)$-action on $X = \underbrace{V \oplus \cdots \oplus V}_{m \text{ copies}}, V = K^n$. We present a proof (cf. [63]) of the Cohen-Macaulayness of $K[X]^{SO_n(K)}$. While *Spec* $K[X]^{O_n(K)}$ being a certain determinantal variety inside Sym $M_m$, the space of symmetric $m \times m$ matrices (cf. chapter 10)) gets identified with the opposite cell in a certain Schubert variety, *Spec* $K[X]^{SO_n(K)}$ does not have a direct relationship to Schubert varieties. Nevertheless, as in Chapter 11, certain "Schubert variety connections" are used in a crucial way (cf. §12.1,§12.2,§12.3 ). As in Chapter 11, we adopt the "deformation technique" for this purpose. To be more precise, the proof of the Cohen-Macaulayness of $K[X]^{SO_n(K)}$ is accomplished in the following steps:

- One first constructs a $K$-subalgebra $S$ of $K[X]^{SO_n(K)}$ by prescribing a set of algebra generators $\{f_\alpha, \alpha \in D\}$, $D$ being a doset associated to a finite partially ordered set $P$ (here, "doset" is as defined in DeConcini-Lakshmibai [21]; see also Definition 12.4.0.1).
- One then constructs a "standard monomial" basis for $S$ by

(i) defining "standard monomials" in the $f_\alpha$'s (cf. Definition 12.2.1.1)

(ii) writing down the straightening relation for a non-standard (degree 2) monomial $f_\alpha f_\beta$ (cf. Proposition 12.3.1.1)

(iii) proving linear independence of standard monomials (by relating the generators of $S$ to certain determinantal varieties inside the space of symmetric matrices) (cf. Proposition 12.2.2.2)

(iv) proving the generation of $S$ (as a vector space) by standard monomials (using (ii)). In fact, to prove the generation for $S$, one first proves generation for a "graded version" $R(D)$ of $S$, where $D$ as above is a doset associated to a finite partially ordered set $P$. One then deduces the generation for $S$. In fact, one constructs a "standard monomial" basis for $R(D)$. While the generation by standard monomials for $S$ is deduced from the generation by standard monomials for $R(D)$, the linear independence of standard monomials in $R(D)$ is deduced from the linear independence of standard monomials in $S$ (cf. (iii) above).

- As a consequence, one obtains a presentation for $S$ as a $K$-algebra (cf. Theorem 12.3.3.4).
- The Cohen-Macaulayness (cf. Proposition 12.4.1.2, Corollary 12.4.1.3) of $R(D)$ is proved by realizing it as a "doset algebra with straightening law" (cf. DeConcini-Lakshmibai [21]), and using the results of [21].
- One then deduces the Cohen-Macaulayness of $S$ from that of $R(D)$ (cf. Theorem 12.4.1.4).
- One proves (cf. Proposition 12.5.0.5) that Spec $S$ is regular in codimension 1 by using the fact that $Spec\, K[X]^{SO_n(K)} \to Spec\, K[X]^{O_n(K)}$ is a 2-sheeted cover.
- One deduces (cf. Proposition 12.5.0.6) the normality of $Spec\, S$ (using Serre's criterion for normality - $Spec\, A$ is normal if and only if $A$ has $S_2$ and $R_1$).
- One then shows that the inclusion $S \subseteq K[X]^{SO_n(K)}$ is in fact an equality by showing that the morphism $Spec\, K[X]^{SO_n(K)} \to Spec\, S$ (induced by the inclusion $S \subseteq K[X]^{SO_n(K)}$) satisfies the hypotheses in Zariski Main Theorem (and hence is an isomorphism (cf. Theorem 12.5.0.7)). One then deduces the Cohen-Macaulayness of $K[X]^{SO_n(K)}$ (cf. Corollary 12.5.0.9)

As the first set of main consequences, we obtain (cf. Theorem 12.5.0.8)

- **First Fundamental Theorem for $SO_n(K)$-invariants**, i.e., describing algebra generators for $K[X]^{SO_n(K)}$.
- **Second Fundamental Theorem for $SO_n(K)$-invariants**, i.e., describing generators for the ideal of relations among these algebra generators for $K[X]^{SO_n(K)}$.
- **A Standard Monomial Basis for $K[X]^{SO_n(K)}$.**

As the second main consequence, we obtain (cf. §12.6) a characteristic-free proof of the Cohen-Macaulayness of the moduli space $M$ of equivalence classes of semi-stable, rank 2 vector bundles on a smooth projective curve of genus $> 2$ by relating it to $K[X]^{SO_3(K)}$. In Mehta-Ramadas [80], the Cohen-Macaulayness for $M$ is deduced by proving the Frobenius-split properties for $M$.

As the third main consequence, we obtain (cf. Theorem 12.7.0.8) a characteristic-free basis for the ring of invariants for the (diagonal) adjoint action of $SL_2(K)$ on $\underbrace{sl_2(K) \oplus \cdots \oplus sl_2(K)}_{m \text{ copies}}$.

Our main goal in this chapter is to prove the Cohen-Macaulayness of $K[X]^{SO_n(K)}$; as mentioned above, this is accomplished by first constructing a "standard monomial" basis for the subalgebra $S$ of $K[X]^{SO_n(K)}$, deducing Cohen-Macaulayness of $S$, and then proving that $S$ in fact equals $K[X]^{SO_n(K)}$. Thus one does not use the results of DeConcini-Procesi [22] (especially, Theorem 5.6 of [22]), we rather give a different proof of Theorem 5.6 of [22].

## 12.1 Preliminaries

For the convenience of the discussion of the results in this chapter, we recollect below the results from chapter 6, specifically the standard monomial basis for the coordinate rings of symmetric determinantal varieties. Since the results of §12.2.1 rely on an explicit description of the straightening relations (of a degree 2 non-standard monomial), in this section we derive such straightening relations (cf. Proposition 12.1.7.3) by relating symmetric determinantal varieties to Schubert varieties in the Lagrangian Grassmannian.

### 12.1.1 The Lagrangian Grassmannian variety

Let $V = K^{2m}$, ($K$ being the base field which we suppose to be algebraically closed of characteristic $\neq 2$) together with a non-degenerate, skew-symmetric bilinear form $\langle , \rangle$. Let $G = Sp(V)$ (the group of linear automorphisms of $V$ preserving $\langle , \rangle$). If $J$ is the matrix of the form, then $G$ may be identified with the fixed point set of the involution $\sigma : SL(V) \to SL(V), \sigma(A) = J^{-1}({}^tA)^{-1}J$. Taking $J$ to be

$$J = \begin{pmatrix} 0 & J_m \\ -J_m & 0 \end{pmatrix}$$

where $J_m$ is the $m \times m$ matrix with 1's along the anti-diagonal, and 0's off the anti-diagonal, $B$ (resp. $T$), the set of upper triangular (resp. diagonal) matrices in $G$ is a Borel sub group (resp. a maximal torus) in $G$ (cf. Steinberg [113]). Let $L_m$ = {maximal totally isotropic sub spaces of $V$}, *the symplectic (or also Lagrangian) Grassmannian Variety*. We have a canonical inclusion $L_m \hookrightarrow G_{m,2m}$, where $G_{m,2m}$ is the Grassmannian variety of $m$-dimensional subspaces of $K^{2m}$.

### 12.1.2 Schubert varieties in $L_m$

Let $I(m, 2m) = \{\underline{i} = (i_1, \ldots, i_m) | 1 \leq i_1 < \cdots < i_m \leq 2m\}$. For $j, 1 \leq j \leq 2m$, let $j' = 2m + 1 - j$. Let

$$I_G(m, 2m) = \{\underline{i} \in I(m, 2m) | \text{ for every } j, 1 \leq j \leq 2m, \text{ precisely one of } \{j, j'\} \text{ occurs in } \underline{i}\}$$

Recall (cf. Chapter 6) that the Schubert varieties in $L_m$ are indexed by $I_G(m, 2m)$; further, the partial order on the set of Schubert varieties in $L_m$ (given by inclusion) induces a partial order on $I_G(m, 2m)$: $\underline{i} \geq \underline{j} \Leftrightarrow i_t \geq j_t, \forall t$. Let $w \in I_G(m, 2m)$, and let $X(w)$ be the associated Schubert variety. For the projective embedding $f_m : L_m \hookrightarrow \mathbb{P}(\wedge^m V)$ (induced by the Plücker embedding $G_{m,2m} \hookrightarrow \mathbb{P}(\wedge^m V)$), let $R(w)$ denote the homogeneous co-ordinate ring of $X(w)$.

**A Standard Monomial Basis for $R(w)$:** We have (cf. Chapter 8) a basis $\{p_{\tau,\varphi}\}$ for $R(w)_1$ indexed by admissible pairs on $X(w)$, i.e., admissible pairs $(\tau, \varphi)$ such that $w \geq \tau$. This basis includes the extremal weight vectors $p_\tau, \tau \in I_G(m, 2m)$ corresponding to the admissible pair $(\tau, \tau)$. Thus denoting by $\mathcal{A}$ the set of all admissible pairs, we have that $\mathcal{A}$ includes the diagonal of $I_G(m, 2m) \times I_G(m, 2m)$. Recall the following (cf. Chapter 8):

**Definition 12.1.2.1** *A monomial of the form*

$$p_{\tau_1,\varphi_1} p_{\tau_2,\varphi_2} \cdots p_{\tau_r,\varphi_r}, \tau_1 \geq \varphi_1 \geq \tau_2 \geq \cdots \geq \varphi_r$$

*is called a* standard monomial *(of degree r). Such a monomial is said to be* standard on $X(w)$*, if in addition* $w \geq \tau_1$.

**Theorem 12.1.2.2** *Standard monomials on* $X(w)$ *of degree r form a basis of* $R(w)_r$.

As a consequence, we have a qualitative description of a typical quadratic relation on a Schubert variety $X(w)$ as given by the following Proposition.

**Proposition 12.1.2.3** *(cf. [21, 62]) Let* $(\tau_1, \phi_1), (\tau_2, \phi_2)$ *be two admissible pairs on* $X(w)$ *such that* $p_{\tau_1,\varphi_1} p_{\tau_2,\varphi_2}$ *is non-standard so that* $\phi_1 \not\geq \tau_2, \phi_2 \not\geq \tau_1$. *Let*

$$p_{\tau_1,\varphi_1} p_{\tau_2,\varphi_2} = \sum_i c_i p_{\alpha_i,\beta_i} p_{\gamma_i,\delta_i}, \quad c_i \in K^* \tag{*}$$

*be the expression for* $p_{\tau_1,\varphi_1} p_{\tau_2,\varphi_2}$ *as a sum of standard monomials on* $X(w)$.

(i) *For every i, we have,* $\alpha_i \geq$ *both* $\tau_1$ *and* $\tau_2$. *Further, for some i, if* $\alpha_i = \tau_1$ *(resp.* $\tau_2$*), then* $\beta_i > \varphi_1$ *(resp.* $\varphi_2$*).*
(ii) *For every i, we have,* $\delta_i \leq$ *both* $\varphi_1$ *and* $\varphi_2$. *Further, for some i, if* $\delta_i = \varphi_1$ *(resp.* $\varphi_2$*), then* $\gamma_i < \tau_1$ *(resp.* $\tau_2$*).*
(iii) *Suppose there exists a permutation* $\sigma$ *of* $\{\tau_1, \varphi_1, \tau_2, \varphi_2\}$ *such that* $\sigma(\tau_1) \geq \sigma(\varphi_1) \geq \sigma(\tau_2) \geq \sigma(\varphi_2)$, *then* $(\sigma(\tau_1), \sigma(\varphi_1)), (\sigma(\tau_2), \sigma(\varphi_2))$ *are both admissible pairs, and* $p_{\sigma(\tau_1),\sigma(\varphi_1)} p_{\sigma(\tau_2),\sigma(\varphi_2)}$ *occurs with coefficient* $\pm 1$ *in (*).*

We shall refer to a relation as in (*) as *a straightening relation*.

**A presentation for** $R(w)$**:** For $w \in I_G(m, 2m)$, let $\mathcal{A}_w = \{(\tau, \varphi) \in \mathcal{A} \mid w \geq \tau\}$. Consider the polynomial algebra $K[x_{\tau,\varphi}, (\tau, \varphi) \in \mathcal{A}_w]$. For two admissible pairs $(\tau_1, \phi_1), (\tau_2, \phi_2)$ in $\mathcal{A}_w$ such that $p_{\tau_1,\varphi_1} p_{\tau_2,\varphi_2}$ is non-standard, denote $F_{(\tau_1,\phi_1),(\tau_2,\phi_2)} = x_{\tau_1,\phi_1}, x_{\tau_2,\phi_2} - \sum_i c_i x_{\alpha_i,\beta_i} x_{\gamma_i,\delta_i}$, $\alpha_i, \beta_i, \gamma_i, \delta_i, c_i$ being as in Proposition 12.1.2.3. Let $I_w$ be the ideal in $K[x_{\tau,\varphi}, (\tau, \varphi) \in \mathcal{A}_w]$ generated by such $F_{(\tau_1,\phi_1),(\tau_2,\phi_2)}$'s. Consider the surjective map $f_w : K[x_{\tau,\varphi}, (\tau, \varphi) \in \mathcal{A}_w] \to R(w)$, $x_{\tau,\varphi} \mapsto p_{\tau,\varphi}$. We have from Chapter 8 the following:

**Proposition 12.1.2.4** $f_w$ *induces an isomorphism* $K[x_{\tau,\varphi}, (\tau, \varphi) \in \mathcal{A}_w]/I_w \cong R(w)$.

### 12.1.3 The opposite big cell in $L_m$

Recall the following from Chapter 6:

**Facts:** 1. We have a natural embedding

$$L_m \hookrightarrow G_{m,2m}$$

$G_{m,2m}$ being the Grassmannian variety of $m$-dimensional sub spaces of $K^{2m}$.

2. Indexing the simple roots of $G$ as in Bourbaki [8], we have an identification

$$G/P \cong L_m$$

$P$ being the maximal parabolic sub group of $G$ corresponding to "omitting" the simple root $\alpha_m$ (the right-end root in the Dynkin diagram of $G$).

3. The opposite big cell $O^-$ in $G_{m,2m}$ may be identified as

$$(*) \qquad O^- = \left\{ \begin{pmatrix} I_m \\ Y \end{pmatrix}, \ Y \in M_{m,m}(K) \right\}$$

where $I_m$ is the identity $m \times m$ matrix, and $M_{m,m}K)$ is the space of $m \times m$ matrices (with entries in $K$). This gives rise to an identification of the opposite big cell $O_G^-$ in $L_m$ as

$$(**) \qquad O_G^- = \left\{ \begin{pmatrix} I_m \\ Z \end{pmatrix}, \ Z \in Sym\ M_m \right\}$$

where $Sym\ M_m$ is the space of symmetric $m \times m$ matrices (with entries in $K$).

### 12.1.4 The functions $f_{\tau,\varphi}$ on $O_G^-$

Let $\underline{j} \in I(m, 2m)$, $p_{\underline{j}}$ the corresponding Plücker co-ordinate on $G_{m,2m}$. Denote by $f_{\underline{j}}$, the restriction of $p_{\underline{j}}$ to $O^-$. As seen in Chapter 6, we have (under the identification (*)) that if $z \in O^-$ corresponds to the matrix $A$, then $f_{\underline{j}}(z)$ is simply the following minor of $A$: let $\underline{j} = (j_1, \ldots, j_m)$, and let $j_r$ be the largest entry $\leq m$. Let $\{k_1, \ldots, k_{m-r}\}$ be the complement of $\{j_1, \ldots, j_r\}$ in $\{1, \ldots, m\}$. Then $f_{\underline{j}}(z)$ is the $(m-r)$-minor of $A$ with column indices $k_1, \ldots k_{m-r}$, and row indices $j_{r+1}, \ldots, j_m$ (here the rows of $A$ are indexed as $m+1, \ldots, 2m$). Conversely, given a minor of $A$, say, with column indices $b_1, \ldots, b_s$, and row indices $j_{m-s+1}, \ldots, j_m$ (again, the rows of $A$ are indexed as $m+1, \ldots, 2m$), it is $f_{\underline{j}}(z)$, where $\underline{j} = (j_1, \ldots, j_m)$ is given as follows: $\{j_1, \ldots, j_{m-s}\}$ is the complement of $\{b_1, \ldots, b_s\}$ in $\{1, \ldots, m\}$, and $j_{d-s+1}, \ldots, j_m$ are simply the row indices.

**The Partial Order $\geq$:** Given $A = (a_1, \cdots, a_s)$, $A' = (a'_1, \cdots, a'_{s'})$, for some $s, s' \geq 1$, recall the partial order: $A \geq A'$ if $s \leq s'$, $a_j \geq a'_j$, $1 \leq j \leq s$.

As seen in Chapter 6, we have that on $G/P$, any $p_{\tau,\varphi}$ ($(\tau, \varphi)$ being an admissible pair) is the restriction of some Plücker co-ordinate on $G_{m,2m}$. Let us denote by $f_{\tau,\varphi}$ the restriction of $p_{\tau,\varphi}$ to $O_G^-$. Given $z \in O_G^-$, let $M$ be the corresponding matrix in Sym $M_m$ (under the identification (**)); then as above, $f_{\tau,\varphi}(z)$ is a certain minor of $A$, say with row (resp. column) indices $A := \{a_1, \cdots, a_s\}$ (resp. $B := \{b_1, \cdots, b_s\}$); we have, $A \geq B$. Denote this minor by $p(A, B)$. Conversely, such a minor corresponds to a unique $f_{\tau,\varphi}$. Thus we have a bijection

$$\{\text{admissible pairs}\} \overset{bij}{\to} \{\text{minors } p(A, B) \text{ of } X, A \geq B\}$$

$X$ being a symmetric $m \times m$ matrix of indeterminates.

**Convention.** If $\tau = \varphi = (1, \ldots, m)$, then $f_{\tau,\varphi}$ evaluated at $z$ is 1; we shall make it correspond to the minor of $X$ with row indices (and column indices) given by the empty set.

### 12.1.5 The opposite cell in $X(w)$

For a Schubert variety $X(w)$ in $L_m$, let us denote $O_G^- \cap X(w)$ by $Y(w)$. We consider $Y(w)$ as a closed subvariety of $O_G^-$. In view of Proposition 12.1.2.4, we obtain that the ideal defining $Y(w)$ in $O_G^-$ is generated by

$$\{f_{\tau,\varphi} \mid, \ w \not\geq \tau\}.$$

### 12.1.6 Symmetric determinantal varieties

Let $\mathcal{Z} = Sym\ M_m$, the space of all symmetric $m \times m$ matrices with entries in $K$. We shall identify $\mathcal{Z}$ with $\mathbb{A}^N$, where $N = \frac{1}{2}m(m+1)$. We have $K[\mathcal{Z}] = K[z_{i,j},\ 1 \leq i \leq j \leq m]$.

**The variety** $D_t(Sym\ M_m)$. Let $X = (x_{ij}),\ 1 \leq i, j \leq m, x_{ij} = x_{ji}$ be a $m \times m$ symmetric matrix of indeterminates. Let $A, B,\ A \subset \{1, \cdots, m\},\ B \subset \{1, \cdots, m\},\ \#A = \#B = s$, where $s \leq m$. Denote by $p(A, B)$ the $s$-minor of $X$ with row indices given by $A$, and column indices given by $B$. For $n,\ 1 \leq n \leq m$, let $I_t$ be the ideal in $K[Z]$ generated by $\{p(A, B),\quad A \subset \{1, \cdots, m\},\ B \subset \{1, \cdots, m\},\ \#A = \#B = t\}$. Let $D_t(Sym\ M_m)$ be the *symmetric determinantal variety* (a closed subvariety of $\mathcal{Z}$), with $I_t$ as the defining ideal. In the discussion below, we also allow $t = m+1$ in which case $D_t(Sym\ M_m) = \mathcal{Z}$.

**Identification of** $D_t(Sym\ M_m)$ **with** $Y(\phi)$. From Chapter 6, we have an identification

$$O_G^- = \left\{ \begin{pmatrix} I_m \\ X \end{pmatrix} \right\}$$

where $X$ is a symmetric $m \times m$ matrix. As mentioned above, we have a bijection:

$$\{f_{\tau,\varphi},\ (\tau, \varphi) \in \mathcal{A}\} \overset{\texttt{bij}}{\to} \{\ \texttt{minors}\ p(A, B), A, B \in I(r, m), A \geq B, 1 \leq r \leq m\ \texttt{of}\ X\}$$

(here, $\mathcal{A}$ is the set of all admissible pairs, and $I(r, m) = \{\underline{i} = (i_1, \ldots, i_r) | 1 \leq i_1 < \cdots < i_r \leq m\}$.

Let $\phi$ be the $m$-tuple, $\phi = (t, t+1, \cdots, m, 2m+2-t, 2m+3-t, \cdots, 2m)(= (t, t+1, \cdots, m, (t-1)', (t-2)', \cdots, 1'))$ (note that $\phi$ consists of the two blocks $[t, m]$, $[2m+2-t, 2m]$ of consecutive integers - here, for $i < j$, $[i, j]$ denotes the set $\{i, i+1, \cdots, j\}$, and for $1 \leq s \leq 2m$, $s' = 2m+1-s$). If $t = m+1$, then we set $\phi = (m', (m-1)', \cdots, 1')$; note then that $Y(\phi) = O_G^- (\cong Sym\ M_m)$. From Chapter 6, we have

**Theorem 12.1.6.1** *$D_t(Sym\ M_m) \cong Y(\phi)$; further, $\dim D_t(Sym\ M_m) = \frac{1}{2}(t-1)(2m+2-t) = \frac{1}{2}(t-1)(t-1)'$.*

**Corollary 12.1.6.2** *$K[D_t(Sym\ M_m)] \cong R(\phi)_{(p_{id})}$, the homogeneous localization of $R(\phi)$ at $p_{id}$ ($id$ being the $m$-tuple $(1, \cdots, m)$).*

**Singular Locus of $D_t(Sym\ M_m)$:** For our discussion in §12.5 (especially, in the proof of Lemma 12.5.0.5), we will be required to know Sing $D_t(Sym\ M_m)$, the singular locus of $D_t(Sym\ M_m)$. Let $\phi = ([t, m], [2m+2-t, 2m])$ as above. From [57], we have

$$Sing\ X(\phi) = X(\phi')$$

where $\phi' = ([t-1, m], [2m+3-t, 2m])$. This together with Theorem 12.1.6.1 implies the following theorem:

**Theorem 12.1.6.3** *Sing* $D_t(Sym\ M_m) = D_{t-1}(Sym\ M_m)$

### 12.1.7 The set $H_m$

Let

$$H_m = \{(A, B) \in \bigcup_{0 \leq s \leq m} I(s, m) \times I(s, m) \mid A \geq B\}$$

where our convention is that $(\emptyset, \emptyset)$ is the element of $H_m$ corresponding to $s = 0$. We define $\succeq$ on $H_m$ as follows:

- We declare $(\emptyset, \emptyset)$ as the largest element of $H_m$.
- For $(A, B), (A', B')$ in $H_m$, say, $A = (a_1, \cdots, a_s)$, $B = (b_1, \cdots, b_s)$, $A' = (a'_1, \cdots, a'_{s'})$, $B' = (b'_1, \cdots, b'_{s'})$ for some $s, s' \geq 1$, we define $(A, B) \succeq (A', B')$ (or also $p(A, B) \succeq p(A', B')$) if $B \geq A'$ (here, the partial order $\geq$ is as in §12.1.4).

**The Bijection $\theta$:** Let $X = (x_{ij})$ be a generic symmetric $m \times m$ matrix. The above partial order $\succeq$ induces a partial order $\succeq$ on the set of all minors of $X$, namely, $p(A, B) \succeq p(A', B')$, if $(A, B) \succeq (A', B')$. Let $(\tau, \varphi) \in \mathcal{A}$, i.e., $(\tau, \varphi)$ is an admissible pair. As elements of $I_G(m, 2m)$, let

$$\tau = (a_1, \cdots, a_r, b'_1, \cdots, b'_s), \varphi = (c_1, \cdots, c_r, d'_1, \cdots, d'_s)$$

where $r + s = m, a_i, b_j, c_i, d_j$ are $\leq n$, and (recall that) for $1 \leq q \leq 2m, q' = 2m + 1 - q$. We would like to remark that the fact that $(\tau, \varphi)$ is an admissible pair implies that number of entries $\leq m$ in $\tau$ and $\varphi$ are the same (see [67], [57] for details). Denote

$$\underline{i} := (i_1, \cdots, i_m) := (a_1, \cdots, a_r, d'_1, \cdots, d'_s) \ (\in I(m, 2m))$$

(here, $I(m, 2m)$ is as in §12.1.2). Set

$$A_{\underline{i}} = \{2m + 1 - i_m, 2m + 1 - i_{m-1}, \cdots, 2m + 1 - i_{r+1}\},$$

$$B_{\underline{i}} = \text{ the complement of } \{i_1, i_2 \cdots, i_r\} \text{ in } \{1, 2, \cdots, m\}.$$

Define $\theta : \mathcal{A} \to$ {all minors of $X$} by setting $\theta(\underline{i}) = p(A_{\underline{i}}, B_{\underline{i}})$ (here, the constant function 1 is considered as the minor of $X$ with row indices (and column indices) given by the empty set). Then $\theta$ is a bijection. Note that $\theta$ reverses the respective (partial) orders, i.e., given $\underline{i}, \underline{i}' \in I(m, 2m)$, corresponding to admissible pairs $(\tau, \varphi), (\tau', \varphi')$ we have, $\underline{i} \leq \underline{i}' \iff \theta(\underline{i}) \succeq \theta(\underline{i}')$. Using $\succeq$, we define *standard monomials* in $p(A, B)$'s for $(A, B) \in H_m$:

**Definition 12.1.7.1** *A monomial* $p(A_1, B_1) \cdots p(A_s, B_s), s \in \mathbb{N}$ *is standard if* $p(A_1, B_1) \succeq \cdots \succeq p(A_s, B_s)$.

Let $X = \underbrace{V \oplus \cdots \oplus V}_{m \text{ copies}}$, $V = K^n$. From Chapter 10, §10.4, we have

**Theorem 12.1.7.2** *(i) Spec* $K[X]^{O(V)}$, *the categorical quotient* $X /\!\!/ O(V)$ *gets identified with* $D_{n+1}(Sym\ M_m)$.
*(ii) Let* $H_n = \{(A, B) \in H_m \mid \#A \leq n\}$. *Standard monomials in* $\{p(A, B), (A, B) \in H_n\}$ *form a basis for* $K[D_{n+1}(Sym\ M_m)]$, *the algebra of regular functions on* $D_{n+1}(Sym\ M_m)$ $(\subset Sym\ M_m)$.

As a direct consequence of Proposition 12.1.2.3, we obtain

**Proposition 12.1.7.3** *Let* $p(A_1, A_2), p(B_1, B_2)$ *(in* $K[D_t(Sym\ M_m)]$*) be non standard. Let*

$$p(A_1, A_2)p(B_1, B_2) = \sum a_i p(C_{i1}, C_{i2})p(D_{i1}, D_{i2}), a_i \in K^* \qquad (*)$$

*be the straightening relation, i.e., R.H.S. is a sum of standard monomials. Then for every* $i$, $C_{i1}, C_{i2}, D_{i1}, D_{i2}$ *have cardinalities* $\leq t-1$; *further,*

*(i)* $C_{i1} \geq$ *both* $A_1$ *and* $B_1$; *further, if for some* $i$, $C_{i1}$ *equals* $A_1$ *(resp.* $B_1$*), then* $C_{i2} > A_2$ *(resp.* $> B_2$*).*
*(ii)* $D_{i2} \leq$ *both* $A_2$ *and* $B_2$; *further, if for some* $i$, $D_{i2}$ *equals* $A_2$ *(resp.* $B_2$*), then* $D_{i1} < A_1$ *(resp.* $< B_1$*).*
*(iii) Suppose there exists a* $\sigma \in S_4$ *(the symmetric group on 4 letters) such that* $\sigma(A_1) \geq \sigma(A_2) \geq \sigma(B_1) \geq \sigma(B_2)$, *then* $(\sigma(A_1), \sigma(A_2)), (\sigma(B_1), \sigma(B_2))$ *are in* $H_{t-1}$, *and* $p(\sigma(A_1), \sigma(A_2))p(\sigma(B_1), \sigma(B_2))$ *occurs with coefficient* $\pm 1$ *in (*).*

**Remark 12.1.7.4.** On the R.H.S. of (*), $C_{i1}, C_{i2}$ could both be the empty set (in which case $p(C_{i1}, C_{i2})$ is understood as 1). For example, with $X$ being a $2 \times 2$ symmetric matrix of indeterminates, we have

$$p_{2,1}^2 = p_{2,2}p_{1,1} - p_{\emptyset,\emptyset}p_{12,12}.$$

**Remark 12.1.7.5.** In the sequel, while writing a straightening relation as in Proposition 12.1.7.3, if for some $i$, $C_{i1}, C_{i2}$ are both the empty set, we keep the corresponding $p(C_{i1}, C_{i2})$ on the right hand side of the straightening relation (even though its value is 1) in order to have homogeneity in the relation.

Taking $t = m + 1$ (in which case $D_t(Sym\ M_m) = \mathcal{Z} = Sym\ M_m$) in Theorem 12.1.7.2 and Proposition 12.1.7.3, we obtain

**Theorem 12.1.7.6** *(i) Standard monomials in* $\{p(A, B), (A, B) \in H_m\}$*'s form a basis for* $K[\mathcal{Z}] (\cong K[x_{ij}, 1 \leq i \leq j \leq m])$.
*(ii) Relations similar to those in Proposition 12.1.7.3 hold on Z.*

**Remark 12.1.7.7.** Note that Theorem 12.1.7.2 recovers Theorem 5.1 of DeConcini-Procesi [22]. But we had taken the above approach of deducing Theorem12.1.7.2 from Theorems 12.1.2.2, 12.1.6.1 in order to derive the straightening relations as given by Proposition 12.1.7.3 (which are crucial for the discussion in §12.2.1).

**A presentation for $K[D_t(Sym\ M_m)]$.** Consider the polynomial algebra $K[x(A,B),(A,B)\in H_{t-1}]$. For two non-comparable pairs $(A_1,A_2),(B_1,B_2)$ in $H_{t-1}$, denote

$$F((A_1,A_2);(B_1,B_2)) = x(A_1,A_2)(B_1,B_2) - \sum a_i x(C_{i1},C_{i2})x(D_{i1},D_{i2})$$

where $C_{i1},C_{i2},D_{i1},D_{i2},a_i$ are as in Proposition 12.1.7.3. Let $J_{t-1}$ be the ideal generated by

$$\{F((A_1,A_2);(B_1,B_2)),(A_1,A_2),(B_1,B_2)\ \texttt{non-comparable}\}$$

Consider the surjective map $f_{t-1}: K[x(A,B),(A,B)\in H_{t-1}]\to K[D_t(Sym\ M_m)]$, $x(A,B)\mapsto p(A,B)$. Then in view of Proposition 12.1.2.4 and Theorem 12.1.6.1, we obtain

**Proposition 12.1.7.8** *$f_{t-1}$ induces an isomorphism*

$$K[x(A,B),(A,B)\in H_{t-1}]/J_{t-1}\cong K[D_t(Sym\ M_m)]$$

## 12.2 The algebra $S$

Let $V=K^n$ together with a non-degenerate bilinear form $\langle,\rangle$. Let $X=V^{\oplus m}$ ($=V\oplus\cdots\oplus V$ ($m$ `copies`)), and $G=SO_n(K)$. Denote $R=K[X]$. We present the proof of Cohen-Macaulayness of $R^G$ (cf. [63]). This is accomplished by proving the Cohen-Macaulayness of a certain subalgebra $S$ of $R^G$, and showing $S$ in fact equals $R^G$.

For $\underline{u}=(u_1,u_2,...,u_m)\in X$, writing $u_i=(u_{i1},\cdots,u_{in})$ (with respect to the standard basis for $K^n$), we shall identify $\underline{u}$ with the $m\times n$ matrix $U:=(u_{ij})_{m\times n}$.

**The Functions $p(A,B)$:** Let $\underline{u}:=(u_1,\cdots,u_m)\in X$. For $A,B$ in $I(r,m)$, where $1\le r\le n$, and $A\ge B$, let $p(A,B)$ denote the regular function on $X$, $p(A,B)((\underline{u}))=$ the $r$-minor of the symmetric $m\times m$ matrix $(\langle u_i,u_j\rangle)$ with row indices given by the entries of $A$, and column indices given by the entries of $B$.

**The Functions $u(I)$:** For $I\in I(n,m)$, let $u(I)$ be the function $u(I):X\to K$, $u(I)(\underline{u}):=$ the $n$-minor of $U$ with the row indices given by the entries of $I$ (here, $U=(u_{ij})$ is as above). We have, $g\cdot u(I)=(det\ g)u(I),\ g\in O(V)$. Hence $u(I)$ is in $K[X]^{SO(V)}$.

**Lemma 12.2.0.1** *For $I,J\in I(n,m)$, we have $u(I)u(J)=p(I,J)$.*

*Proof.* Let $M, N$ denote the $n \times n$ submatrices of the $m \times n$ matrix $U = (u_{ij})$ with row indices given by $I, J$ respectively. We have

$$u(I)u(J) = (det\ M)\ (det\ N) = (\det\ M)\ (det\ {}^tN) = det(M\ {}^tN) = p(I, J)$$

(note that $M\ {}^tN$ is the submatrix of $(\langle u_i, u_j \rangle)$ with row indices given by the entries of $I$, and column indices given by the entries of $J$).

**The Algebra** $S$**:** Let $S$ be the subalgebra of $R^G$ generated by $p(A, B), A, B \in I(r, m), r \leq n-1, A \geq B,\ u(I), I \in I(n, m)$.

**Remark 12.2.0.2.** The $K$-algebra $S$ could have been simply defined as the $K$-subalgebra of $R^G$ generated by $\{\langle u_i, u_j \rangle\}$ (i.e., by $\{p(A, B), \#A = \#B = 1\}$) $\cup \{u(I), I \in I(n, m)\}$. But we have a purpose in defining it as above, namely, the standard monomials (in $S$) will be built out of the $p(A, B)$'s with $\#A \leq n-1$, and the $u(I)$'s (cf. Definition 12.2.1.1 below).

### 12.2.1 Standard monomials and their linear independence

In this subsection, we define standard monomial in $S$, and prove their linear independence.

**The Set** $H$**:** Let

$$H_u := I(n, m)$$

(note that $H_u$ indexes the $u(I)$'s). Let

$$H_p = \{(A, B) \in I(r, m) \times I(r, m), 1 \leq r \leq n-1, A \geq B\}$$

(note that $H_p$ indexes the $p(A, B)$'s).

Let

$$H = H_p \cup H_u$$

We define a partial order $\geq$ on $H$ as follows:

(i) For elements $H_p$, it is just the partial order on $H_p$.

(ii) For elements $H_u$, it is the partial order $\geq$ as defined in §12.1.4.

(iii) No element of $H_u$ is greater than any element of $H_p$.

(iv) For $(A, B) \in H_p$, $I \in H_u$, $(A, B) \geq I$, if $B \geq I$ (again, $\geq$ being as in §12.1.4).

Thus $S$ has a set of algebra generators $\{p(A, B), (A, B) \in H_p, u(I), I \in H_u\}$ indexed by the partially-ordered set $H$.

**Definition 12.2.1.1** *A monomial*

$$F = p(A_1, B_1) \cdots p(A_r, B_r)u(I_1) \cdots u(I_s)$$

*in* $\{p(A, B), u(I), (A, B) \in H_p, I \in H_u\}$ *is said to be* standard *if*

$$(A_1, B_1) \geq \cdots \geq (A_1, B_1) \geq I_1 \geq \cdots \geq I_s$$

*i.e,* $A_1 \geq B_1 \geq A_2 \geq \cdots \geq B_r \geq I_1 \geq \cdots \geq I_s$.

### 12.2.2 Linear independence of standard monomials

In this subsection, we prove the linear independence of standard monomials.

**Lemma 12.2.2.1** *Let* $(A, B) \in H_p$, $I \in H_u$.

*(i) The set of standard monomials in the* $p(A, B)$*'s is linearly independent.*
*(ii) The set of standard monomials in the* $u(I)$*'s is linearly independent.*

*Proof.* The subalgebra generated by $\{p(A, B),\ A, B \in H_n\}$ being $R^{O(V)}$, gets identified with $K[D_{n+1}(Sym(M_m)]$, and hence (i) follows from Theorem 12.1.7.2. (ii) follows from [72], Theorem 1.6.6,(1) applied to $K[u_{ij}, 1 \le i \le m, 1 \le j \le n]$.

**Proposition 12.2.2.2** *Standard monomials (in S) are linearly independent.*

*Proof.* Let

$$(*) \qquad F = G + H = 0,$$

be a relation among standard monomials, where $G = \sum c_i G_i$, $H = \sum d_j H_j$ and for each $p(A_1, B_1) \cdots p(A_r, B_r)u(I_1) \cdots u(I_s)$ in $G$ (resp. $H$), $s$ is even - including 0 - (resp. odd). Consider a $g$ in $O_n(K)$, with $\det g = -1$. Then noting that $u(I)u(J) = p(I, J)$, and using the facts that $g \cdot p(A, B) = p(A, B)$, $g \cdot u(I) = (det\ g)u(I)$, we have, $F - gF = \sum 2d_j H_j = 0$. Hence, if we show that $\sum d_j H_j = 0$, then it would follow (in view of Theorem 12.1.7.2) that (*) is the trivial sum. Thus we may suppose

$$(**) \qquad F = \sum d_j H_j$$

where each $H_j$ is a standard monomial of the form:

$$H_j = p(R_1, S_1) \cdots p(R_l, S_l)u(I)$$

Now multiplying (**) by $u(I_n)$, $I_n$ being $(1, \cdots, n)$ (and noting as above that $u(I)u(I_n) = p(I, I_n)$), we obtain

$$\sum d_i P_i = 0$$

where each $P_i$ is a standard monomial of the form $p(U_1, Q_1) \cdots p(U_m, Q_m)$ (note that for each standard monomial $H_j$ appearing in (**), $H_j u(I_n)$ is again standard). Now the required result follows from the linear independence of $p(A, B)$'s in $K[X]^{O(V)}$(cf. Theorem 12.1.7.2).

## 12.3 The algebra $S(D)$

Let $S$ be as in the previous section.

### 12.3.1 Quadratic relations

In this subsection, we describe certain straightening relations (i.e., expressions for non-standard monomials as linear sums of standard monomials) to be used while proving the generation (of $S$ as a $K$-vector space) by standard monomials.

**Proposition 12.3.1.1** *(i) Let $I, I' \in H_u$ be not comparable. We have,*

$$u(I)u(I') = \sum_r b_r u(I_r)u(I'_r),\ b_r \in K^*$$

*where for all r, $I_r \geq$ both $I, I'$, and $I'_r \leq$ both $I, I'$; in fact, $I_r >$ both $I, I'$ (for, if $I_r = I$ or $I'$, then $I, I'$ would be comparable).*

*(ii) Let $(A_1, A_2), (B_1, B_2) \in H_p$ be not comparable. Then we have*

$$p(A_1, A_2)p(B_1, B_2) = \sum a_i p(C_{i1}, C_{i2})p(D_{i1}, D_{i2}), a_i \in K^*,$$

*where $(C_{i1}, C_{i2}), (D_{i1}, D_{i2})$ belong to $H_p$, and $C_{i2} \geq D_{i1}$; further, for every i, we have*

*a) $C_{i1} \geq$ both $A_1$ and $B_1$; if $C_{i1} = A_1$ (resp. $B_1$), then $C_{i2} > A_2$ (resp. $B_2$).*
*b) $D_{i2} \leq$ both $A_2$ and $B_2$; if $D_{i2} = A_2$ (resp. $B_2$), then $D_{i1} < A_1$ (resp. $B_1$).*

*(iii) Let $I \in H_u$, $(A, B) \in H_p$ be such that $B \not\geq I$. We have,*

$$p(A, B)u(I) = \sum_t d_t p(A_t, B_t)u(I_t),\ d_t \in K^*$$

*where for every t, we have, $(A_t, B_t) \in H_p$, $B_t \geq I_t$. Further, $A_t \geq$ both $A$ and $I$; if $A_t = A$, then $B_t > B$.*

*Proof.* In the course of the proof, we will be repeatedly using the fact that the sub algebra generated by $\{p(A, B), A, B \in H_n\}$ (where recall that $H_n = \{(A, B) \in H_m, \#A \leq n\}$, $H_m$ being as in §12.1.7) being $R^{O(V)}$ (cf. Chapter 10, Theorem 10.4.0.3), the relations given by Proposition 12.1.7.3 hold in $K[X]^{O(V)}$.
Assertion (i) follows from [72], Theorem 4.1.1,(2).
Assertion (ii) follows from Proposition 12.1.7.3.
(iii). We have, $p(A, B)u(I)u(I_n) = p(A, B)p(I, I_n)$ ($I_n$ being $(1, 2, \cdots, n)$). The hypothesis that $B \not\geq I$ implies that $p(A, B)p(I, I_n)$ is not standard. Hence Proposition 12.1.7.3 implies that in $K[D_{n+1}(Sym(M_m))]$

$$p(A, B)p(I, I_n) = \sum a_i p(C_{i1}, C_{i2})p(D_{i1}, D_{i2}), a_i \in K^*$$

where R.H.S. is a standard sum, i.e., $C_{i1} \geq C_{i2} \geq D_{i1} \geq D_{i2}, \forall i$. Further, for every $i$, $C_{i1} \geq$ both $A$ and $I$; if $C_{i1} = A$, then $C_{i2} > B$; $C_{i2} \geq$ both $B$ and $I_n$; $D_{i2} \leq$ both $B$ and $I_n$ which forces $D_{i2} = I_n$ (note that in view of Proposition 12.1.7.3, all minors in the above relation have size $\leq n$); and hence $\#D_{i1} = n$, for all $i$. Hence $p(D_{i1}, D_{i2}) = u(D_{i1})u(I_n)$, for all $i$. Hence cancelling $u(I_n)$, we obtain

$$p(A, B)u(I) = \sum a_i p(C_{i1}, C_{i2})u(D_{i1}),$$

where $C_{i1} \geq$ both $A$ and $I$; if $C_{i1} = A$, then $C_{i2} > B$. This proves (iii).

To prove the generation of $S$ (as a $K$-vector space) by standard monomials, we define a $K$-algebra $S(D)$, construct a standard monomial basis for $S(D)$ and deduce the results for $S$ (in fact, it will turn out that $S(D) \cong S$). We first define the $K$-algebra $R(D)$ as follows:

Let

$$D = H \cup \{\mathbf{1}\}$$

$H$ being as in §12.2.1. Extend the partial order on $H$ to $D$ by declaring $\{\mathbf{1}\}$ as the largest element. Let $P(D)$ be the polynomial algebra

$$P(D) := K[X(A, B), Y(I), X(\mathbf{1}), (A, B) \in H_p, I \in H_u]$$

Let $\mathfrak{a}(D)$ be the homogeneous ideal in the polynomial algebra $P(D)$ generated by the relations (i)-(iii) of Proposition 12.3.1.1 ($X(A, B), Y(I)$ replacing $p(A, B), u(I)$ respectively), with relation (ii) homogenized as

$$X(A_1, A_2)X(B_1, B_2) = \sum a_i X(C_{i1}, C_{i2})X(D_{i1}, D_{i2})$$

where $X(C_{i1}, C_{i2})$ is to be understood as $X(\mathbf{1})$ if both $C_{i1}, C_{i2}$ equal the empty set (cf. Remark 12.1.7.5). Let

$$R(D) = P(D)/\mathfrak{a}(D)$$

We shall denote the classes of $X(A, B), Y(I), X(\mathbf{1})$ in $R(D)$ by $x(A, B), y(I), x(\mathbf{1})$ respectively.

**The Algebra $S(D)$:** Set $S(D) = R(D)_{(x(\mathbf{1}))}$, the homogeneous localization of $R(D)$ at $x(\mathbf{1})$. We shall denote $\frac{x(A,B)}{x(\mathbf{1})}, \frac{y(I)}{x(\mathbf{1})}$ (in $S(D)$) by $c(A, B), d(I)$ respectively.

Let $\varphi_D : S(D) \to S$ be the map $\varphi_D(c(A, B)) = p(A, B), \varphi_D(d(I)) = u(I)$. Let

$$\theta_D : R(D) \to S(D)$$

be the canonical map. Denote $\gamma_D : R(D) \to S$ as the composite $\gamma_D = \varphi_D \circ \theta_D$.

### 12.3.2 A standard monomial basis for $R(D)$

We define a monomial in $x(A, B), y(I), x(\mathbf{1}))$ (in $R(D)$) to be standard in exactly the same way as in Definition 12.2.1.1.

**Proposition 12.3.2.1** *The standard monomials in the $x(A, B), y(I), x(\mathbf{1})$ are linearly independent.*

*Proof.* The result follows by considering $\gamma_D : R(D) \to S$, and using the linear independence of standard monomials in $S$ (cf. Proposition 12.2.2.2).

**Generation of $R(D)$ by Standard Monomials:** We shall now show that any non-standard monomial $F$ in $R(D)$ is a linear sum of standard monomials. Observe that if $M$ is a standard monomial, then $x(\mathbf{1})^l M$ is again standard; hence we may suppose $F$ to be:

$$F = x(A_1, B_1) \cdots x(A_r, B_r) y(I_1) \cdots y(I_s)$$

Fix an integer $N$ sufficiently large. To each element $A \in \cup_{r=1}^{n} I(r, m)$, we associate an $(n+1)$-tuple as follows: Let $A \in I(r, m)$, for some $r$, $A = (a_1, \cdots, a_r)$. To $A$, we associate the $n+1$-tuple $\overline{A} = (a_1, \cdots, a_r, m, m, \cdots, m, 1)$.

To $F$, we associate the integer $n_F$ (and call it the *weight of* $F$) which has the entries of $\overline{A_1}, \overline{B_1}, \overline{A_2}, \overline{B_2}, \cdots, \overline{A_r}, \overline{B_r}, \overline{I_1}, \cdots, \overline{I_s}$ as digits (in the $N$-ary presentation). The hypothesis that $F$ is non-standard implies that either $x(A_i, B_i)x(A_{i+1}, B_{i+1})$ is non-standard for some $i \leq r-1$, or $x(A_r, B_r)y(I_1)$ is non-standard, or $u(I_t)u(I_{t+1})$ is non-standard for some $t \leq s-1$ is non-standard. Straightening these using Proposition 12.3.1.1, we obtain that $F = \sum a_i F_i$ where $n_{F_i} > n_F$, for all $i$, and the result follows by decreasing induction on $n_F$ (note that while straightening a degree 2 relation using Proposition 12.3.1.1, if $x(\mathbf{1})$ occurs in a monomial $G$, then the digits in $n_G$ corresponding to $x(\mathbf{1})$ are taken to be $m, m, \cdots, m$ ($2n$+2-times)). Also note that the largest $F$ of degree $r$ in $x(A, B)$'s and degree $s$ in the $y(I)$'s is $x(\{m\}, \{m\})^r u(I_0)^s$ (where $I_0$ is the $n$-tuple $(m+1-n, m+2-n, \cdots, m)$) which is clearly standard.

Hence we obtain

**Proposition 12.3.2.2** *Standard monomials in $x(A, B), y(I), x(\mathbf{1}))$ generate $R(D)$ as a $K$-vector space.*

Combining Propositions 12.3.2.1, 12.3.2.2, we obtain

**Theorem 12.3.2.3** *Standard monomials in $x(A, B), y(I), x(\mathbf{1})$ give a basis for the $K$-vector space $R(D)$.*

### 12.3.3 Standard monomial bases for $S(D)$

Standard monomials in $c(A, B), d(I)$ in $S(D)$ $S(D))$ are defined in exactly the same way as in Definition 12.2.1.1.

**Theorem 12.3.3.1** *Standard monomials in $c(A, B), d(I)$ give a basis for the $K$-vector space $S(D)$.*

*Proof.* The linear independence of standard monomials follows as in the proof of Proposition 12.3.2.1 by considering $\varphi_D : S(D) \to S$, and using the linear independence of standard monomials in $S$ (cf. Proposition 12.2.2.2).

To see the generation of $S(D)$ by standard monomials, consider a non-standard monomial $F$ in $S(D)$, say,

$$F = c(A_1, B_1) \cdots c(A_i, B_i) d(I_1) \cdots d(I_k)$$

Then $F = \theta_D(G)$, where $G = x(A_1, B_1) \cdots x(A_i, B_i) y(I_1) \cdots y(I_k)$. The required result follows from Proposition 12.3.2.2.

**Theorem 12.3.3.2** *Standard monomials in $p(A, B), u(I)$ form a basis for the $K$-vector space $S$.*

*Proof.* We already have established the linear independence of standard monomials (cf. Proposition 12.2.2.2). The generation by standard monomials follows by considering the surjective map $\varphi_D : S(D) \to S$ and using the generation of $S(D)$ by standard monomials (cf. Theorem 12.3.3.1).

**Theorem 12.3.3.3** *The map $\varphi_D : S(D) \to S$ is an isomorphism of $K$-algebras.*

*Proof.* Under $\varphi_D$, the standard monomials in $S(D)$ are mapped bijectively onto the standard monomials in $S$. The result follows from Theorems 12.3.3.1 and 12.3.3.2.

**Theorem 12.3.3.4** ***A Presentation for** $S$:*

*(i) The $K$-algebra $S$ is generated by $\{p(A,B), u(I), (A,B) \in H_p, I \in H_u\}$.*
*(ii) The ideal of relations among the generators $\{p(A,B), u(I)\}$ is generated by relations (i)-(iii) of Proposition 12.3.1.1.*
*(iii) (**Standard Monomial Basis**) Standard monomials in $\{p(A,B), u(I), (A,B) \in H_p, I \in H_u\}$ form a basis for $S$.*

*Proof.* The result follows from Theorem 12.3.3.3 (and the definition of $S(D)$)

**Remark 12.3.3.5.** We shall see (cf. Theorem 12.5.0.7 below) that the inclusion $S \hookrightarrow R^G$ is in fact an equality.

## 12.4 Cohen-Macaulayness of $S$

In this section, we prove the Cohen-Macaulayness of $S$ in the following steps:

- (i) We prove that $S$ is a doset algebra with straightening law over a doset $D$ contained in $P \times P$, for a certain partially ordered set $P$.
- (ii) $P$ is a wonderful poset.
- (iii) Conclude the Cohen-Macaulayness of $S$ using [21].

Let $P$ be a partially ordered set. Recall

**Definition 12.4.0.1** *(cf. [21]) A doset of $P$ is a subset $D$ of $P \times P$ such that*

*(i) The diagonal $\Delta(P) \subset D$.*
*(ii) If $(a,b) \in D$, then $a \geq b$.*
*(iii) Let $a \geq b \geq c$ in $P$.*
*(a) Let $(a,b), (b,c)$ be in $D$. Then $(a,c) \in D$.*
*(b) Let $(a,c) \in D$. Then $(a,b), (b,c)$ are in $D$.*

**Remark 12.4.0.2.** (i) We shall refer to an element $(\alpha, \beta)$ of $D$ as an *admissible pair*; $(\alpha, \alpha)$ will be called a *trivial admissible pair*.
(ii) As a consequence of (iii) above, we have the following:
If $a \geq b \geq c \geq d$ in $P$ are such that $(a,d)$ is in $D$, then $(b,c) \in D$.

**Definition 12.4.0.3** *(cf. [21]) A doset algebra with straightening law over the doset $D$ is a $K$-algebra $E$ with a set of algebra generators $\{x(A,B), (A,B) \in D\}$ such that* *(i)* *$E$ is graded with $E_0 = K$, and $E_r$ equals the $K$-span of monomials of degree $r$ in $x(A, B)$'s.*

*(ii)* *"Standard monomials" $x(A_1, B_1) \cdots x(A_r, B_r)$ (i.e., $A_1 \geq B_1 \geq A_2 \geq B_2 \cdots A_r \geq B_r$) is a $K$-basis for $E_r$.*

*(iii)* *Given a non-standard monomial $F := x(A_1, B_1) \cdots x(A_r, B_r)$, in the straightening relation*

$$(*) \qquad F = \sum a_i F_i$$

*expressing $F$ as a sum of standard monomials, writing $F_i = x(A_{i1}, B_{i1}) \cdots x(A_{ir}, B_{ir})$ (a standard monomial) we have the following:*

*(a) For every permutation $\sigma$ of $\{A_1, B_1, A_2, B_2, \cdots, A_r, B_r\}$, we have, $\{A_{i1}, B_{i1}, A_{i2}, B_{i2}, \cdots, A_{ir}, B_{ir}\}$ is lexicographically greater than or equal to $\{\sigma(A_1), \sigma(B_1), \sigma(A_2), \sigma(B_2), \cdots, \sigma(A_r), \sigma(B_r)\}$.*

*(b) If there exists a permutation $\tau$ of $\{A_1, B_1, A_2, B_2, \cdots, A_r, B_r\}$ such that $\tau(A_1) \geq \tau(B_1) \geq \tau(A_2) \geq \tau(B_2) \geq \cdots \geq \tau(A_r) \geq \tau(B_r)$, then the monomial $x(\tau(A_1), \tau(B_1)) \cdots x(\tau(A_r), \tau(B_r))$ occurs on the R.H.S. of (*) with coefficient $\pm 1$. (We shall refer to this situation as "$\{A_1, B_1, A_2, B_2, \cdots, A_r, B_r\}$ is totally ordered up to a reshuffle")*
*(Note that in (iii)(b), we have (in view of (iii) in Definition 12.4.0.1)) that $(\tau(A_i), \tau(B_i))$'s are admissible pairs.*

**The Discrete Doset Algebra $K\{D\}$:** The discrete doset algebra $K\{D\}$ is the doset algebra with straightening relations given as follows: Let $F := x(A_1, B_1) \cdots x(A_r, B_r)$ be a non-standard monomial. Then

$$F = \begin{cases} x(\tau(A_1), \tau(B_1)) \cdots x(\tau(A_r), \tau(B_r)), & \text{if } \tau(A_1) \geq \tau(B_1) \geq \cdots \geq \tau(A_r) \geq \tau(B_r) \\ 0, & \text{otherwise} \end{cases}$$

for some $\tau$ as in (iii),(b) above (if such a $\tau$ exists).

**Remark 12.4.0.4.** The conditions in (iii) are equivalent to the corresponding conditions for $r = 2$.

Let $K\{P\}$ be *the discrete algebra* over $P$, namely, *the Stanley-Reisner algebra*, defined as the quotient of the polynomial algebra $K[x_\alpha, \alpha \in P]$ by the ideal generated by $\{x_\alpha x_\beta, \ \alpha, \beta \in P \text{ non-comparable}\}$.

Recall

**Proposition 12.4.0.5** *(cf. [21], Theorem 3.5) Let $E$ be a doset algebra with straightening law over a doset $D$ inside $P \times P$.*

*(i)* *There exists a sequence $E = B_1, B_2, \cdots, B_r = K\{D\}$ of doset algebras with straightening law over the doset $D$ such that there exists a flat family $\mathcal{B}_j$, $1 \leq j \leq r-1$ (over Spec $K[t]$) with generic fiber Spec $B_j$, and special fiber Spec $B_{j+1}$.*

*(ii)* *$K\{D\}$ is Cohen-Macaulay if $K\{P\}$ is.*

### 12.4.1 A doset algebra structure for $R(D)$

We first define

**The Partially Ordered Set $P$:** Let

$$P := \cup_{r=1}^{n} I(r, m) \cup \{\mathbf{1}\}$$

Extend the partial order on $\cup_{r=1}^{n} I(r, m)$ to $P$ by declaring $\mathbf{1}$ to be the largest element.

Let $D$ be as in §12.3, namely,

$$D = H_p \cup H_u \cup \{\mathbf{1}\}$$

**Lemma 12.4.1.1** *$D$ is a doset inside $P \times P$.*

*Proof.* Clearly, $D \subset P \times P$, and contains $\Delta(P)$. Note that $H_u$ (as a subset of $P \times P$) is identified with the diagonal $\Delta(H_u)$ (inside $P \times P$); similarly, $\mathbf{1}$ is identified with $(\mathbf{1}, \mathbf{1})$. Also note that the non-trivial admissible pairs in $D$ are among the $(A, B)$'s $((A, B) \in H_p)$. The remaining conditions in Definition 12.4.0.1 hold in view of the results in §12.1.6 and the results of [62, 69].

**Proposition 12.4.1.2** *$R(D)$ is a doset algebra with straightening laws over the doset $D$.*

*Proof.* Condition (1) in Definition 12.4.0.3 follows from the definition of the $K$-algebra $R(D)$; note that $R(D)$ has algebra generators given by $\{x(A, B), (A, B) \in H_p\}, y(I), I \in H_u, x(\mathbf{1})$, indexed by $D$. Note that the generators $\{y(I), I \in H_u, x(\mathbf{1})\}$ are indexed by trivial admissible pairs, and the generators indexed by non-trivial admissible pairs are among $\{x(A, B), (A, B) \in H_p\}$. Condition (2) in Definition 12.4.0.3 follows from Theorem 12.3.2.3.

**Verification of Condition (3):** As in the proof of Theorem 4.1 of [21], in view of Proposition 12.3.1.1, it suffices to verify condition (3) in Definition 12.4.0.3 for a degree two non-standard monomial $F$ (cf. Remark 12.4.0.4). We divide the verification into the following cases:

**Case 1:** Let $F = y(I)y(J)$. In this case, the situation of 3(b) (in Definition 12.4.0.3) does not exist. The condition 3(a) follows from Proposition 12.3.1.1,(1).

**Case 2:** Let $F = x(A, B)y(I)$. In this case again, the situation of 3(b) (in Definition 12.4.0.3) does not exist (since $B \not\geq I$, and $I$ can not be $>$ $A$ or $B$ - note that no element of $H_u$ is greater than any element of $H_p$). Condition 3(a) follows from Proposition 12.3.1.1,(3).

**Case 3:** Let $F = x(A_1, A_2)x(B_1, B_2)$. Condition 3(a) follows from Proposition 12.3.1.1,(2). Condition 3(b) follows from Theorem 12.1.6.1, and DeConcini-Lakshmibai [21], Theorem 4.1 (especially, its proof); here, we should remark that one first concludes such relations for $Y(\phi)$ ($Y(\phi)$ as in Theorem 12.1.6.1), then for $S$, and hence for $R(D)$ (note that $S$ being $R(D)_{(x(\mathbf{1}))}$ such relations in $S$ imply similar relations in $R(D)$, since $x(\mathbf{1})^l$ is the largest monomial in any given degree $l$).

**Corollary 12.4.1.3** *$R(D)$ is Cohen-Macaulay.*

*Proof.* This follows from Propositions 12.4.0.5, 12.4.1.2. Note that $P$ is a wonderful poset (in the sense of DeConcini-Eisenbud-Procesi [20]), in fact, a distributive lattice, and hence the discrete algebra $K\{P\}$ is Cohen-Macaulay (cf. [20], Theorem 8.1).

The above Corollary together with the fact that $S$ is a homogeneous localization of $R(D)$ implies

**Theorem 12.4.1.4** *$S$ is Cohen-Macaulay.*

## 12.5 The equality $R^{SO_n(K)} = S$

We preserve the notation from the previous sections. In this section, we shall first prove that the morphism $q : Spec\, R^{SO_n(K)} \longrightarrow Spec\, S$ induced by the inclusion $S \subseteq R^{SO_n(K)}$, is finite, surjective, and birational. Then, we shall prove that $Spec\, S$ is normal, and deduce (using Zariski's Main Theorem) that $q$ is an isomorphism thus proving that the inclusion $S \subseteq R^{SO_n(K)}$ is an equality, i.e., $R^{SO_n(K)} = S$. As a consequence, we will obtain (in view of Theorem 12.4.1.4) that $R^{SO_n(K)}$ is Cohen-Macaulay.

**Notation:** In this section, for an integral domain $A, \kappa(A)$ will denote the quotient field of $A$.

**Lemma 12.5.0.1** *Let $\widetilde{A}$ be a finitely generated $K$-algebra ($K$ being an algebraically closed field of characteristic different from 2). Further, let $\widetilde{A}$ be a domain. Let $\gamma$ be an involutive $K$-algebra automorphism of $\widetilde{A}$. Let $A = \widetilde{A}^{\Gamma}$ where $\Gamma \cong \mathbb{Z}/2\mathbb{Z}$ is the group generated by $\gamma$. We have,*

*(i) The canonical map $p : Spec\, \widetilde{A} \longrightarrow Spec\, A$ induced by the inclusion $A \subset \widetilde{A}$ is a finite, surjective morphism.*
*(ii) $\kappa(\widetilde{A})$ is a quadratic extension of $\kappa(A)$.*

*Proof.* (i) It is easy to see that a finite set $\mathcal{B}$ of $A$-algebra generators of $\widetilde{A}$ can be chosen so that they are all eigenvectors of $\gamma$ corresponding to the eigenvalue $-1$. (This is because for any $f \in \widetilde{A}$, one has $f = 1/2(f + \gamma(f)) + 1/2(f - \gamma(f))$.)

Since $a := f.\gamma(f) \in A$ for any $f \in \widetilde{A}$, each generator satisfies the equation $x^2 + a = 0$ (over $A$). Also, since product of any two generators is an eigenvector corresponding to the eigenvalue 1, it follows that $\widetilde{A}$ is generated as an $A$-module by the set of algebra generators $\mathcal{B}$ together with 1. Therefore $\widetilde{A}$ is a finite module over $A$. Hence $p$ is a finite morphism; surjectivity of $p$ follows from the fact (cf. Mumford [86], Ch I.7, Proposition 3) that a finite morphism $f : Spec\, B \longrightarrow Spec\, A$ of affine varieties is surjective if and only if $f^* : A \longrightarrow B$ is injective. Assertion (1) follows.

(ii) From the discussion in (i), we obtain that the $A$-algebra $\widetilde{A}$ has algebra generators, say, $\{f_1, \cdots, f_r\}$ such that

- $f_i^2 \in A, 1 \leq i \leq r$
- $f_i f_j \in A, \forall i, j$

Hence we obtain that for every $\alpha \in \widetilde{A}$, $\alpha^2 \in A$. This implies that every $s \in \kappa(\widetilde{A})$ satisfies a quadratic equation $x^2 + a$ over $\kappa(A)$; further, $\kappa(\widetilde{A})$ is a (finite) separable extension of $\kappa(A)$ (since char $K \neq 2$). Assertion (2) follows from this.

**Corollary 12.5.0.2** *Let $A = R^{O_n(K)}$, $\widetilde{A} = R^{SO_n(K)}$. We have,*

*(i) The canonical map $p : Spec\,\widetilde{A} \longrightarrow Spec\,A$ induced by the inclusion $A \subset \widetilde{A}$ is a finite, surjective morphism.*
*(ii) $\kappa(\widetilde{A})$ is a quadratic extension of $\kappa(A)$.*

*Proof.* Taking $\gamma \in O_n(K)$ to be an order 2 element which projects onto the generator of $O_n(K))/SO_n(K) =: \Gamma(= \mathbb{Z}/2\mathbb{Z})$, we have, $\gamma$ defines an involutive $K$-algebra automorphism of $\widetilde{A}$, with $\widetilde{A}^{\Gamma} = A$. The result follows from Lemma 12.5.0.1.

**Proposition 12.5.0.3** *The morphism $q : Spec\,R^{SO_n(K)} \longrightarrow Spec\,S$ induced by the inclusion $S \subseteq R^{SO_n(K)}$ is surjective, finite, and birational.*

*Proof.* Denote $\widetilde{Y} = Spec(R^{SO_n(K)})$, $Y = Spec(R^{O_n(K)})$, $Z = Spec\,S$. Consider the inclusions

$$(*) \qquad R^{O_n(K)} \subset S \subseteq R^{SO_n(K)}$$

Note that the first inclusion is a strict inclusion, since, $S = R^{O_n(K)}[u(I), I \in I(n,m)]$ (and $u(I) \notin R^{O_n(K)}$ for $I \in I(n,m)$). This induces the following commutative diagram

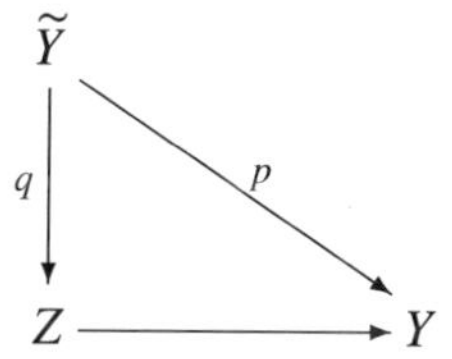


The finiteness of $q$ follows from the finiteness of $p$ (cf. Corollary 12.5.0.2, (1)); this together with the inclusion $q^* : S \hookrightarrow R^{SO_n(K)}$ implies the surjectivity of $q$ (cf. Mumford [86], ch I.7, Proposition 3).
Now the inclusions given by (*) give the following inclusions of the respective quotient fields:

$$\kappa(R^{O_n(K)}) \subset \kappa(S) \subseteq \kappa(R^{SO_n(K)})$$

(with the first inclusion being a strict inclusion). This together with the fact that $\kappa(R^{SO_n(K)})$ is a quadratic extension of $\kappa(R^{O_n(K)})$ (cf. Corollary 12.5.0.2, (2)) implies that $\kappa(S)$ is a quadratic extension of $\kappa(R^{O_n(K)})$ as well. Hence we obtain that $\kappa(S) = \kappa(R^{SO_n(K)})$ proving the birationality of $q$.

Finally, to verify the hypotheses in Zariski's Main Theorem for the morphism $q$, it remains to show that $S$ is a normal domain. Again in view of Theorem 12.4.1.4 and Serre's criterion for normality (namely, *Spec A* is normal if and only if $A$ has $S_2$ and $R_1$), to prove the normality of $S$, it suffices to show that *Spec S* is regular in codimension 1 (i.e., singular locus of *Spec S* has codimension at least 2). Towards proving this, we first obtain a criterion for the branch locus of a finite morphism to have codimension at least 2.

Let $\pi : Z \longrightarrow Y$ be a finite morphism where $Y$ is a reduced and irreducible affine scheme over an algebraically closed field $K$ of arbitrary characteristic. Then, there exists an open subscheme $Y_1 \subset Y$ (namely, the set of unramified points for $\pi$) such that $res(\pi) : \pi^{-1}(Y_1) \rightarrow Y_1$ is an étale morphism (see Mumford [86], Ch.III.10; here, $res(\pi)$ denotes the restriction of $\pi$.).

Let $Y, Z$ be affine, say $Y = Spec\, A$, $Z = Spec\, B$, where $A, B$ are finitely generated $K$-algebras. Further, let $A$ be a domain, and $B$ an integral extension of $A$ which is finitely generated as an $A$ module (so that $\pi : Z \longrightarrow Y$ is a finite morphism).

Suppose that $\kappa(B)$ is a finite separable extension of $\kappa(A)$. Then, there exists a $s \in B$ such that $\kappa(B) = \kappa(A)[s]$. Indeed if $\frac{b}{b'}$ is a primitive element for the extension $\kappa(B)$ of $\kappa(A)$ with $b, b' \in B$, then there exists a $\lambda \in \kappa(A)$ such that $\kappa(B) = \kappa(A)[b + \lambda b']$. To see this, observe that since there are only finitely many intermediate fields between $\kappa(A)$ and $\kappa(B)$ (while $\kappa(A)$ is infinite), we have that not all $\kappa(A)[b + \lambda b'], \lambda \in \kappa(A)$ can be distinct. Hence, for some $\lambda, \mu \in \kappa(A), \lambda \neq \mu$, $b + \mu b'$ is in $\kappa(A)[b + \lambda b']$. From this it follows that $\frac{b}{b'}$ is in $\kappa(A)[b + \lambda b']$.

**Lemma 12.5.0.4** *With the above notation, let $Y = Spec\, A$ and let $Z = Spec\, B$. Consider the finite morphism $\pi : Z \longrightarrow Y$ induced by the inclusion $A \subset B$. As above, let $b$ be a primitive element such that $\kappa(B) = \kappa(A)[b]$. Let $a \in A$ be such that $A[1/a][b] = B[1/a]$. Further, let the discriminant of $b$ be invertible in $A[1/a]$. Then $res(\pi) : Spec\, B[1/a] \rightarrow Spec\, A[1/a]$ is étale.*
*(Here, $res(\pi)$ denotes the restriction of $\pi$.)*

*Proof.* Let $U = Spec\,(A[1/a]), V = \pi^{-1}(U) = Spec\,(B[1/a])$. Since $B[1/a] = A[1/a][b]$ (by Hypothesis), we obtain that $B[1/a]$ is a free $A[1/a]$-module with basis $\{1, b, \cdots, b^{N-1}\}$ (here, $N = [\kappa(B) : \kappa(A)]$). Further, by Hypothesis, discriminant $(b)$ is invertible in $A[1/a]$. Hence we obtain that $res(\pi) : Spec\,(B[1/a]) \rightarrow Spec\,(A[1/a])$ is étale.

**Proposition 12.5.0.5** *The variety Spec S is regular in codimension* 1.

*Proof.* Taking $A = R^{O_n(K)}$, $B = S$, as seen in the proof of Proposition 12.5.0.3, we have that $\kappa(B)$ is a quadratic extension of $\kappa(A)$. Further, the generators $u_I$ (of the $A$-algebra $B = A[u(I), I \in I(n, m)]$) satisfy the relation $u(I)^2 - p(I, I) = 0$ over $A$ (note that $A = K[p(I, J), I, J \in I(r, m), r \leq n]$ (cf. Theorem 12.1.7.2)). Hence, $\kappa(B)$ is also separable over $\kappa(A)$, since $char(K) \neq 2$. Now we take $Y = Spec\, A$, $Z = Spec\, B$, and $\pi : Z \rightarrow Y$ the morphism induced by the inclusion $R^{O_n(K)} \subseteq S$, in Lemma 12.5.0.4. Fixing a particular $I \in I(n, m)$, we have that the relation $u(I)u(J) = p(I, J)$ implies that $u(J) = p(I, J)u(I)/p(I, I)$ and hence $u(J) \in S[1/p(I, I)]$ for all $J$. In particular $S[1/p(I, I)]$ is a free $R^{O_n(K)}[1/p(I, I)]$-module with basis $\{1, u(I)\}$. Also the discriminant $\delta(u_I)$ is seen to be $4p(I, I)$ which is invertible in $R^{O_n(K)}[1/p(I, I)]$. Hence, taking $a = p(I, I), b = u(I)$, the hypothesis of Lemma 12.5.0.4 are satisfied. Hence

$$res(\pi) : \bigcup_{I \in I(n,m)} S[1/p(I, I)] \rightarrow \bigcup_{I \in I(n,m)} R^{O_n(K)}[1/p(I, I)]$$

is étale. Let $\mathfrak{a}$ be the ideal generated by $\{p(I, I), I \in I(n, m)\}$. Note that in view of the relations

$$u(I)u(J) = p(I, J), I, J \in I(n, m)$$

(cf. Lemma 12.2.0.1), we have

$$p(I, J)^2 = p(I, I)p(J, J)$$

Thus we have, $p(I, J) \in \sqrt{\mathfrak{a}}, \forall I, J \in I(n, m)$. Let $Y_0 = V(\sqrt{\mathfrak{a}})$. Then $Y_0$ is simply $D_n(Sym\ M_m)$. Also, $Y$ being $D_{n+1}(Sym\ M_m)$ (cf. Theorem 12.1.7.2), we have $Y_0$ is the singular locus of $Y$ (cf. Theorem 12.1.6.3). Now the branch locus $Y_b$ of $\pi : Z \longrightarrow Y$ is contained in $Y_0$. Hence the fact that $codim_Y Y_0 \geq 2$ (since $Y$ is normal) implies that

$$(*) \qquad codim_Y Y_b \geq 2$$

Denote

$$Y_e := \{\texttt{unramified points for } \pi\}$$

We have, $res(\pi) : \pi^{-1}(Y_e) \to Y_e$ is étale, and $Y_b = Y \setminus Y_e$. Denote

$$Z_e := \pi^{-1}(Y_e),\ Z_b := \pi^{-1}(Y_b)$$

Denoting by $Z_{\texttt{sing}}$ the singular locus of $Z$, we have (cf. (*))

$$(**) \qquad codim_Z(Z_{\texttt{sing}} \cap Z_b)\ (\geq codim_Z Z_b) \geq 2$$

On the other hand, $res(\pi) : Z_e \to Y_e$ being étale, we have,

$$\pi^{-1}(Y_{\texttt{sing}} \cap Y_e) = Z_{\texttt{sing}} \cap Z_e$$

Hence we obtain,

$$(***) \quad codim_Z(Z_{\texttt{sing}} \cap Z_e) = codim_Y(Y_{\texttt{sing}} \cap Y_e) \geq codim_Y(Y_{\texttt{sing}}) \geq 2$$

(**) and (***) imply that $codim_Z(Z_{\texttt{sing}}) \geq 2$, and the result follows.

The above Proposition together with Theorem 12.4.1.4 implies the following

**Proposition 12.5.0.6** *Spec S is normal.*

The result follows from Serre's criterion for normality: *Spec A* is normal if and only if $A$ has $S_2$ and $R_1$.

**Theorem 12.5.0.7** *The inclusion* $S \subseteq R^{SO_n(K)}$ *is an equality, i.e.,* $R^{SO_n(K)} = S$.

*Proof.* Propositions 12.5.0.3, 12.5.0.6 imply (in view of Zariski Main Theorem (cf. Mumford [86], ch III.9)) that the morphism $q : Spec\ R^{SO_n(K)} \longrightarrow Spec\ S$ is in fact an isomorphism. The result follows from this.

Combining the above Theorem with Theorems 12.3.3.4, 12.4.1.4 we obtain the following theorems.

**Theorem 12.5.0.8** ***A presentation for*** $R^G$:

*(i)* ***(First fundamental Theorem)*** *The K-algebra* $R^G(= S)$ *is generated by* $\{p(A, B), u(I), (A, B) \in H_p, I \in H_u\}$.
*(ii)* ***(Second fundamental Theorem)*** *The ideal of relations among the generators* $\{p(A, B), u(I)\}$ *is generated by relations (1)–(3) of Proposition 12.3.1.1.*
*(iii)* ***(Standard Monomial Basis)*** *Standard monomials in* $\{p(A, B), u(I), (A, B) \in H_p, I \in H_u\}$ *form a basis for* $R^G$.

**Theorem 12.5.0.9** $R^{SO_n(K)}$ *is Cohen-Macaulay.*

The result follows from the above Theorem and Theorem 12.4.1.4.

## 12.6 Application to moduli problem

In this section, using Theorem 12.5.0.9, we give a characteristic-free proof of the Cohen-Macaulayness of the moduli space $\mathcal{M}_2$ of equivalence classes of semi-stable rank 2, degree 0 vector bundles on a smooth projective curve of genus $> 2$ by relating it to $K[X]^{SO_3(K)}$. It is known (cf. [95],§7, Theorem 3) that $\mathcal{M}_2$ is smooth when the genus is 2.

Assume for the moment that the characteristic of the field $K$ is zero. Consider the moduli space $\mathcal{M}_n$ of equivalence classes of semi-stable, rank $n$, degree 0 vector bundles on a smooth projective curve $C$ of genus $m > 2$. Let $V$ be the trivial vector bundle on $C$. The automorphism group of $V$ ($\cong H^0(C, \text{Aut } V)$) can be identified with $GL_n(K)$. The tangent space at $V$ of the versal deformation space (cf. [108]) of $V$ is $H^1(C, \text{End } V)$ and hence it can be identified with $m$ copies of the space $M_n(K)$ of $n \times n$ matrices, $V$ being identified with the "origin". The canonical action of Aut $V$ on this tangent space gets identified with the diagonal adjoint action of $GL_n(K)$ on $m$ copies of $M_n(K)$. Now the moduli space $\mathcal{M}_n$ is a GIT quotient $Z /\!\!/ H$, for a suitable $Z$ and $H$ a projective linear group of suitable rank. The versal deformation space of $V$ gets embedded in $Z$ and by "Luna slice" type of arguments (cf. [87], Appendix to chapter 1, D), the analytic local ring of $\mathcal{M}_n$ at $V$ gets identified with the analytic local ring of $H^1(C, \text{End } V) /\!\!/ PGL_n(K)$ at the "origin".

Suppose now that $n = 2$. If $V$ is a stable vector bundle, then the point in $\mathcal{M}_2$ that it defines is smooth. If $V$ is not stable, then the point in $\mathcal{M}_2$ that it defines can be represented by $L_1 \oplus L_2$, where $L_1$ and $L_2$ are line bundles of degree zero. If $L_1 \cong L_2$, then $End\ (V) \cong M_2(K)$, and the considerations are the same as for the case when $V$ is trivial. If $L_1$ is not isomorphic to $L_2$, then $Aut\ V$ is the torus $T := \mathbb{G}_m \times \mathbb{G}_m$, and the analytic local ring of $\mathcal{M}_2$ at $V$ is isomorphic to the analytic local ring of $H^1(C, End\ V) /\!\!/ T$ at the origin (in fact the action of $T$ is trivial). This is certainly Cohen-Macaulay.

Although we are only interested in the rank 2 case, let us remark that considerations of the previous paragraph hold when the rank of $V$ is arbitrary, and the analytical local ring of $\mathcal{M}_n$ is isomorphic to $Z /\!\!/ G$, where $Z = \prod Z_i$, $G = \prod G_i$, where $Z_i$ is $g$ copies of $M_{n_i}(K)$ and $G_i = PGL_{n_i}(K)$.

Although we have assumed that the characteristic of the field to be zero, the above considerations remain valid in positive characteristic but with certain restrictions; for example, if $n = 2$, then the characteristic should not be 2 or 3.

Thus, in order to prove that $\mathcal{M}_2$ is Cohen-Macaulay, it suffices to show that the point corresponding to the trivial bundle is so. Further, since the analytic local ring at the point corresponding to the trivial bundle gets identified with the completion of the local ring at the origin of $M_2(K)^{\oplus m} /\!\!/ PGL_2(K)$, it suffices to prove that $M_2(K)^{\oplus m} /\!\!/ PGL_2(K)$ is Cohen-Macaulay.

Let then $n = 2$, $M_2 := M_{2,2}(K)$, $M_2^0 := sl_2(K)$. Let

$$Z = M_2 \oplus \cdots \oplus M_2 (g \text{ copies}),\ Z_0 = M_2^0 \oplus \cdots \oplus M_2^0 (g \text{ copies})$$

Let $A = K[Z]$, $A_0 = K[Z_0]$. Let $G = SL_2(K)$. Consider the diagonal action of $G$ on $Z$, $Z_0$ induced by the action of $G$ on $M_2$, $M_2^0$ by conjugation.

From the above discussion, we have that the completion of the local ring at the point in $\mathcal{M}$ corresponding to the trivial rank 2 vector bundle is isomorphic to the completion of $A^G$ at the point which is the image of the origin (in $Z(= \mathbf{A}^{4g})$) under $Z \to \operatorname{Spec} A^G$. On the other hand, we have, $A^G = A_0^G[x_1, \cdots, x_g]$, $x_i$'s being indeterminates (since, $M_2 \cong M_2^0 \oplus K$); further, we have that the adjoint action of $SL_2(K)$ on $sl_2(k)$ is isomorphic to the natural representation of $SO_3(K)$ on $K^3$ (note that the Lie algebras $sl_2(K)$ and $so_3(K)$ are isomorphic). Hence the ring $A_0^G$ gets identified with $K[X]^{SO_3(K)}$, $X$ being $V \oplus \cdots \oplus V(g \text{ copies})$, $V = K^3$. Hence we obtain

**Theorem 12.6.0.1** *(i) The GIT quotients $Z /\!\!/ G$, $Z_0 /\!\!/ G$ are Cohen-Macaulay.*
*(ii) The moduli space $\mathcal{M}$ is Cohen-Macaulay.*
*(iii) We have standard monomial bases for the co-ordinate rings of these spaces.*
*(iv) We have first & second fundamental theorems for these spaces, i.e., algebra generators, and generators for the ideal of relations among the generators.*

**Remark 12.6.0.2.** The results in Theorem 12.6.0.1 being characteristic-free, we may deduce from Theorem 12.6.0.1 that the moduli space $\mathcal{M}$ behaves well under specializations; for instance, if the curve $C$ is defined over $\mathbb{Z}$, and if $\mathcal{M}_{\mathbb{Z}}$ is the corresponding moduli space, then for any algebraically closed field $K$ of characteristic $\neq 2, 3$, the base change of $\mathcal{M}_{\mathbb{Z}}$ by $K$ gives the moduli space $\mathcal{M}_K$ over $K$.

**Remark 12.6.0.3.** This section is motivated by Mehta-Ramadas [80]. In loc.cit, the Cohen-Macaulayness for $Z /\!\!/ G$, $Z_0 /\!\!/ G$, $\mathcal{M}$ are deduced by proving the Frobenius-split properties for these spaces.

## 12.7 Results for the adjoint action of $SL_2(K)$

Consider $G = SL_2(K)$, $ch\ K \neq 2, 3$. Let

$$Z = \underbrace{sl_2(K) \oplus \cdots \oplus sl_2(K)}_{m \text{ copies}} = Spec\, R, \quad \texttt{say}$$

Using the results of §12.4, §12.5, we shall describe a "standard monomial basis" for $R^G$; the elements of this basis will be certain monomials in $tr\,(A_iA_j)$'s, and $tr\,(A_iA_jA_k)$'s.

**Identification of $sl_2(K)$ and $so_3(K)$:** Let

$$X = \begin{pmatrix} 0 & 1 \\ 0 & 0 \end{pmatrix}, H = \begin{pmatrix} 1 & 0 \\ 0 & -1 \end{pmatrix}, Y = \begin{pmatrix} 0 & 0 \\ 1 & 0 \end{pmatrix}$$

be the Chevalley basis of $sl_2(K)$.

Let $\langle , \rangle$ be a symmetric non-degenerate bilinear form on $V = K^3$. Taking the matrix of the form $\langle , \rangle$ to be $J =$ `anti-diagonal`$(1, 1, 1)$, we have,

$$SO_3(K) = \{A \in SL_3(K) \mid J^{-1}({}^tA)^{-1}J = A\}$$
$$so_3(K) = \{A \in sl_3(K) \mid J^{-1}({}^tA)J = -A\}$$

The Chevalley basis of $so_3(K)$ is given by

$$X' = \begin{pmatrix} 0 & \sqrt{2} & 0 \\ 0 & 0 & -\sqrt{2} \\ 0 & 0 & 0 \end{pmatrix}, H' = \begin{pmatrix} 2 & 0 & 0 \\ 0 & 0 & 0 \\ 0 & 0 & -2 \end{pmatrix}, Y' = \begin{pmatrix} 0 & 0 & 0 \\ \sqrt{2} & 0 & 0 \\ 0 & -\sqrt{2} & 0 \end{pmatrix}$$

The map $X \mapsto X', H \mapsto H', Y \mapsto Y'$ gives an isomorphism $sl_2(K) \cong so_3(K)$ of Lie algebras. Further, the map

$$\theta : sl_2(K) \to K^3, \begin{pmatrix} a & b \\ c & -a \end{pmatrix} \mapsto \left(\frac{b}{\sqrt{2}}, -a, \frac{c}{\sqrt{2}}\right)$$

identifies the adjoint action of $SL_2(K)$ on $sl_2(K)$ with the natural action of $SO_3(K)$ on $K^3$; and the induced map

$$\theta_m : Z \longrightarrow \underbrace{V \oplus \cdots \oplus V}_{m \text{ copies}},\ V = K^3$$

identifies the diagonal (adjoint) action of $SL_2(K)$ on $Z$ with the diagonal (adjoint) action of $SO_3(K)$ on $\underbrace{V \oplus \cdots \oplus V}$.

**Lemma 12.7.0.1** *Let $z \in Z$, say $z = (A_1, \cdots, A_m)$. Let $\theta_m(z) = (u_1, \cdots, u_m)$. We have*

*(i) $tr\,(A_iA_j) = 2\langle u_i, u_j\rangle$.*
*(ii) Let $I \in I(3, m)$, say, $I = (i, j, k)$. Then $tr\,(A_iA_jA_k) = 2u(I)$.*

*(here, $u(I)$ is as in §12.2)*

The proof is an easy verification.

**Notation:** In the sequel we shall denote

$$U(i, j) := tr\,(A_iA_j), i \geq j,\ U(i, j, k) := tr\,(A_iA_jA_k), (i, j, k) \in I(3, m)$$

**Remark 12.7.0.2.** Note that if $i, j, k$ are not distinct, then $tr\,(A_iA_jA_k) = 0$; for, say, $i = j$, then $A_i$ being a $2 \times 2$ traceless matrix, we have (in view of Cayley-Hamilton theorem), $A_i^2 = -|A_i|$, and hence $tr\,(A_i^2A_k) = -|A_i|tr(A_k) = 0$ (since $A_k \in sl_2(K)$).

In view of the identification given by $\theta_m$, we obtain (cf. Theorems 12.5.0.7, 12.3.3.2) a "standard monomial basis" for $S := R^{SL_2(K)}$. By Theorem 12.3.3.2, monomials

$$p(\underline{\alpha})p(\underline{A}, \underline{B})u(\underline{I})$$

where

$$\begin{aligned} p(\underline{\alpha}) &:= p(\alpha_1, \alpha_2) \cdots p(\alpha_{2r-1}, \alpha_{2r}), \text{ for some } r,\ \alpha_i \in [1, m] \\ p(\underline{A}, \underline{B}) &:= p(A_1, B_1) \cdots p(A_s, B_s), \text{ for some } s,\ A_i, B_i \in I(2, m) \\ u(\underline{I}) &:= u(I_1) \cdots u(I_t), \text{ for some } t,\ I_\ell \in I(3, m) \\ \alpha_1 \geq \cdots \geq \alpha_{2r} &\geq A_1 \geq B_1 \geq A_2 \geq \cdots \geq B_s \geq I_1 \geq \cdots \geq I_\ell \end{aligned}$$

give a basis for $R^{SO_3(K)}$ (here, $[1, m]$ denotes the set $\{1, \cdots, m\}$).
We shall refer to $p(\underline{\alpha})p(\underline{A}, \underline{B})u(\underline{I})$ as a *standard monomial of multi-degree* $(r, s, t)$. Denote

$$\mathcal{M}_{r,s,t} := \{\texttt{standard monomials of multi-degree } (r, s, t)\}$$

**Standard Monomials of Type I,II,III:** Let notation be as above. We shall refer to

$$\begin{gathered} p(\alpha_1, \alpha_2) \cdots p(\alpha_{2r-1}, \alpha_{2r}),\ \alpha_1 \geq \cdots \geq \alpha_{2r} \\ p(A_1, B_1) \cdots p(A_s, B_s),\ A_1 \geq B_1 \geq A_2 \geq \cdots \geq B_s \\ u(I_1) \cdots u(I_t),\ I_1 \geq \cdots \geq I_\ell \end{gathered}$$

as *standard monomials of Type I,II,III* (and of degree $r, s, t$) respectively.

We now define three types of standard monomials in $U(i, j)(= tr\,(A_iA_j)), i \geq j$, $U(i, j, k)(= tr\,(A_iA_jA_k)), (i, j, k) \in I(3, m)$ analogous to the above three types. Type I and Type III are direct extensions to traces:

**Definition 12.7.0.3** *A monomial of the form*

$$U(\alpha_1, \alpha_2) \cdots U(\alpha_{2r-1}, \alpha_{2r}),\ \alpha_1 \geq \cdots \geq \alpha_{2r}$$

*will be called a Type I standard monomial.*

**Definition 12.7.0.4** *A monomial of the form*

$$U(I_1) \cdots U(I_t),\ I_1 \geq \cdots \geq I_\ell$$

*will be called a Type III standard monomial.*

For defining Type II standard monomials, we first define a bijection between $\{p(A,B), A, B \in I(2,m), A \geq B\}$ and $\{U(j,i)U(l,k), j \geq i, l \geq k, i \not\geq l\}$. Note that given $U(j,i)U(l,k), j \geq i, l \geq k$, we may suppose that $j$ is the greatest among $\{i, j, k, l\}$ (if $l$ is the greatest, then we may write $U(j,i)U(l,k)$ as $U(l,k)U(j,i)$); then the latter set is simply

$$\{\texttt{all non-standard Type I, degree 2 monomials in the } U(a,b)\texttt{'s}\}$$

Let us denote the two sets by $\mathcal{A}, \mathcal{B}$ respectively. Let $p(A,B) \in \mathcal{A}$, say, $A = (a,b), b > a, B = (c,d), d > c$. Define $\omega : \mathcal{A} \to \mathcal{B}$ as follows:

$$\omega(p(A,B)) = \begin{cases} U(b,c)U(d,a), & \text{if } d \geq a \\ U(b,d)U(a,c), & \text{if } d < a \end{cases}$$

Given a standard monomial $p(A_1, B_1) \cdots p(A_s, B_s), A_i, B_i \in I(2,m)$, we shall define

$$\omega(p(A_1,B_1) \cdots p(A_s,B_s)) := \omega(p(A_1,B_1)) \cdots \omega(p(A_s,B_s))$$

**Definition 12.7.0.5** *A monomial*

$$U(j_1,i_1)U(l_1,k_1) \cdots U(j_s,i_s)U(l_s,k_s), \; j_t > i_t, l_t > k_t, \; 1 \leq t \leq s$$

*is called a Type II standard monomial if it equals $\omega(p(A_1,B_1) \cdots p(A_s,B_s))$ for some (Type II) standard monomial $p(A_1,B_1) \cdots p(A_s,B_s)$.*

**Remark 12.7.0.6.** Note in particular that for $s = 1$, the Type II standard monomials (in the $U(j,i)$'s) are precisely the non-standard Type I, degree 2 monomials in the $U(a,b)$'s.

Let us extend $\omega$ to $\mathcal{M}_{r,s,t}$, the definition of $\omega(\mathcal{F})$, for $\mathcal{F}$ a Type I or III standard monomial being obvious, namely,

$$\omega(p(\alpha_1,\alpha_2) \cdots p(\alpha_{2r-1},\alpha_{2r})) = U(\alpha_1,\alpha_2) \cdots U(\alpha_{2r-1},\alpha_{2r})$$
$$u(I_1) \cdots u(I_t) = U(I_1) \cdots U(I_t)$$

Define

$$\omega(p(\underline{\alpha})p(\underline{A},\underline{B})u(\underline{I})) = \omega(p(\underline{\alpha}))\omega(p(\underline{A},\underline{B}))\omega(u(\underline{I})), \; p(\underline{\alpha})p(\underline{A},\underline{B})u(\underline{I}) \in \mathcal{M}_{r,s,t}$$

**Definition 12.7.0.7** *A monomial in the $U(j,i)$'s, $j \geq i$ and $U(I)$'s, $I \in I(3,m)$ of the form*

$$\omega(p(\underline{\alpha})p(\underline{A},\underline{B})u(\underline{I})), \; p(\underline{\alpha})p(\underline{A},\underline{B})u(\underline{I}) \in \mathcal{M}_{r,s,t}$$

*will be called a standard monomial of type $(r,s,t)$.*

**Notation:** We shall denote

$$\mathcal{N}_{r,s,t} = \omega(\mathcal{M}_{r,s,t})$$

**Theorem 12.7.0.8** *Let* $Z = \underbrace{sl_2(K) \oplus \cdots \oplus sl_2(K)}_{m \text{ copies}} = Spec\, R$, *say, where,* $m > 3$. *Standard monomials (in the traces) of type* $(r, s, t)$ *where* $r, s, t$ *are positive integers form a basis for* $R^{SL_2(K)}$ *for the adjoint action of* $SL_2(K)$ *on* $Z$

*Proof.* Denote $S := R^{SL_2(K)}$. Write

$$S = \bigoplus_{(r,s,t)} S_{r,s,t}$$

where $r, s, t$ are positive integers, and $S_{r,s,t}$ is the $K$-span of $\mathcal{M}_{r,s,t}$. In fact, we have (in view of linear independence of standard monomials (cf. Theorem 12.3.3.2)) that $\mathcal{M}_{r,s,t}$ is a basis for $S_{r,s,t}$. Using the bijection $\omega$, we shall show that $\mathcal{N}_{r,s,t}$ is also a basis for $S_{r,s,t}$. Clearly this requires a proof only in the case $s \neq 0$. Let $N_{r,s,t} = \# \mathcal{M}_{r,s,t}$. The relations (cf. Lemma 12.7.0.1; note that $p(i, j) = \langle u_i, u_j \rangle$)

$$p(i, j) = \frac{1}{2} U(i, j)$$

$$p(A, B) = \frac{1}{4}(U(a, c)U(b, d) - U(b, c)U(a, d))$$

$$u(I) = \frac{1}{2} U(I)$$

give raise to the transition matrix, say, $M$. Then it is easy to see that for a suitable indexing of the elements of $\mathcal{M}_{r,s,t}$, and $\mathcal{N}_{r,s,t}$, the matrix $M$ takes the upper triangular form with the diagonal entries being non-zero. To be very precise, to $p(\underline{\alpha})p(\underline{A}, \underline{B})u(\underline{I}) \in \mathcal{M}_{r,s,t}$, we associate a $(4s + 2r + 3t)$-tuple $n(\underline{\alpha}, \underline{A}, \underline{B}, \underline{I})$ as follows. Let

$$p(\underline{\alpha}) = p(\alpha_1, \alpha_2) \cdots p(\alpha_{2r-1}, \alpha_{2r}),\ \alpha_1 \geq \cdots \geq \alpha_{2r}$$

$$p(\underline{A}, \underline{B}) = p(A_1, B_1) \cdots p(A_s, B_s),\ A_1 \geq B_1 \geq A_2 \geq \cdots \geq B_s$$

$$u(\underline{I}) = u(I_1) \cdots u(I_t),\ I_1 \geq \cdots \geq I_\ell$$

$$A_i = (a_{i1}, a_{i2}), a_{i1} < a_{i2},\ B_i = (b_{i1}, b_{i2}), b_{i1} < b_{i2},\ A_i \geq B_i\ 1 \leq i \leq s$$

Set

$$n(\underline{\alpha}, \underline{A}, \underline{B}, \underline{I}) = (a_{12}, a_{11}, b_{12}, b_{11}, a_{22}, a_{21} \cdots, b_{s2}, b_{s1}, \underline{\alpha}, \underline{I})$$

where $\underline{\alpha} = (\alpha_1, \cdots, \alpha_{2r})$, and $\underline{I} = (I_1, \cdots, I_t)$. We take an indexing on $\mathcal{M}_{r,s,t}$ induced by the lexicographic order on the $n(\underline{\alpha}, \underline{A}, \underline{B}, \underline{I})$'s, and take the induced indexing on $\mathcal{N}_{r,s,t}$ (via the bijection $\omega$). With respect to these indexings on $\mathcal{M}_{r,s,t}, \mathcal{N}_{r,s,t}$, it is easily seen that the transition matrix $M$ is upper triangular with the diagonal entries being non-zero. It follows that $\mathcal{N}_{r,s,t}$ is a basis for $S_{r,s,t}$.

**Remark 12.7.0.9.** Note that the standard monomial basis (in the traces) as given by Theorem 12.7.0.8 is characteristic-free. Also, note that Theorem 12.7.0.8 recovers the result of Procesi [100], Theorem 3.4(a), for the case of $SL_2(K)$, namely, $tr\,(A_iA_j), tr\,(A_iA_jA_k), i, j, k \in [1, m]$ generate $R^G$ as a $K$-algebra (in a characteristic-free way).

# 13

# Applications of standard monomial theory

In this chapter, we present some applications of SMT. One of the most important applications of SMT is the determination of the singular loci of Schubert varieties. We present results for Schubert varieties in the flag variety $SL_n(K)/B$. Using the standard monomial basis for Schubert varieties, Lakshmibai and Seshadri determined ([68]) the singular loci of Schubert varieties in the flag variety. It might be said that this work is the beginning of the study of singularities of Schubert varieties. After this work, over the past two decades, a lot of work has been done (by several authors) on the singularities of Schubert varieties. For a complete account, see [4]. After presenting the results on singularities of Schubert varieties in the flag variety, we discuss some important algebraic varieties - ladder determinantal varieties, quiver varieties, varieties of complexes - for which geometric properties such as normality, Cohen-Macaulayness etc., are concluded by relating them to Schubert varieties. We also present (as yet another application of SMT), the degeneration of Schubert varieties (in the Grassmannian) to toric varieties (cf. [33]). Here again, though many more results have been proved in recent times (cf. [12, 17]), we have presented the results for Schubert varieties in the Grassmannian because of the simplicity of the proof in this case!

## 13.1 Tangent space and smoothness

### 13.1.1 The Zariski tangent space

Let $x$ be a point on a variety $X$. Let $\mathfrak{m}_x$ be the maximal ideal of the local ring $\mathcal{O}_{X,x}$ with residue field $K(x)(=\mathcal{O}_{X,x}/\mathfrak{m}_x)$. Note that $K(x) = K$ (since $K$ is algebraically closed). The *Zariski tangent space* to $X$ at $x$ is defined as

$$\begin{aligned} T_x(X) &= \mathrm{Der}_K(\mathcal{O}_{X,x}, K(x)) \\ &= \{D : \mathcal{O}_{X,x} \to K(x),\ K\text{-linear such that } D(ab) = D(a)b + aD(b)\} \end{aligned}$$

(here $K(x)$ is regarded as an $\mathcal{O}_{X,x}$-module). It can be seen easily that $T_x(X)$ is canonically isomorphic to $\mathrm{Hom}_{K\text{-mod}}(\mathfrak{m}_x/\mathfrak{m}_x^2, K)$.

### 13.1.2 Smooth and non-smooth points

A point $x$ on a variety $X$ is said to be a *simple* or *smooth* or *nonsingular point of* $X$ if $\mathcal{O}_{X,x}$ is a regular local ring. A point $x$ which is not simple is called a *multiple* or *non-smooth* or *singular point* of $X$. The set $\operatorname{Sing} X = \{x \in X \mid x \text{ is a singular point}\}$ is called the *singular locus of* $X$. A variety $X$ is said to be *smooth* if $\operatorname{Sing} X = \emptyset$.

**Theorem 13.1.2.1** *Let* $x \in X$. *Then* $\dim_K T_x(X) \geq \dim \mathcal{O}_{X,x}$ *(*$\dim \mathcal{O}_{X,x}$ *is also denoted* $\dim_x X$*) with equality if and only if* $x$ *is a simple point of* $X$.

*Proof.* We have

$$\dim_K T_x(X) = \dim_K(\mathfrak{m}_x/\mathfrak{m}_x^2) \geq \dim \mathcal{O}_{X,x},$$

with equality if and only if $\mathcal{O}_{X,x}$ is regular. The result now follows.

### 13.1.3 The space $T(w, \tau)$

Let $G$ be a semisimple algebraic group.
Let $T, B, W, R, S$ etc., be as in Chapter 3. For $\alpha \in R$, let $X_\alpha$ be the element of the Chevalley basis for $\mathfrak{g}(= \operatorname{Lie} G)$, corresponding to $\alpha$. We follow [8] for denoting elements of $R$, $R^+$, $S$ etc.

For $\tau \leq w$, let $T(w, \tau)$ be the Zariski tangent space to $X(w)$ at $e_\tau$. Let $w_0$ be the element of largest length in $W$. Now the tangent space to $G$ at $e_{id}$ is $\mathfrak{g}$, and hence the tangent space to $G/B$ at $e_{id}$ is $\oplus_{\beta \in R^+} \mathfrak{g}_{-\beta}$. For $\tau \in W$, identifying $G/B$ with $G/{}^\tau B$ (where ${}^\tau B = \tau B \tau^{-1}$) via the map $gB \mapsto (n_\tau g n_\tau^{-1})\, {}^\tau B$, $n_\tau$ being a fixed lift of $\tau$ in $N_G(T)$, we have, the tangent space to $G/B$ at $e_\tau$ is $\oplus_{\beta \in \tau(R^+)} \mathfrak{g}_{-\beta}$, i.e.,

$$T(w_0, \tau) = \oplus_{\beta \in \tau(R^+)} \mathfrak{g}_{-\beta}.$$

Set

$$N(w, \tau) = \{\beta \in \tau(R^+) \mid X_{-\beta} \in T(w, \tau)\}.$$

Since $T(w, \tau)$ is a $T$-stable subspace of $T(w_0, \tau)$, we have

$$T(w, \tau) = \text{ the span of } \{X_{-\beta},\ \beta \in N(w, \tau)\}.$$

### 13.1.4 A canonical affine neighborhood of a $T$-fixed point

Let $\tau \in W$. Let $U_\tau^-$ be the unipotent subgroup of $G$ generated by the root subgroups $U_{-\beta}$, $\beta \in \tau(R^+)$ (note that $U_\tau^-$ is the unipotent part of the Borel sub group ${}^\tau B^-$, opposite to ${}^\tau B$ $(= \tau B \tau^{-1})$). We have

$$U_{-\beta} \cong \mathbb{G}_a,\ U_\tau^- \cong \Pi_{\beta \in \tau(R^+)} U_{-\beta}.$$

Now, $U_\tau^- e_\tau$ gets identified with $\mathbb{A}^N$, where $N = \# R^+$ via the above identification and we shall denote the induced coordinate system on $U_\tau^- e_\tau$ by $\{x_{-\beta},\ \beta \in \tau(R^+)\}$. In the sequel, we shall denote $U_\tau^- e_\tau$ by $\mathcal{O}_\tau^-$. Thus we obtain that $\mathcal{O}_\tau^-$ is an affine neighborhood of $e_\tau$ in $G/B$.

### 13.1.5 The affine variety $Y(w, \tau)$

For $w \in W,\ w \geq \tau$, let us denote $Y(w, \tau) := \mathcal{O}_\tau^- \cap X(w)$. It is a nonempty affine open subvariety of $X(w)$, and a closed subvariety of the affine space $\mathcal{O}_\tau^-$.

### 13.1.6 Equations defining $Y(w, \tau)$ in $O_\tau^-$

Let $w \in W$. For $1 \leq i \leq l(= \text{rank}G)$, fix a basis $D_i$ for the kernel of the surjective map $H^0(G/B, L_i) \to H^0(X(w), L_i)$ (given by restriction), $L_i$ being the ample generator of $\text{Pic}(G/P_i)$. We have (Chapter 3, §3.3.3) that the ideal sheaf of $X(w)$ in $G/B$ is generated by $\{D_i,\ 1 \leq i \leq l\}$. Let $\tau \in W$ be such that $\tau \leq w$. Let $Y(w, \tau)$ be as in §13.1.5, $I(w, \tau)$ the ideal defining $Y(w, \tau)$ as a closed subvariety of $\mathcal{O}_\tau^-$. We have

$$I(w, \tau) \text{ is generated by } \{f|_{\mathcal{O}_\tau^-},\ f \in D_i,\ 1 \leq i \leq l\}.$$

For example, take $G = SL(4)$. The ideal of $Y((3412), id)$ in $\mathcal{O}_{Id}^-$ is generated by $\{p_4, p_{234}\}$, and the ideal of $Y((2413), id)$ is generated by $\{p_3, p_4, p_{34}, p_{134}, p_{234}\}$.

### 13.1.7 Jacobian criterion for smoothness

Let $Y$ be an affine variety in $\mathbb{A}^n$, and let $I(Y)$ be the ideal defining $Y$ in $\mathbb{A}^n$. Let $I(Y)$ be generated by $\{f_1, f_2, \ldots, f_r\}$. Let $J$ be the Jacobian matrix $(\frac{\partial f_i}{\partial x_j})$. Then we have, $\text{rank}\, J_P \leq \text{codim}_{\mathbb{A}^n} Y$ with equality if and only if $P$ is a smooth point of $Y$ (here $J_P$ denotes $J$ evaluated at $P$).

**Example 1:** Consider the Schubert variety $X(3412)$ in $SL_4/B$. As stated in §13.1.6, the equations defining $Y((3412), id)$ in $\mathcal{O}_{id}^-$ are $\{p_4, p_{234}\}$, and the equations defining $Y((2413), Id)$ are $\{p_3, p_4, p_{34}, p_{134}, p_{234}\}$. Identifying $\mathcal{O}_{id}^-$ with the group of unipotent lower triangular matrices in $SL_4$, we have

$$\begin{aligned}
p_3 &= x_{31} \\
p_4 &= x_{41} \\
p_{34} &= x_{42}x_{31} - x_{41}x_{32} \\
p_{134} &= x_{43}x_{32} - x_{42} \\
p_{234} &= x_{21}(x_{43}x_{32} - x_{42}) - (x_{43}x_{31} - x_{41})
\end{aligned}$$

Hence, the equations defining $Y((3412), Id)$ are $\{x_{41},\ x_{21}(x_{43}x_{32} - x_{42}) - (x_{43}x_{31} - x_{41})\}$. The Jacobian matrix is

$$\begin{bmatrix} 0 & 0 & 0 & 1 & 0 & 0 \\ (x_{43}x_{32} - x_{42}) & -x_{43} & x_{21}x_{43} & 1 & -x_{21} & -x_{31} \end{bmatrix}.$$

Here, the columns are indexed by $x_{21}, x_{31}, x_{32}, x_{41}, x_{42}, x_{43}$. The rank of this matrix at $P$ =id (*i.e.* all $x_{ij} = 0$) is $1 < 2 = 6 - l(3412) = \text{codim}Y((3412), Id)$. Therefore, $X(3412)$ is singular at $e_{id}$.

**Example 2:** Consider the Schubert variety $X(2413)$ in $SL_4/B$; the equations defining $Y((2413), Id)$ are $\{x_{31},\ x_{41},\ x_{31}x_{42}-x_{41}x_{32}, x_{42}-x_{43}x_{32},\ x_{21}(x_{43}x_{32}-x_{42})-(x_{43}x_{31}-x_{41})\}$. The Jacobian matrix is

$$\begin{bmatrix} 0 & 1 & 0 & 0 & 0 & 0 \\ 0 & 0 & 0 & 1 & 0 & 0 \\ 0 & x_{42} & -x_{41} & -x_{32} & x_{31} & 0 \\ 0 & 0 & -x_{43} & 0 & 1 & -x_{32} \\ (x_{43}x_{32}-x_{42}) & -x_{43} & x_{21}x_{43} & 1 & -x_{21} & (x_{32}x_{21}-x_{31}) \end{bmatrix}.$$

This matrix is easily seen to have rank 3 when all variables are set to zero; $3 = 6 - l(2413) = \text{codim} Y((2413), Id)$ which implies $X(2413)$ smooth at $e_{id}$.

### 13.1.8 $T$-stable curves

For a root $\alpha > 0$, let $Z_\alpha$ denote the $SL(2)$-copy in $G$ corresponding to $\alpha$; note that $Z_\alpha$ is simply the subgroup of $G$ generated by $U_\alpha$ and $U_{-\alpha}$. Given $x \in W$, precisely one of $\{U_\alpha, U_{-\alpha}\}$ fixes the point $e_x$. Thus $Z_\alpha \cdot e_x$ is a $T$-stable curve in $G/B$ (note that $Z_\alpha \cdot e_x \cong \mathbf{P}^1$), and conversely any $T$-stable curve in $G/B$ is of this form (cf. [13]). Corresponding to each $\alpha > 0$, the number of $T$-stable curves in $G/B$ is $\frac{1}{2}(\#W)$, and thus the number of $T$-stable curves in $G/B$ is $\frac{1}{2}(\#W)l(w_0)$, $w_0$ being the element of largest length in $W$. Now a $T$-stable curve $Z_\alpha \cdot e_x$ is contained in a Schubert variety $X(w)$ if and only if $e_x, e_{s_\alpha x}$ are both in $X(w)$. Given $y \le w$, let $\mathcal{R}(y, w) = \{\alpha > 0 \mid s_\alpha y \le w\}$. Then there are precisely $r(y, w)$ $T$-stable curves in $X(w)$ passing through $e_y$, where $r(y, w) = \#\mathcal{R}(y, w)$.

**Deodhar's Inequality:** (see [13] for example) For $x \le y \le w$, $\#\{\alpha \in R^+ \mid x \le s_\alpha y \le w\} \ge l(w) - l(x)$.

In view of Deodhar's Inequality, we have $r(y, w) \ge l(w)$ (taking $x = id$ in Deodhar's Inequality). Thus there are at least $l(w)$ $T$-stable curves in $X(w)$ passing through $e_y$.

**Lemma 13.1.8.1** *Let $w, \tau \in W$, $w \ge \tau$. Let $\beta \in \tau(R^+)$. If $w \ge s_\beta \tau$, then $X_{-\beta} \in T(w, \tau)$.*

*Proof.* The hypothesis that $w \ge s_\beta \tau$ implies that the curve $Z_\beta \cdot e_\tau$ is contained in $X(w)$. Now the tangent space to $Z_\beta \cdot e_\tau$ at $e_\tau$ is the one-dimensional span of $X_{-\beta}$. The required result now follows.

We shall show in the following section that $w, \tau$ being as above, $T(w, \tau)$ is precisely the span of $\{X_{-\beta},\ \beta \in \tau(R^+) \mid w \ge s_\beta \tau\}$, $G$ being $SL_n(K)$.

## 13.2 Singularities of Schubert varieties in the flag variety

In this section, we shall describe the singular locus of a Schubert variety in the flag variety.

Let $G = SL_n(K)$, $T$ the maximal torus in $G$ consisting of diagonal matrices, and $B$ the Borel subgroup consisting of upper triangular matrices.

### 13.2.1 Ideal of $Y(w, \tau)$

Let $w, \tau \in W,\ w \geq \tau$. Let $Y(w,\tau)$ be as in §13.1.5, and $I(w,\tau)$ the ideal defining $Y(w,\tau)$ as a closed subvariety of $\mathcal{O}_\tau^-$. For $1 \leq d \leq l$, we have $H^0(G/B, L(\omega_d)) = (\wedge^d K^n)^*$, the linear dual of $\wedge^d K^n$. Further, the Plücker coordinates $p_\theta,\ \theta \in I_{d,n}$ form a basis of $H^0(G/B, L(\omega_d))$. The kernel of $H^0(G/B, L(\omega_d)) \to H^0(X(w), L(\omega_d))$ has a basis consisting of $\{p_\theta,\ \theta \in I_{d,n}\ |w^{(d)} \not\geq \theta\}$, where $w^{(d)}$ is the element of $I_{d,n}(= W^{P_d}$, $P_d$ being as in Chapter 4, §4.1.6) given by $\pi_d(X(w)) = X_{P_d}(w^{(d)})$, $\pi_d$ being the canonical projection $G/B \to G/P_d$. Hence we obtain the following Theorem:

**Theorem 13.2.1.1** *The ideal $I(w,\tau)$ is generated by*

$$\{p_\theta \,|_{\mathcal{O}_\tau^-}, \theta \in I_{d,n},\ 1 \leq d \leq n-1 \mid w^{(d)} \not\geq \theta\}.$$

Let $w, \tau$ be as above. We shall now give a criterion for $X(w)$ to be smooth at $e_\tau$. We first remark that

(1) For $\theta, \mu \in W^d (= W^{P_d})$, $1 \leq d \leq n-1$, $p_\theta(e_\mu) \neq 0 \iff \theta = \mu$, where, recall that for $\theta = (i_1 \cdots i_d) \in W^d\ (= I_{d,n})$, $e_\theta$ denotes the vector $e_{i_1} \wedge \cdots \wedge e_{i_d}$ in $\wedge^d K^n$, and $p_\theta$ denotes the Plücker coordinate associated to $\theta$.

(2) Let $X_\alpha$ be the element of the Chevalley basis of $\mathfrak{g}$, corresponding to $\alpha \in R$. If $X_\alpha p_\mu \neq 0$, $\mu \in W^d$, then $X_\alpha p_\mu = \pm p_{s_\alpha \mu}$, where $s_\alpha$ is the reflection corresponding to the root $\alpha$.

(3) For $\alpha \neq \beta$, if $X_\alpha p_\mu,\ X_\beta p_\mu$ are non-zero, then $X_\alpha p_\mu \neq X_\beta p_\mu$.

The first remark is obvious, since $\{p_\theta \mid \theta \in W^d\}$ is the basis of $(\wedge^d k^n)^* (= H^0(G/P_d, L(\omega_d)))$, dual to the basis $\{e_\phi,\ \phi \in I_{d,n}\}$ of $\wedge^d k^n$.

The second remark is a consequence of $SL_2$ theory, using the following facts

(a) $|\langle \chi, \alpha^* \rangle| = \left| \frac{2(\chi,\alpha)}{(\alpha,\alpha)} \right| = 0$ or $1$, $\chi$ being the weight of $p_\mu$.

(b) $p_\mu$ is the lowest weight vector for the Borel subgroup ${}^\mu B = \mu B \mu^{-1}$.

The third remark follows from weight considerations (note that if $X_\alpha p_\mu \neq 0$, then $X_\alpha p_\mu$ is a weight vector (for the $T$-action) of weight $\chi + \alpha$, $\chi$ being the weight of $p_\mu$).

**Theorem 13.2.1.2** *(cf. [68]) Let $\tau, w \in W,\ \tau \leq w$. Then*

$$\dim T(w,\tau) = \#\{\gamma \in \tau(R^+) \mid w \geq s_\gamma \tau\}.$$

*Proof.* By Theorem 13.2.1.1, we have, $I(w,\tau)$ is generated by $\{p_\theta \,|_{\mathcal{O}_\tau^-}, \theta \in I_{d,n}, 1 \leq d \leq n-1 \mid w^{(d)} \not\geq \theta\}$. Denoting the affine coordinates on $\mathcal{O}_\tau^-$ by $x_\beta,\ \beta \in \tau(R^-)$, we have that the evaluations of $\frac{\partial p_\theta}{\partial x_\beta}$ and $X_\beta p_\theta$ at $e_\tau$ coincide. Let $J_w$ denote the Jacobian matrix of $Y(w,\tau)$ (considered as a subvariety of the affine space $\mathcal{O}_\tau^-$). We shall index the rows of $J_w$ by $\{p_\theta \,|_{\mathcal{O}_\tau^-}, \theta \in I_{d,n},\ 1 \leq d \leq n-1 \mid w^{(d)} \not\geq \theta\}$

and the columns by $x_\beta,\ \beta \in \tau(R^-)$. Let $J_w(\tau)$ denote $J_w$ evaluated at $\tau$. Now in view of (1) & (2) above, the $(p_\theta, x_\beta)$-th entry in $J_w(\tau)$ is non-zero if and only if $X_\beta p_\theta = \pm p_\tau$. Hence in view of (3) above, we obtain that in each row of $J_w(\tau)$, there is at most one non-zero entry. Hence rank$J_w(\tau)$ = the number of non-zero columns of $J_w(\tau)$. Now, $X_\beta p_\theta = \pm p_\tau$ if and only if $\theta \equiv s_\beta\tau(\bmod W_{P_d})$. Thus the column of $J_w(\tau)$ indexed by $x_\beta$ is non-zero if and only if $w \not\geq s_\beta\tau$. Hence rank$J_w(\tau) = \#\{\gamma \in \tau(R^+) \mid w \not\geq s_\gamma\}$ and thus we obtain

$$\dim T(w,\tau) = \#\{\gamma \in \tau(R^+) \mid w \geq s_\gamma\tau\}.$$

**Theorem 13.2.1.3** *Let $w, \tau$ be as in Theorem 13.2.1.2. Then $\{X_\beta,\ \beta \in \tau(R^-) \mid w \geq s_\beta\tau\}$ is a basis for $T(w,\tau)$.*

*Proof.* Let $\beta \in \tau(R^-)$ be such that $w \geq s_\beta\tau$. We have (by Lemma 13.1.8.1), $X_\beta \in T(w,\tau)$, and hence $\dim T(w,\tau) \geq \#\{\gamma \in \tau(R^+) \mid w \geq s_\gamma\tau\}$. On the other hand, by Theorem 13.2.1.2, $\dim T(w,\tau) = \#\{\gamma \in \tau(R^+) \mid w \geq s_\gamma\tau\}$. The result follows from this.

**Corollary 13.2.1.4** *$X(w)$ is smooth at $e_\tau$ if and only if $l(w) = \#\{\alpha \in R^+ \mid w \geq \tau s_\alpha\}$.*

*Proof.* We have, $X(w)$ is smooth at $e_\tau$ if and only if $\dim T(w,\tau) = \dim X(w)$, and the result follows in view of Theorem 13.2.1.2 (note that if $\beta = \tau(\alpha)$, then $s_\beta\tau = \tau s_\alpha$).

**Corollary 13.2.1.5** *$X(w)$ is smooth if and only if $l(w) = \#\{\alpha \in R^+ \mid w \geq s_\alpha\}$.*

*Proof.* If Sing$X(w)$ (the singular locus of $X(w)$) is not empty, then it is a closed $B$-stable subset of $X(w)$, and hence is a union of Schubert subvarieties of $X(w)$. In particular, we obtain that Sing$X(w)$ is non-empty if and only if $e_{\text{id}} \in$ Sing$X(w)$. Thus $X(\tau)$ is smooth if and only if it is smooth at $e_{\text{id}}$, and the result follows from Theorem 13.2.1.2.

**Remark 13.2.1.6.** The proof of Theorem 13.2.1.3 uses the fact (and just this fact alone) that the extremal weight vectors in $H^0(G/P_d, L(\omega_d))$ form a basis for $H^0(G/P_d, L(\omega_d))$.

We make an abstraction of this fact and deduce the following

**Theorem 13.2.1.7** *Let $\omega$ be a minuscule fundamental weight with $P$ as the associated maximal parabolic subgroup. Let $\tau, w \in W^P, \tau \leq w$. Then the tangent space $T(w,\tau)$ has a basis given by*

$$\{X_\beta,\ \beta \in \tau(R^- \setminus R_P^-) \mid w \geq s_\beta\tau\}$$

The proof follows from the fact that $\omega$ being minuscule, the extremal weight vectors in $H^0(G/P, L(\omega))$ form a basis for $H^0(G/P, L(\omega))$ (cf. Chapter 8).

**Application to orthogonal Grassmannian:** Let $H = SL_{2n}(K), G = SO_{2n}(K)$. Let $Q_n, P_n$ be as in Chapter 7, §7.2. Let $\tau, w \in W^{P_n}, \tau \leq w$. Let $T_H(w,\tau)$ (resp. $T_G(w,\tau)$) denote the tangent space to $X_{Q_n}(w)$ (resp. $X_{P_n}(w)$) at $\tau Q_n$ (resp. $\tau P_n$), notation being as in Chapter 7, §7.2.5. Let $\sigma$ be as in Chapter 7.

**Theorem 13.2.1.8** *$T_H(w,\tau)$ is $\sigma$-stable. Further*

$$T_G(w,\tau) = T_H(w,\tau)^\sigma$$

The proof is immediate from Theorem 13.2.1.7 and the description of $R_G$ (Chapter 7, §7.1) (note that the fundamental weight $\omega_n$ (for $G$) is minuscule).

**Remark 13.2.1.9.** One has similar "root-system" description as in Theorem 13.2.1.3 for the singular locus of a Schubert variety for other classical groups (see [4, 57, 58, 65] for details).

We also would like to remark that even though one has now a "type-free" description of singular points on a Schubert variety (thanks to [9, 14, 54]), SMT-theoretic approach seems to be the only approach which gives an explicit determination of the tangent space at a singular point on a Schubert variety!

**Application to symplectic Grassmannian:** Using the "root-system" description for the tangent space $T(w,\tau)$ for the symplectic group case, we have results similar to Theorem 13.2.1.8. In fact more is true as given by the following

**Theorem 13.2.1.10** *(cf. [57]) Let $H = SL_{2n}(K)$, $G = Sp_{2n}(K)$. Let $B_H$, $B_G$ be the Borel subgroups of $H, G$ respectively as in Chapter 6. Let $\tau, w \in W_G, \tau \le w$. Let $T_H(w,\tau)$ (resp. $T_G(w,\tau)$) denote the tangent space to $X_{B_H}(w)$ (resp. $X_{B_G}(w)$) at $\tau B_H$ (resp. $\tau B_G$). Let $\sigma$ be as in Chapter 6. We have $T_H(w,\tau)$ is $\sigma$-stable. Further*

$$T_G(w,\tau) = T_H(w,\tau)^\sigma.$$

### 13.2.2 A criterion for smoothness

In this sub section, we give a criterion for smoothness of Schubert varieties in $SL_n(K)/B$, in terms of "pattern avoidance" (cf. [66]). We first observe that for $n = 4$, $X(3412)$, $X(4231)$ are the only two singular Schubert varieties. In [66], it is shown that the situation for a general $SL(n)$ is quite analogous to that of $SL(4)$ as given by the following Theorem:

**Theorem 13.2.2.1** *(cf. [66]) Let $\eta = (a_1 \ldots a_n) \in S_n$. Then $X(\eta)$ is singular if and only if there exist $i, j, k, l$, $1 \le i < j < k < l \le n$ such that either (1) or (2) below holds.*

*(1) $a_k < a_l < a_i < a_j$*

*(2) $a_l < a_j < a_k < a_i$.*

### 13.2.3 Components of the singular locus

In this subsection, we give an explicit description of the irreducible components of the singular locus of a Schubert variety in the flag variety. In $G = SL_4(K)$, we have that $Sing\, X(3412) = X(1324)$ and $Sing\, X(4231) = X(2143)$. It turns out that for a general $n$, the situation is "nothing more than this".

Let then $X(w) \subset G/B$ be singular ($G$ being $SL(n)$). Let $w = (a_1 \cdots a_n)$. Since $X(w)$ is singular, there exist $i, j, k, l$, $1 \le i < j < k < l \le n$ such that

$$\text{either } a_k < a_l < a_i < a_j \quad \text{or } a_l < a_j < a_k < a_i. \tag{13.1}$$

It is shown in [66] that in the former case, if $w'$ is obtained from $w$ by replacing $a_i, a_j, a_k, a_l$ respectively by $a_k, a_i, a_l, a_j$, then $e_{w'} \in \text{Sing}X(w)$; and in the latter case, if $w'$ is obtained from $w$ by replacing $a_i, a_j, a_k, a_l$ respectively by $a_j, a_l, a_i, a_k$, then $e_{w'} \in \text{Sing}X(w)$.

For $w \in W$, let $P_w$ (resp. $Q_w$) be the maximal element of the set of parabolic subgroups which leave $\overline{BwB}$ (in $G$) stable under multiplication on the left (resp. right).

**Definition 13.2.3.1** *Given parabolic subgroups $P$, $Q$, we say that $\overline{BwB}$ is $P$-$Q$ stable if $P \subset P_w$ and $Q \subset Q_w$.*

**Theorem 13.2.3.2** *Let $w = (a_1 \ldots a_n)$. Then*

$$S_{Q_w} = \{\epsilon_i - \epsilon_{i+1} | a_i > a_{i+1}\}$$
$$S_{P_w} = \{\epsilon_i - \epsilon_{i+1} | i + 1 \textit{ appears before } i \textit{ in } \{a_1, a_2, \ldots, a_n\}\}.$$

*Proof.* Let $\alpha = \epsilon_i - \epsilon_{i+1}$. Then $\alpha \in P_w$ (resp. $Q_w$) if and only if $w^{-1}(\alpha)$ (resp. $w(\alpha)$) is $< 0$. The result now follows from this (note that for $1 \le t \le n$, $w(t) = a_t$, $w^{-1}(t) = m$, where $m$ is given by $a_m = t$).

**The set $F_\eta$**

Let $\eta = (a_1 \ldots a_n) \in S_n$ (the symmetric group on $n$ letters). Let $E_\eta$ be the set of all $\tau' \le \eta$ such that either 1) or 2) below holds.

1) There exist $i, j, k, l$, $1 \le i < j < k < l \le n$, such that

(a) $a_k < a_l < a_i < a_j$

(b) if $\tau' = (b_1 \ldots b_n)$, then there exist $i', j', k', l'$, $1 \le i' < j' < k' < l' \le n$ such that $b_{i'} = a_k, b_{j'} = a_i, b_{k'} = a_l, b_{l'} = a_j$

(c) if $\tau$ (resp. $\eta'$) is the element obtained from $\eta$ (resp. $\tau'$) by replacing $a_i, a_j, a_k, a_l$ resp. by $a_k, a_i, a_l, a_j$ (respectively $b_{i'}, b_{j'}, b_{k'}, b_{l'}$ resp. by $b_{j'}, b_{l'}, b_{i'}, b_{k'}$), then $\tau' \ge \tau$ and $\eta' \le \eta$.

2) There exist $i, j, k, l$, $1 \le i < j < k < l \le n$, such that

(a) $a_l < a_j < a_k < a_i$

(b) if $\tau' = (b_1 \ldots b_n)$, then there exist $i', j', k', l'$, $1 \le i' < j' < k' < l' \le n$ such that $b_{i'} = a_j, b_{j'} = a_l, b_{k'} = a_i, b_{l'} = a_k$

(c) if $\tau$ (resp. $\eta'$) is the element obtained from $\eta$ (resp. $\tau'$) by replacing $a_i, a_j, a_k, a_l$ resp. by $a_j, a_l, a_i, a_k$ (respectively $b_{i'}, b_{j'}, b_{k'}, b_{l'}$ resp. by $b_{k'}, b_{i'}, b_{l'}, b_{j'}$), then $\tau' \ge \tau$ and $\eta' \le \eta$.

Let $F_\eta = \{\tau \in E_\eta \mid \overline{B\tau B} \text{ is } P_\eta\text{-}Q_\eta \text{ stable}\}$.

**Theorem 13.2.3.3** *For $\eta \in S_n$, the singular locus of $X(\eta)$ is equal to $\cup_\lambda X(\lambda)$, where $\lambda$ runs over the maximal (under the Bruhat order) elements of $F_\eta$.*

This was first conjectured by Lakshmibai-Sandhya (cf. [66]); later it was proved by several authors (cf. [5, 48, 78]).

## 13.3 Singular loci of Schubert varieties in the Grassmannian

In this section, we describe the components of Sing $X$, the singular locus of a Schubert variety $X$ in $G_{d,n}$. We also describe a recursive formula for the multiplicity at a singular point on $X$.

As seen in chapter 4, the Schubert varieties in $G_{d,n}$ are indexed by $W^{P_d} = \{(a_1, \ldots, a_d) \mid 1 \leq a_1 < \cdots < a_d \leq n\}$, $P_d$ being as in Chapter 4, §4.1.6. For our discussion, it is more convenient to use the language of partitions (or Young diagrams). To $(a_1, \ldots, a_d) \in W^{P_d}$ we associate the Young diagram $\mathbf{a} := (\mathbf{a_d} \geq \cdots \geq \mathbf{a_1})$, where $\mathbf{a_i} = a_{d-i+1} - (d - i + 1)$. For any partition $\mathbf{a} = (\mathbf{a_d} \geq \cdots \geq \mathbf{a_1})$, we will denote by $X_\mathbf{a}$ the Schubert variety corresponding to $(a_1, \ldots, a_d)$.

The dimension of $X_\mathbf{a}$ in this setup equals $|\mathbf{a}| := \mathbf{a_1} + \ldots \mathbf{a_d}$. It is clear that $\mathbf{a_i}$ cannot be larger than $n - d$.

Now we describe Sing$X_\mathbf{a}$, the singular locus of $X_\mathbf{a}$. To describe Sing $X_\mathbf{a}$, we use the "rectangle" notation. Let $\mathbf{a}$ be a partition. we write

$$\mathbf{a} = (\mathbf{p_1^{q_1}}, \ldots, \mathbf{p_r^{q_r}}) = (\underbrace{\mathbf{p_1}, \ldots, \mathbf{p_1}}_{\mathbf{q_1}}, \underbrace{\mathbf{p_2}, \ldots, \mathbf{p_2}}_{\mathbf{q_2}}, \cdots \underbrace{\mathbf{p_r}, \ldots, \mathbf{p_r}}_{\mathbf{q_r}}),$$

where $\mathbf{p_i}$ denotes the Young diagram consisting of $p_i$ boxes arranged in one row. With this notation, we say that $\mathbf{a}$ consists of $r$ rectangles: $\mathbf{p_1} \times \mathbf{q_1}, \ldots, \mathbf{p_r} \times \mathbf{q_r}$.

**Theorem 13.3.0.1** *(cf. [73]) Let* $\mathbf{a}$ *consist of r rectangles. Then*

*(1) Sing* $X_\mathbf{a}$ *has* $r - 1$ *components* $X_{\mathbf{a_1}}, \ldots, X_{\mathbf{a_{r-1}}}$, *where*

$$\mathbf{a_i} = (\mathbf{p_1^{q_1}}, \ldots, \mathbf{p_{i-1}^{q_{i-1}}}, \mathbf{p_i^{q_i-1}}, (\mathbf{p_{i+1}} - \mathbf{1})^{\mathbf{q_{i+1}+1}}, \mathbf{p_{i+2}^{q_{i+2}}}, \ldots, \mathbf{p_r^{q_r}}).$$

*(Note that* $\mathbf{a}/\mathbf{a}_i$, $1 \leq i \leq r - 1$ *are simply the hooks in the Young diagram* $\mathbf{a}$. *For example, if* $\mathbf{a} = (7, 5^4, 3^2, 1^3)$, *then* $\mathbf{a}_2 = (7, 5^3, 2^3, 1^3)$*)*

*(2) For a rectangular partition* $\mathbf{a}$, $X_\mathbf{a}$ *is nonsingular.*

**Remark 13.3.0.2.** Note that the Young diagram of a component of Sing $X_\mathbf{a}$ is obtained by deleting a hook from the Young diagram of $X_\mathbf{a}$. In particular, we have the following

**Theorem 13.3.0.3** *Sing* $D_t = D_{t-1}$.

(Here, $D_t$ is the determinantal variety in $M_{r,d}$ (cf. Chapter 5))

*Proof.* Set $n = r + d$. We have (cf. Chapter 5, Theorem 5.2.2.7) $D_t \cong Y_\phi$ where $\phi = (t, t+1, \cdots, d, n+2-t, n+3-t, \cdots, n)$, and $Y_\phi$ is the opposite cell in $X(\phi)$ (note that $\phi$ consists of the two blocks $[t, d]$, $[n+2-t, n]$ of consecutive integers; here, for $a, b \in \mathbb{N}$, $a < b$, $[a, b]$ denotes the set $\{a, a+1, \cdots, b\}$). The partition determined by $\phi$ is given by $\mathbf{a} = (\underbrace{n-d, \cdots, n-d}_{t-1 \text{ times}}, \underbrace{t-1, \cdots, t-1}_{d+1-t \text{ times}})$.

Now by Theorem 13.3.0.1,

Sing $X_{\mathbf{a}} = X_{\mathbf{b}}$, where $\mathbf{b}$ is the partition $\mathbf{b} = (\underbrace{n-d, \cdots, n-d}_{t-2 \text{ times}}, \underbrace{t-2, \cdots, t-2}_{d+2-t \text{ times}})$.

Now the $d$-tuple $\tau$ associated to $\mathbf{b}$ is given by $\tau = (t-1, t, \cdots, d, n+3-t, n+4-t, \cdots, n) = ([t-1, d], [n+3-t, n])$. Hence Sing$Y_{\phi} = Y_{\tau}$. Hence we obtain Sing$D_t = D_{t-1}$.

### 13.3.1 Multiplicity at a singular point

In this subsection, we give a recursive formula for the multiplicity at a singular point on a Schubert variety $X$ in the Grassmannian variety.

#### Multiplicity of a local ring

Let $A$ be a finitely generated $K$-algebra. Further let $A$ be local with $\mathfrak{m}$ as the unique maximal ideal. For $l \geq 0$, let $\psi_A(l) = \text{length}(A/\mathfrak{m}^l)\ (= \dim_k(A/\mathfrak{m}^l))$. $\psi_A$ is called the *Hilbert-Samuel* function of $A$. Recall (see [27] for example) the following:

**Theorem 13.3.1.1** *There exists a polynomial $P_A(x) \in \mathbb{Q}[x]$, called the Hilbert-Samuel polynomial of $A$ such that*

*(i) $\psi_A(l) = P_A(l)$, $l \gg 0$.*
*(ii) $\deg P_A(x) = \dim A$.*
*(iii) The leading coefficient of $P_A(x)$ is of the form $e_A/n!$, where $e_A \in \mathbb{Z}^+$ and $n = \dim A$.*

**Definition 13.3.1.2** *With notations as in Theorem 13.3.1.1, the number $e_A$ is called the* multiplicity *of $A$.*

**Definition 13.3.1.3** *For a point $P$ on an algebraic variety $X$,* the multiplicity of $X$ *at $P$ is defined to be the number $e_A$, where $A = \mathcal{O}_{X,P}$, the stalk at $P$, and is denoted $\mathrm{mult}_P X$.*

**Proposition 13.3.1.4** *Let $P$ be a point on an algebraic variety $X$. Then $P$ is a smooth point of $X$ if and only if $\mathrm{mult}_P X = 1$.*

For a proof, see [86].

Recall (see [27] for example) the following:

**Theorem 13.3.1.5** *Let $gr(A, \mathfrak{m}) = \bigoplus_{l \geq 0} \mathfrak{m}^l/\mathfrak{m}^{l+1}$. There exists a polynomial $Q_A(x) \in \mathbb{Q}[x]$, such that*

*(i) $Q_A(l) = \dim_k \mathfrak{m}^l/\mathfrak{m}^{l+1}$, $l \gg 0$.*
*(ii) $\deg Q_A(x) = \dim A - 1\ (= n-1)$.*
*(iii) The leading coefficient of $Q_A(x)$ is of the form $\frac{f_A}{(n-1)!}$, where $f_A \in \mathbb{Z}^+$.*

**Remark 13.3.1.6.** With notations as above, by considering the exact sequence

$$0 \to \mathfrak{m}^l/\mathfrak{m}^{l+1} \to A/\mathfrak{m}^{l+1} \to A/\mathfrak{m}^l \to 0,$$

we have for $l \gg 0$,

$$\begin{aligned} Q_A(l) &= P_A(l+1) - P_A(l) \\ &= [\frac{e_A}{n!}(l+1)^n + ()(l+1)^{n-1} + \cdots] - [\frac{e_A}{n!}(l)^n + ()(l)^{n-1} + \cdots] \\ &= \frac{e_A}{(n-1)!}(l)^{n-1} + \cdots . \end{aligned}$$

Hence we obtain

$$e_A = f_A.$$

**Example.** Let $R$ be the polynomial algebra $K[x_1, \cdots x_n]$, $\mathfrak{a}$ the maximal ideal generated by $\{x_1, \cdots x_n\}$, and $A = R_{\mathfrak{a}}$ (the localization at $\mathfrak{a}$). Let $\mathfrak{m} = \mathfrak{a}R_{\mathfrak{a}}$. We have, $\mathfrak{m}^l/\mathfrak{m}^{l+1}$ is the span of monomials of total degree $l$ and has dimension $\binom{l+n-1}{n-1}$. Hence $Q_A(l) = \binom{l+n-1}{n-1}$ and is a polynomial in $l$ of degree $n-1$; further, the leading coefficient of $Q_A(l)$ is equal to $\frac{1}{(n-1)!}$. Therefore $e_A = 1$.

### Multiplicity of a graded affine $K$-algebra

For this subsection we refer the reader to [37], [27]. Let $B$ be a graded, finitely generated $K$-algebra. The function $f_B(n) = \dim_k B_n, n \in \mathbb{Z}^+$ is called the *Hilbert function* of $B$, and $\sum_{n\in\mathbb{Z}^+} f_B(n)t^n$ is called the *Hilbert series* of $B$.

**Theorem 13.3.1.7** *[37] There exists a polynomial $P_B(x) \in \mathbb{Q}[x]$ of degree $\dim(X)$, where $X = Proj(B)$ such that $f_B(n) = P_B(n)$, for $n \gg 0$. Further, the leading coefficient of $P_B(n)$ is of the form $c_B/r!$, where $c_B \in \mathbb{N}$ and $r = deg P_B(x)$.*

### A formula for the multiplicity

In this subsection, we give a recursive formula for the multiplicity at a singular point on a Schubert variety $X$ in the Grassmannian.

Let $X = X(w)$ be a Schubert variety in the Grassmannian $G_{d,n}$ ($\cong SL_n(K)/P_d$, $P_d$ being as in Chapter 4, §4.1.6). Let us denote $P_d$ by just $P$. Given $x \in X$, we shall denote $\text{mult}_x X$ by just $m_x X$. For $x \in B \cdot e_\tau$, we have $m_x(X) = m_{e_\tau}(X)$. Thus it suffices to determine $m_{e_\tau}(X)$, $\tau \le w$. For $e_\tau \in X$, let us denote $m_{e_\tau}(X)$ by just $m_\tau(w)$. Consider a $\tau \in W^P$ such that $\tau \le w$.

**Evaluation of Plücker coordinates on $U_\tau^- e_\tau$**

I. Let us first consider the case $\tau = \text{id}$. We identify $U^- e_{\text{id}}$ with

$$\left\{ \begin{pmatrix} & \text{Id}_{d\times d} & \\ x_{d+1\,1} & \cdots & x_{d+1\,d} \\ \vdots & & \vdots \\ x_{n\,1} & \cdots & x_{n\,d} \end{pmatrix}, \quad x_{ij} \in k, \quad d+1 \le i \le n, 1 \le j \le d \right\}.$$

Let $A$ be the affine algebra of $U^-e_{\text{id}}$. Let us identify $A$ with the polynomial algebra $K[x_{-\beta}, \beta \in R^+ \setminus R_P^+]$. To be very precise, we have $R^+ \setminus R_P^+ = \{\epsilon_j - \epsilon_i, 1 \le j \le d, d+1 \le i \le n\}$; given $\beta \in R^+ \setminus R_P^+$, say $\beta = \epsilon_j - \epsilon_i$, we identify $x_{-\beta}$ with $x_{ij}$. Hence we obtain that the expression for $p_\theta|_{U^-e_{\text{id}}}$ in the local coordinates $x_{-\beta}$'s is homogeneous.

**Example 13.3.1.8** *Consider $G_{2,4}$. Then*

$$U^-e_{\text{id}} = \left\{ \begin{pmatrix} 1 & 0 \\ 0 & 1 \\ x_{31} & x_{32} \\ x_{41} & x_{42} \end{pmatrix}, \quad x_{ij} \in k \right\}.$$

*On $U^-e_{\text{id}}$, we have $p_{12} = 1$, $p_{13} = x_{32}$, $p_{14} = x_{42}$, $p_{23} = x_{31}$, $p_{24} = x_{41}$, $p_{34} = x_{31}x_{42} - x_{41}x_{32}$.*

Thus a Plücker coordinate is homogeneous in the local coordinates $x_{ij}$, $d+1 \le i \le n$, $1 \le j \le d$. This phenomenon does not hold for $G/P$, if $P$ is not maximal as seen in the following example.

**Example 13.3.1.9** *Consider $SL(4)/B$. We have*

$$U^-e_{\text{id}} = \left\{ \begin{pmatrix} 1 & 0 & 0 & 0 \\ x_{21} & 1 & 0 & 0 \\ x_{31} & x_{32} & 1 & 0 \\ x_{41} & x_{42} & x_{43} & 1 \end{pmatrix}, \quad x_{ij} \in k \right\}.$$

*Then on $U^-e_{\text{id}}$, $p_{23} = -x_{31} + x_{32}x_{21}$ is non-homogeneous!*

II. Returning to $G_{d,n}$, let now $\tau$ be any other element in $W^P$, say $\tau = (a_1, \dots, a_n)$. Then $U_\tau^- e_\tau$ consists of $\{N_{d,n}\}$, where $N_{d,n}$ is obtained from $\begin{pmatrix} \text{Id} \\ X \end{pmatrix}_{n \times d}$ (with notations as above) by permuting the rows by $\tau^{-1}$. (Note that $U_\tau^- e_\tau = \tau U^- e_{\text{id}}$.)

**Example 13.3.1.10** *Consider $G_{2,4}$, and let $\tau = (2314)$. Then $\tau^{-1} = (3124)$, and*

$$U_\tau^- e_\tau = \left\{ \begin{pmatrix} x_{31} & x_{32} \\ 1 & 0 \\ 0 & 1 \\ x_{41} & x_{42} \end{pmatrix}, \quad x_{ij} \in k \right\}.$$

*We have on $U_\tau^- e_\tau$, $p_{12} = -x_{32}$, $p_{13} = x_{31}$, $p_{14} = x_{31}x_{42} - x_{41}x_{32}$, $p_{23} = 1$, $p_{24} = x_{42}$, $p_{34} = -x_{41}$.*

As in the case $\tau = \text{id}$, we find that for $\theta \in W^P$, $p_\theta|_{U_\tau^- e_\tau}$ is homogeneous in local coordinates. In fact we have

**Proposition 13.3.1.11** *Let $\theta \in W^P$. We have a natural isomorphism*

$$K[x_{-\beta}, \beta \in R^+ \setminus R_P^+] \cong k[x_{-\tau(\beta)}, \beta \in R^+ \setminus R_P^+],$$

*given by*

$$p_\theta \mapsto p_{\tau\theta}$$

*(here we have denoted $p_\theta|_{U^-\text{id}}$, $p_{\tau\theta}|_{U_\tau^- e_\tau}$ by just $p_\theta$, $p_{\tau\theta}$ respectively).*

As a consequence, we have

**Corollary 13.3.1.12** *Let $\theta \in W^P$. Then the polynomial expression for $p_\theta|_{U_\tau^- e_\tau}$ in the local coordinates $\{x_{-\tau(\beta)}, \beta \in R^+ \setminus R_P^+\}$ is homogeneous.*

### The integer $d_\tau(\theta)$

Let $\theta \in W^P$. Let $f_\tau(\theta)$ be the polynomial expressing $p_\theta|_{U_\tau^- e_\tau}$ in terms of the local coordinates $\{x_{-\beta}, \beta \in \tau(R^+ \setminus R_P^+)\}$. We define $d_\tau(\theta)$ (also denoted by $\deg_\tau(\theta)$) by

$$d_\tau(\theta) := \deg f_\tau(\theta)$$

(note that $f_\tau(\theta)$ is homogeneous (cf. Corollary 13.3.1.12)). In fact, we have an explicit expression for $d_\tau(\theta)$, as follows:

**Proposition 13.3.1.13** *Let $\theta \in W^P$. Let $\tau = (a_1, \ldots, a_n)$, $\theta = (b_1, \ldots, b_n)$, $\tau, \theta$ being considered as permutations. Let $r = \#\{a_1, \ldots, a_d\} \cap \{b_1, \ldots, b_d\}$. Then $d_\tau(\theta) = d - r$.*

*Proof.* Let $s = d - r$, and

$$\{c_1, \ldots, c_s\} = \{a_1, \ldots, a_d\} \setminus \{a_1, \ldots, a_d\} \cap \{b_1, \ldots, b_d\}$$
$$\{f_1, \ldots, f_s\} = \{b_1, \ldots, b_d\} \setminus \{a_1, \ldots, a_d\} \cap \{b_1, \ldots, b_d\}.$$

For $1 \leq t \leq s$, let $\beta_t = \tau(\epsilon_i - \epsilon_j)(= \epsilon_{a_i} - \epsilon_{a_j})$, where $i$ and $j$ are given by $a_i = c_t$, $a_j = f_t$ (note that $i \leq d < j$, and hence $\beta_t \in \tau(R^+ \setminus R_P^+)$). For $\sigma = (\sigma_1, \ldots, \sigma_n) \in W^P$, let $q_\sigma = e_{\sigma_1} \wedge \ldots \wedge e_{\sigma_d}$, the extremal weight vector in $V_k(\omega_d)$ of weight $\sigma(\omega_d)$. For $\beta \in R$, say $\beta = \epsilon_i - \epsilon_j$, we have

$$X_\beta q_\tau = \begin{cases} \pm q_{s_\beta \tau}, \text{ if } j \in \{a_1, \ldots, a_d\}, i \notin \{a_1, \ldots, a_d\} \\ 0, \text{ otherwise} \end{cases}$$

(note that in the case when $j \in \{a_1, \ldots, a_d\}$ and $i \notin \{a_1, \ldots, a_d\}$, $(s_\beta \tau)^{(d)}$ is the $d$-tuple obtained from $(a_1, \ldots, a_d)$ by replacing $j$ by $i$). Setting $D' = X_{-\beta_1} \ldots X_{-\beta_s}$, we have $D' q_\tau = \pm q_\theta$. This implies

$$\langle X_{-\beta_1} \ldots X_{-\beta_s} q_\tau, p_\theta \rangle = \pm 1,$$

where $\langle , \rangle$ is the canonical $\mathfrak{g}$-invariant pairing on $\wedge^d V \times (\wedge^d V)^*$, $V = K^n$ (in view of the fact that the bases $\{p_\theta, \theta \in W^P\}$, $\{q_\theta, \theta \in W^P\}$ are dual to each other).

Now using the $\mathfrak{g}$-invariance of $\langle , \rangle$, we obtain $\langle q_\tau, X_{-\beta_s} \dots X_{-\beta_1} p_\theta \rangle = \pm 1$. Thus we obtain $X_{-\beta_s} \dots X_{-\beta_1} p_\theta = \pm p_\tau$, and hence the monomial $x_{-\beta_s} \dots x_{-\beta_1}$ appears in $f_\tau(\theta)$ (the polynomial expression for $p_\theta|_{U_\tau^- e_\tau}$ in terms of $x_{-\beta}$'s, $\beta \in \tau(R^+ \setminus R_P^+)$). The required result now follows.

**Theorem 13.3.1.14** *(cf. [73])* ***A recursive formula for*** $m_\tau(w)$***:*** *With notations as above, we have*

$$m_\tau(w) \deg_\tau w = \sum m_\tau(w'),$$

*where on the right we sum over all divisors $X(w')$ in $X(w)$ such that $e(\tau) \in X(w')$.*

For other formulas for the multiplicity at a singular point, we refer the reader to [4].

## 13.4 Results for Schubert varieties in a minuscule $G/P$

Let $G$ be semisimple and $P$ a maximal parabolic subgroup of minuscule type, i.e., the associated fundamental weight $\omega$ satisfies $(\omega, \beta^*) \leq 1$, for all $\beta \in R^+$; we refer to such a $G/P$ as a *minuscule* $G/P$. For example, any maximal parabolic in $SL_n(K)$ is minuscule, the corresponding $G/P$ being a Grassmannian. Other examples of maximal parabolics which are minuscule are $G = Sp_{2n}(K)$, $P = P_1$, $G = SO_{2n}(K)$, $P = P_{n-1}, P_n$, $G = SO_{2n+1}(K)$, $P = P_n$ (note that for $G = Sp_{2n}(K)$, the corresponding $G/P$ is a projective space, and for $G = SO_{2n}(K)$, $G = SO_{2n+1}(K)$, the corresponding $G/P$'s are orthogonal Grassmannians consisting of maximal totally isotropic subspaces in $K^N$, $N = 2n, 2n+1$ equipped with a non-degenerate symmetric bilinear form). In addition, for $G = E_6$, $P_1, P_6$ are minuscule maximal parabolic subgroups, while for $G = E_7$, $P_1$ is the only minuscule maximal parabolic subgroup. The above list in fact exhausts all minuscule maximal parabolic subgroups for all simple $G$ (note that there are no minuscule maximal parabolic subgroups for $G$ of type $E_8, F_4, G_2$). See the table in Figure A.1 in the Appendix.

For $\theta \in W$, and $P = P_d$ (the maximal parabolic subgroup with $S \setminus \{\alpha_d\}$ as the associated set of simple roots), we shall denote by $\theta^{(d)}$, the element in $W^{P_d}$ representing $\theta W_{P_d}$. For the rest of this section, we shall suppose that $P$ is minuscule. The geometry of a minuscule $G/P$ is very similar to that of the Grassmannian; for example, similar to the Grassmannian case, for a minuscule $G/P$, we have that the extremal weight vectors (namely, the Weyl group translates of a highest weight vector) in $H^0(G/P, L(\omega))$ form a $K$-basis for $H^0(G/P, L(\omega))$. We have results for Sing $X$ ($X$ being a Schubert variety in $G/P$) similar to those for a Schubert variety in the Grassmannian. Below we describe briefly these results. For details, we refer the readers to [73].

Let $P = P_d$ and $\tau, w \in W^P$, $w \geq \tau$. Set $R_P(w, \tau) = \{\beta \in \tau(R^+ \setminus R_P^+) \mid w \geq (s_\beta \tau)^{(d)}\}$.

**Theorem 13.4.0.1** *$T(w, \tau)$, the tangent space to $X_P(w)$ at $e_\tau$ is spanned by $\{X_{-\beta}, \beta \in R_P(w, \tau)\}$.*

The proof is similar to that of Theorem 13.2.1.3.

Let $U_\tau^-$ be the unipotent subgroup of $G$ generated by the root subgroups $U_{-\beta}$, $\beta \in \tau(R^+ \setminus R_P^+)$. We have

$$U_{-\beta} \simeq \mathbb{G}_a, \ U_\tau^- \simeq \Pi_{\beta \in \tau(R^+ \setminus R_P^+)} U_{-\beta}.$$

We shall denote the coordinate system on $U_\tau^- e_\tau$ induced by the above identification by $\{x_{-\beta}, \ \beta \in \tau(R^+ \setminus R_P^+)\}$. We shall denote by $A_\tau$ the polynomial algebra $K[x_{-\beta}, \ \beta \in \tau(R^+ \setminus R_P^+)]$. Let $A_{\tau,w} = A_\tau/\mathfrak{I}_P(w)$, where $\mathfrak{I}_P(w)$ is the ideal of elements of $A_\tau$ that vanish on $X_P(w) \cap U_\tau^- e_\tau$. Then $A_{\tau,w}$ is generated as an algebra by $\{x_{-\beta}, \ \beta \in R_P(w, \tau)\}$ (in view of Theorem 13.4.0.1).

### 13.4.1 Homogeneity of $\mathfrak{I}_P(w)$

Let $\omega$ be the fundamental weight associated to $P$. We shall denote the extremal weight vectors in $H^0(G/P, L_\omega)$ by $\{p_\tau, \ \tau \in W^P\}$. Recall (cf. Chapter 8, [111]) that $\{p_\tau, \ \tau \in W^P\}$ is a basis for $H^0(G/P, L_\omega)$. Given $y \in \mathfrak{g}$, we identify $y$ with the corresponding right invariant vector field $D_y$ on $G$. Considering $H^0(G/P, L_\omega)$ as a $\mathfrak{g}$-module, we have

$$D_y f = yf, \ f \in H^0(G/P, L_\omega).$$

Further, the evaluations of $\frac{\partial f}{\partial x_{-\beta}}$ and $X_{-\beta} f$ at $e_\tau$ coincide for $\beta \in \tau(R^+ - R_P^+)$ (recall that for a root $\gamma$, $X_\gamma$ denotes the element in the Chevalley basis of $\mathfrak{g}$ associated to $\gamma$). Take an indexing of $\tau(R^+ - R_P^+)$.

Given $f \in H^0(G/P, L_\omega)$, let

$$\mathcal{D}_f = \{D = X_{-\beta_s} \cdots X_{-\beta_1} \mid Df = a_D p_\tau, \ a_D \in k^*\}$$

(here, $X_{-\beta_s} \cdots X_{-\beta_1}$ is considered as an element of $U(\mathfrak{g})$, the universal enveloping algebra of $\mathfrak{g}$). If $\mathcal{D}_f \neq \emptyset$, then $f$ is a weight vector for the action of $T$, and in $A_\tau$, $f$ gets identified with $\sum_{D \in \mathcal{D}_f} a_D x_D$, where if $D = X_{-\beta_s} \cdots X_{-\beta_1}$, then $x_D$ is the monomial $x_{-\beta_s} \cdots x_{-\beta_1}$.

**Lemma 13.4.1.1** *(cf. [73]) Let $\tau, \theta \in W^P$. Suppose $D_1 = X_{-\beta_s} \cdots X_{-\beta_1}$, $D_2 = X_{-\beta'_t} \cdots X_{-\beta'_1}$ are such that $D_i p_\theta = a_{D_i} p_\tau$, $a_{D_i} \in k^*$, $i = 1, 2$. Then $s = t$, i.e., ord $D_1 =$ ord $D_2$.*

Hence we obtain that for $f = p_\theta$, if $\mathcal{D}_f \neq \emptyset$, then for $D_1, D_2 \in \mathcal{D}_f$, ord $D_1 =$ ord $D_2$; thus in $A_\tau$, $p_\theta$ gets identified with a homogeneous polynomial. Now $I_P(X(w))$, the ideal of $X_P(w)$ in $G/P$ is generated by $\{p_\theta \mid w \not\geq \theta\}$ (cf. [111]). Hence we obtain that $I_P(w)$ (= the ideal of elements of $A_\tau$ that vanish on $X_P(w) \cap U_\tau^- e_\tau$) is homogeneous. Thus we get

$$\mathrm{gr}(A_{\tau,w}, M_{\tau,w}) = A_{\tau,w}$$

where $M_{\tau,w}$ is the maximal ideal in $A_{\tau,w}$ that corresponds to the point $e_\tau$.

### 13.4.2 A basis for $(M_{\tau,w})^r/(M_{\tau,w})^{r+1}$

Let us fix $\tau, w \in W^P, w \geq \tau$. Let $\theta \in W^P$, $\theta \leq w$ be such that $\theta$ and $\tau$ are comparable i.e. either $\tau \geq \theta$ or $\theta \geq \tau$.

**Lemma 13.4.2.1** *(cf. [73]) Let $w, \tau, \theta$ be as above. Then there exists a $D = X_{-\beta_s} \cdots X_{-\beta_1}$, $\beta_i \in R_P(w, \tau)$, $1 \leq i \leq s$ such that $Dp_\theta = a_D p_\tau, a_D \in k^*$.*

Set

$$N_P(\theta) = \{D = X_{-\beta_s} \cdots X_{-\beta_1},\ \beta_i \in R_P(w, \tau),\ 1 \leq i \leq s \mid Dp_\theta = a_D p_\tau,\ a_D \in k^*\}.$$

Note that $N_P(\theta) \neq \emptyset$ (in view of the above Lemma).

**Definition 13.4.2.2** *With notations as above, define $d_\theta := \text{ord}\, D$, where $D \in N_P(\theta)$ (note that the right hand side is independent of $D$ (cf. Lemma 13.4.1.1), so the definition makes sense).*

**Remark 13.4.2.3.** If $\theta = w$, then we refer to $d_w$ as the *degree* of $X_P(w)$ at $e_\tau$ and denote it by $\deg_\tau w$.

Similar to Theorem 13.3.1.14, we have

**Theorem 13.4.2.4** *(cf. [73]) Denoting $\text{mult}_{e_\tau} X$ by $m_\tau(w)$, we have*

$$m_\tau(w)\, deg_\tau w = \sum m_\tau(w')$$

*where the sum on the right hand side runs over all Schubert divisors $X_P(w')$ in $X_P(w)$ such that $e_\tau \in X_P(w')$.*

## 13.5 Applications to other varieties

In this section, we introduce two classes of affine varieties - ladder determinantal varieties (cf. §13.5.1) and quiver varieties (cf. §13.5.3 ) - and we conclude (cf. [34], [60] ) that these varieties are normal and Cohen-Macaulay by identifying them with $Y_Q(w)$ for suitable Schubert varieties $X_Q(w)$ in suitable $SL(n)/Q$ (note that $Y_Q(w)$ is normal and Cohen-Macaulay, since $X_Q(w)$ is).

### 13.5.1 Ladder determinantal varieties

Let $X = (x_{ba})$, $1 \leq b \leq m$, $1 \leq a \leq n$ be a $m \times n$ matrix of indeterminates.

Given $1 \leq b_1 < \cdots < b_h < m$, $1 < a_1 < \cdots < a_h \leq n$, we consider the subset of $X$, defined by

$$L = \{x_{ba} \mid \text{ there exists } 1 \leq i \leq h \text{ such that } b_i \leq b \leq m, 1 \leq a \leq a_i\}.$$

We call $L$ an *one-sided ladder* in $X$, defined by the *outside corners* $\omega_i = x_{b_i a_i}$, $1 \leq i \leq h$. For simplicity of notation, we identify the variable $x_{ba}$ with just $(b, a)$.

Let $\mathbf{s} = (s_1, s_2 \dots, s_l) \in \mathbb{Z}_+^l$, $\mathbf{t} = (t_1, t_2 \dots, t_l) \in \mathbb{Z}_+^l$ such that

$$\begin{gathered} b_1 = s_1 < s_2 < \cdots < s_l \le m, \\ t_1 \ge t_2 \ge \cdots \ge t_l,\ 1 \le t_i \le \min\{m - s_i + 1, a_{i^*}\} \text{ for } 1 \le i \le l, \text{ and} \\ s_i - s_{i-1} > t_{i-1} - t_i \text{ for } 1 < i \le l, \end{gathered} \tag{L1}$$

where for $1 \le i \le l$, we let $i^*$ be the largest integer such that $b_{i^*} \le s_i$.

For $1 \le i \le l$, let

$$L_i = \{x_{ba} \in L \mid s_i \le b \le m\}.$$

Let $K[L]$ denote the polynomial ring $K[x_{ba} \mid x_{ba} \in L]$, and let $\mathbb{A}(L) = \mathbb{A}^{|L|}$ be the associated affine space. Let $I_{\mathbf{s},\mathbf{t}}(L)$ be the ideal in $k[L]$ generated by all the $t_i$-minors contained in $L_i$, $1 \le i \le l$, and $D_{\mathbf{s},\mathbf{t}}(L) \subset \mathbb{A}(L)$ the variety defined by the ideal $I_{\mathbf{s},\mathbf{t}}(L)$. We call $D_{\mathbf{s},\mathbf{t}}(L)$ a *ladder determinantal variety* (associated to an one-sided ladder).

Let $\Omega = \{\omega_1, \dots, \omega_h\}$. For each $1 < j \le l$, let

$$\Omega_j = \{\omega_i \mid 1 \le i \le h \text{ such that } s_{j-1} < b_i < s_j \text{ and } s_j - b_i \le t_{j-1} - t_j\}.$$

Let

$$\Omega' = \left(\Omega \setminus \bigcup_{j=2}^{l} \Omega_j\right) \bigcup_{\Omega_j \neq \emptyset} \{(s_j, a_{j^*})\}.$$

Let $L'$ be the one-sided ladder in $X$ defined by the set of outside corners $\Omega'$. Then it is easily seen that $D_{\mathbf{s},\mathbf{t}}(L) \simeq D_{\mathbf{s},\mathbf{t}}(L') \times \mathbb{A}^d$, where $d = |L| - |L'|$.

Let $\omega'_k = (b'_k, a'_k) \in \Omega'$, for some $k$, $1 \le k \le h'$, where $h' = |\Omega'|$. If $b'_k \notin \{s_1, \dots, s_l\}$, then $b'_k = b_i$ for some $i$, $1 \le i \le h$, and we define $s_{j^-} = b_i$, $t_{j^-} = t_{j-1}$, $s_{j^+} = s_j$, $t_{j^+} = t_j$, where $j$ is the unique integer such that $s_j < b_i < s_{j+1}$. Let $\mathbf{s}'$ (resp. $\mathbf{t}'$) be the sequence obtained from $\mathbf{s}$ (resp. $\mathbf{t}$) by replacing $s_j$ (resp. $t_j$) with $s_{j^-}$ and $s_{j^+}$ (resp. $t_{j^-}$ and $t_{j^+}$) for all $k$ such that $b'_k \notin \{s_1, \dots, s_l\}$, $j$ being the unique integer such that $s_{j-1} < b_i < s_j$, and $i$ being given by $b'_k = b_i$. Let $l' = |\mathbf{s}'|$. Then $\mathbf{s}'$ and $\mathbf{t}'$ satisfy (L1), and in addition we have $\{b'_1, \dots, b'_{h'}\} \subset \{s'_1, \dots, s'_{l'}\}$. It is easily seen that $D_{\mathbf{s},\mathbf{t}}(L') = D_{\mathbf{s}',\mathbf{t}'}(L')$, and hence

$$D_{\mathbf{s},\mathbf{t}}(L) \simeq D_{\mathbf{s}',\mathbf{t}'}(L') \times \mathbb{A}^d.$$

Therefore it is enough to study $D_{\mathbf{s},\mathbf{t}}(L)$ with $\mathbf{s}, \mathbf{t} \in \mathbb{Z}_+^l$ such that

$$\{s_1, \dots, s_l\} \supset \{b_1, \dots, b_h\}. \tag{L2}$$

Without loss of generality, we can also assume that

$$t_l \ge 2, \text{ and } t_{i-1} > t_i \text{ if } s_i \notin \{b_1, \dots, b_h\}, 1 < i \le l. \tag{L3}$$

For $1 \le i \le l$, let

$$L(i) = \{x_{ba} \mid s_i \le b \le m, 1 \le a \le a_{i^*}\}.$$

Note that the ideal $I_{\mathbf{s},\mathbf{t}}(L)$ is generated by the $t_i$-minors of $X$ contained in $L(i)$, $1 \leq i \leq l$.

The ladder determinantal varieties (associated to one-sided ladders) get related to Schubert varieties(cf. [34]). We describe below the main results of [34].

**The Varieties $Z$ and $X_Q(w)$**

Let $G = SL(n)$, $Q = P_{a_1} \cap \cdots \cap P_{a_h}$. Let $\mathcal{O}^-$ be the opposite big cell in $G/Q$. Let $H$ be the one-sided ladder defined by the outside corners $(a_i + 1, a_i)$, $1 \leq i \leq h$. Let $\mathbf{s}, \mathbf{t} \in \mathbb{Z}_+^l$ satisfy (L1), (L2), (L3) above. For each $1 \leq i \leq l$, let $L(i) = \{x_{ba} \mid s_j \leq b \leq n, 1 \leq a \leq a_{i^*}\}$. Let $Z$ be the variety in $\mathbb{A}(H) \simeq \mathcal{O}^-$ defined by the vanishing of the $t_i$-minors in $L(i)$, $1 \leq i \leq l$. Note that $Z \simeq D_{\mathbf{s},\mathbf{t}}(L) \times \mathbb{A}(H \setminus L) \simeq D_{\mathbf{s},\mathbf{t}}(L) \times \mathbb{A}^r$, where $r = \dim SL(n)/Q - |L|$.

We shall now define an element $w \in W_Q^{\min}$, such that the variety $Z$ identifies with the opposite cell in the Schubert variety $X_Q(w)$ in $G/Q$. We define $w \in W_Q^{\min}$ by specifying $w^{(a_i)} \in W^{a_i}$ $1 \leq i \leq h$, where $\pi_i(X(w)) = X(w^{(a_i)})$ under the projection $\pi_i : G/Q \to G/P_{a_i}$.

Define $w^{(a_i)}$, $1 \leq i \leq h$, inductively, as the (unique) maximal element in $W^{a_i}$ such that

(1) $w^{(a_i)}(a_i - t_j + 1) = s_j - 1$ for all $j \in \{1, \ldots, l\}$ such that $s_j \geq b_i$, and $t_j \neq t_{j-1}$ if $j > 1$.

(2) if $i > 1$, then $w^{(a_{i-1})} \subset w^{(a_i)}$.

Note that $w^{(a_i)}$, $1 \leq i \leq h$, is well defined in $W^i$, and $w$ is well defined as an element in $W_Q^{\min}$.

**Theorem 13.5.1.1** *(cf. [34]) The variety $Z$ $(= D_{\mathbf{s},\mathbf{t}}(L) \times \mathbb{A}^r)$ identifies with the opposite cell in $X_Q(w)$, i.e. $Z = X_Q(w) \cap \mathcal{O}^-$ (scheme theoretically).*

As a consequence of the above Theorem, we obtain (cf. [34])

**Theorem 13.5.1.2** *The variety $D_{\mathbf{s},\mathbf{t}}(L)$ is irreducible, normal, Cohen-Macaulay.*

### 13.5.2 The varieties $V_i$, $1 \leq i \leq l$

Let $V_i$, $1 \leq i \leq l$ be the subvariety of $D_{\mathbf{s},\mathbf{t}}(L)$ defined by the vanishing of all $(t_i - 1)$-minors in $L(i)$, where $L(i)$ is as in §13.5.1.

In [34] the singular locus of $D_{\mathbf{s},\mathbf{t}}(L)$ has also been determined, as described below.

**Theorem 13.5.2.1** *Sing $D_{\mathbf{s},\mathbf{t}}(L) = \cup_{i=1}^l V_i$.*

**Remark 13.5.2.2.** Ladder determinantal varieties were first introduced by Abhyankar (cf. [2]).

**Remark 13.5.2.3.** A similar identification as in Theorem 13.5.1.1 for the case $t_1 = \cdots = t_l$ has also been obtained by Mulay (cf. [85]).

### 13.5.3 Quiver varieties

Fulton [31] and Buch-Fulton [11] have recently given a theory of "universal degeneracy loci", characteristic classes associated to maps among vector bundles, in which the role of Schubert varieties is taken by certain degeneracy schemes. The underlying varieties of these schemes arise in the theory of quivers: they are the closures of orbits in the space of representations of the equi-oriented quiver $A_h$. Many other classical varieties also appear as quiver varieties, such as determinantal varieties and the variety of complexes (see [23], [40], [90]).

In [60], the quiver varieties (corresponding to the equi-oriented type A quiver) are shown to be normal and Cohen-Macaulay (in arbitrary characteristic) by identifying them with $Y_Q(w)$ for suitable Schubert varieties $X_Q(w)$ in suitable $SL_n(K)/Q$.

Fix an $h$-tuple of non-negative integers $\mathbf{n} = (n_1, \ldots, n_h)$ and a list of vector spaces $K$-vector spaces $V_1, \ldots, V_h$ with respective dimensions $n_1, \ldots, n_h$. Define $Z$, the *variety of quiver representations* (of dimension $\mathbf{n}$, of the equi-oriented quiver of type $A_h$) to be the affine space of all $(h-1)$-tuples of linear maps $(f_1, \ldots, f_{h-1})$ :

$$V_1 \overset{f_1}{\to} V_2 \overset{f_2}{\to} \cdots \overset{f_{h-2}}{\to} V_{h-1} \overset{f_{h-1}}{\to} V_h.$$

If we endow each $V_i$ with a basis, we get $V_i \cong K^{n_i}$ and

$$Z \cong M_{n_2,n_1} \times \cdots \times M_{n_h,n_{h-1}},$$

where $M_{l,m}$ denotes the affine space of matrices over $K$ with $l$ rows and $m$ columns. The group

$$G_{\mathbf{n}} := GL(n_1) \times \cdots \times GL(n_h)$$

acts on $Z$ by

$$(g_1, g_2, \cdots, g_h) \cdot (f_1, f_2, \cdots, f_{h-1}) = (g_2 f_1 g_1^{-1}, g_3 f_2 g_2^{-1}, \cdots, g_h f_{h-1} g_{h-1}^{-1})$$

Now, let $\mathbf{r} = (r_{ij})_{1 \le i \le j \le h}$ be an array of non-negative integers with $r_{ii} = n_i$, and define $r_{ij} = 0$ for any indices other than $1 \le i \le j \le h$. Define the set

$$Z^\circ(\mathbf{r}) = \{(f_1, \cdots, f_{h-1}) \in Z \mid \forall i < j, \ \mathrm{rank}(f_{j-1} \cdots f_i : V_i \to V_j) = r_{ij}\}.$$

(This set might be empty for a bad choice of $\mathbf{r}$.)

**Proposition 13.5.3.1** *(cf. [24]) The $G_{\mathbf{n}}$-orbits of $Z$ are exactly the sets $Z^\circ(\mathbf{r})$ for $\mathbf{r} = (r_{ij})$ with*

$$r_{ij} - r_{i,j+1} - r_{i-1,j} + r_{i-1,j+1} \ge 0, \quad \forall\, 1 \le i < j \le h.$$

**Definition 13.5.3.2** *We define the quiver variety as the algebraic set*

$$Z(\mathbf{r}) = \{(f_1, \cdots, f_{h-1}) \in Z \mid \forall i, j, \ \mathrm{rank}(f_{j-1} \cdots f_i : V_i \to V_j) \le r_{ij}\}.$$

**Remark 13.5.3.3.** The variety $Z(\mathbf{r})$ is simply the Zariski closure of $Z^\circ(\mathbf{r})$ (cf. [1], [60]).

Given $\mathbf{n} = (n_1, \cdots, n_h)$, for $1 \leq i \leq h$, let

$$a_i = n_1 + n_2 + \cdots + n_i, \qquad a_0 = 0, \qquad \text{and} \qquad n = n_1 + \cdots + n_h.$$

For positive integers $i \leq j$, we shall frequently use the notations

$$[i, j] = \{i, i+1, \ldots, j\}, \qquad [i] = [1, i], \qquad [0] = \{\}.$$

Let $K^n \cong V_1 \oplus \cdots \oplus V_h$ have basis $e_1, \ldots, e_n$ compatible with the $V_i$. Consider its general linear group $GL_n(K)$, the subgroup $B$ of upper-triangular matrices, and the parabolic subgroup $Q$ of block upper-triangular matrices

$$Q = \{(a_{ij}) \in GL(n) \mid a_{ij} = 0 \text{ whenever } j \leq a_k < i \text{ for some } k\}.$$

In this section, we look at $G/Q$ as the space of partial flags as follows: a *partial flag of type* $(a_1 < a_2 < \cdots < a_h = n)$ (or simply a *flag*) is a sequence of subspaces $U_\bullet = (U_1 \subset U_2 \subset \cdots \subset U_h = K^n)$ with $\dim U_i = a_i$. Let $E_i = V_1 \oplus \cdots \oplus V_i = \langle e_1, \ldots, e_{a_i} \rangle$, and $E'_i = V_{i+1} \oplus \cdots \oplus V_h = \langle e_{a_i+1}, \ldots, e_n \rangle$, so that $E_i \oplus E'_i = K^n$. The *flag variety* Fl is the set of all flags $U_\bullet$ as above. Fl has a transitive $GL_n(K)$-action induced from $K^n$, and $Q = \mathrm{Stab}_{GL_n(K)}(E_\bullet)$, so we have the identification $\mathrm{Fl} \cong GL_n(K)/Q,\ g{\cdot}E_\bullet \leftrightarrow gQ$. The *Schubert varieties* are the closures of $B$-orbits on Fl. Such orbits are usually indexed by certain permutations of $[n]$, but we prefer to use *flags of subsets* of $[n]$, of the form

$$\tau = (\tau_1 \subset \tau_2 \subset \cdots \subset \tau_h = [n]), \qquad \#\tau_i = a_i.$$

A permutation $w : [n] \to [n]$ corresponds to the subset-flag with

$$\tau_i = w[a_i] = \{w(1), w(2), \ldots, w(a_i)\}.$$

This gives a one-to-one correspondence between cosets of the symmetric group $W = \mathcal{S}_n$ modulo the Young subgroup $W_{\mathbf{n}} = \mathcal{S}_{n_1} \times \cdots \times \mathcal{S}_{n_h}$, and subset-flags.

Given such $\tau$, let $E_i(\tau) = \langle e_j \mid j \in \tau_i \rangle$ be a coordinate subspace of $K^n$, and $E_\bullet(\tau) = (E_1(\tau) \subset E_2(\tau) \subset \cdots) \in \mathrm{Fl}$. Then we may define the *Schubert cell*

$$\begin{aligned} X_Q^\circ(\tau) &= B \cdot E(\tau) \\ &= \left\{ (U_1 \subset U_2 \subset \cdots) \in \mathrm{Fl} \;\middle|\; \begin{array}{c} \dim U_i \cap K^j = \#\tau_i \cap [j] \\ 1 \leq i \leq h,\ 1 \leq j \leq n \end{array} \right\} \end{aligned}$$

and the *Schubert variety*

$$\begin{aligned} X_Q(\tau) &= \overline{X_Q^\circ(\tau)} \\ &= \left\{ (U_1 \subset U_2 \subset \cdots) \in \mathrm{Fl} \;\middle|\; \begin{array}{c} \dim U_i \cap K^j \geq \#\tau_i \cap [j] \\ 1 \leq i \leq h,\ 1 \leq j \leq n \end{array} \right\} \end{aligned}$$

where $K^j = \langle e_1, \ldots, e_j \rangle \subset K^n$.

Under the identification of $G/Q$ with Fl, the opposite cell $\mathcal{O}^-$ in $G/Q$ gets identified with the set of flags in general position with respect to the spaces $E'_1 \supset \cdots \supset E'_{h-1}$:

$$\mathcal{O}^- = \{(U_1 \subset U_2 \subset \cdots) \in \mathrm{Fl} \mid U_i \cap E'_i = 0\}.$$

Let $Y_Q(\tau) = X_Q(\tau) \cap \mathcal{O}^-$, the opposite cell of $X(\tau)$.

We define a special subset-flag $\tau^{\max} = (\tau_1^{\max} \subset \cdots \subset \tau_h^{\max} = [n])$ corresponding to $\mathbf{n} = (n_1, \ldots, n_h)$. We want each $\tau_i^{\max}$ to contain numbers as large as possible given the constraints $[a_{j-1}] \subset \tau_j^{\max}$ for all $j$. Namely, we define $\tau_i^{\max}$ recursively by

$$\tau_h^{\max} = [n]; \quad \tau_i^{\max} = [a_{i-1}] \cup \{\text{largest } n_i \text{ elements of } \tau_{i+1}^{\max}\}.$$

Furthermore, given $\mathbf{r} = (r_{ij})_{1 \le i \le j \le h}$ indexing a quiver variety, define a subset-flag $\tau^{\mathbf{r}}$ to contain numbers as large as possible given the constraints

$$\#\tau_i^{\mathbf{r}} \cap [a_j] = \begin{cases} a_i - r_{i,j+1} & \text{for } i \le j \\ a_j & \text{for } i > j \end{cases}$$

Namely,

$$\tau_i^{\mathbf{r}} = \{\underbrace{1 \ldots a_{i-1}}_{a_{i-1}}\ \underbrace{\cdots\cdots a_i}_{r_{ii}-r_{i,i+1}}\ \underbrace{\cdots\cdots a_{i+1}}_{r_{i,i+1}-r_{i,i+2}}\ \underbrace{\cdots\cdots a_{i+2}}_{r_{i,i+2}-r_{i,i+3}}\ \cdots\ \underbrace{\cdots\cdots n}_{r_{i,h}}\}$$

where we use the visual notation

$$\underbrace{\cdots\cdots a}_{b} = [a-b+1, a].$$

Recall that $a_j = a_{j-1} + n_j$ and $0 \le r_{ij} - r_{i,j+1} \le n_j$, so that each $\tau_i^{\mathbf{r}}$ is an increasing list of integers. Also $r_{ij} - r_{i,j+1} \le r_{i+1,j} - r_{i+1,j+1}$, so that $\tau_i^{\mathbf{r}} \subset \tau_{i+1}^{\mathbf{r}}$. Thus, $\tau^{\mathbf{r}}$ are indeed subset-flags.

## Examples

We give below four examples.

**Example 1** *A small generic case.*
Let $h = 4$, $\mathbf{n} = (2, 3, 2, 2)$,

$$\mathbf{r} = \begin{array}{|cccc|} 2 & 2 & 0 & 0 \\ & 3 & 1 & 1 \\ & & 2 & 2 \\ & & & 2 \end{array}$$

where $r_{ij}$ are written in the usual matrix positions.

Then we get $(a_1, a_2, a_3, a_4) = (2, 5, 7, 9)$, $n = 9$, and

$$\tau^{\max} = (89 \subset 12589 \subset 1234589 \subset [9]), \qquad \tau^{\mathbf{r}} = (45 \subset 12459 \subset 1234589 \subset [9]),$$

which correspond to the cosets in $W/W_{\mathbf{n}}$

$$w^{\max} = 89|125|34|67, \qquad w^{\mathbf{r}} = 45|129|38|67.$$

(The minimal-length representatives of these cosets are the permutations as written; the other elements are obtained by permuting numbers within each block.) The partial flag variety is $\mathrm{Fl} = \{U_1 \subset U_2 \subset U_3 \subset K^9 \mid \dim U_i = a_i\}$, and the Schubert varieties are:

$$X_Q(\tau^{\max}) = \left\{U_\bullet \,\middle|\, \begin{matrix} K^2 \subset U_2 \\ K^5 \subset U_3 \end{matrix}\right\}, \quad X_Q(\tau^{\mathbf{r}}) = \left\{U_\bullet \,\middle|\, \begin{matrix} U_1 \subset \mathbf{k}^5 \subset U_3,\ K^2 \subset U_2 \\ \dim U_2 \cap K^5 \geq 4 \end{matrix}\right\}.$$

The opposite cells $Y_Q(\tau)$ are defined by the extra conditions $U_i \cap E_i' = 0$.

**Example 2.** *Fulton's universal degeneracy schemes* (cf. [31]).
Given $m > 0$, let $Z$ be the affine space associated to the quiver data $h = 2m$, $\mathbf{n} = (1, 2, \ldots, m, m, \ldots, 2, 1)$. For each $w \in \mathcal{S}_{m+1}$, Fulton defines a "degeneracy scheme" $\Omega_w = Z(\mathbf{r})$ as follows. Denote $\bar{i} = 2m+1-i$, and define $\mathbf{r} = \mathbf{r}(w) = (r_{ij})$ by:

$$r_{ij} = r_{\bar{j}\bar{i}} = i$$
$$r_{i\bar{j}} = \#[i] \cap w[j]$$

for $1 \leq i, j \leq m$. The associated Schubert varieties $X_Q(\tau^{\mathbf{r}})$ are given by $\tau^{\mathbf{r}} = (\tau_1^{\mathbf{r}} \subset \cdots \subset \tau_{\bar{1}}^{\mathbf{r}})$ or by cosets $\widetilde{w} = \widetilde{w}_1 | \cdots | \widetilde{w}_{\bar{1}} \in W/W_{\mathbf{n}}$

$$\tau_i^{\mathbf{r}} = [a_{i-1}] \cup \{a_{\overline{w^{-1}(1)}}, a_{\overline{w^{-1}(2)}}, \ldots, a_{\overline{w^{-1}(i)}}\},$$
$$\tau_{\bar{i}}^{\mathbf{r}} = [a_{\bar{i}} - 1] \cup \{a_{\bar{1}}, a_{\bar{2}}, \ldots, a_{\bar{m}}\}$$

$$\widetilde{w}_i = [a_{i-2}+1, a_{i-1}] \cup \{a_{\overline{w^{-1}(i)}}\}$$
$$\widetilde{w}_{\bar{m}} = [a_{m-1}+1, a_m - 1] \cup \{a_{\overline{w^{-1}(m+1)}}\}$$
$$\widetilde{w}_{\bar{j}} = [a_{\bar{j}-2}+1, a_{\bar{j}-1}]$$

for $1 \leq i \leq m$, $1 \leq j \leq m-1$. Furthermore $\tau^{\max} = \tau^{\mathbf{r}(w)}$ and $\widetilde{w}^{\max} = \widetilde{w}^{\mathbf{r}(w)}$ for $w = e \in \mathcal{S}_{m+1}$, the identity permutation.

**Example 3.** *The variety of complexes.*
For a given $h$ and $\mathbf{n}$, the *variety of complexes* is defined as the union $\mathcal{C} = \cup_{\mathbf{r}} Z(\mathbf{r})$ over all $\mathbf{r} = (r_{ij})$ with $r_{i,i+2} = 0$ for all $i$. The subvarieties $Z(\mathbf{r})$ correspond to the multiplicity matrices $\mathbf{m} = (m_{ij})$ with $m_{ij} = 0$ for all $i + 2 \leq j$, and $m_{ii} + m_{i-1,i} + m_{i,i+1} = n_i$ for all $i$. In [90], Musili-Seshadri have shown that each component of $\mathcal{C}$, is isomorphic to the opposite cell in a Schubert variety.

**Example 4.** *The Classical Determinantal Variety.*
The classical *determinantal variety* of $k \times l$ matrices of rank $\leq t$ is $\mathcal{D} = Z(r)$ for $\mathbf{r} = \begin{pmatrix} l & m \\ 0 & k \end{pmatrix}$ and $\mathbf{m} = \begin{pmatrix} l-m & m \\ 0 & k-m \end{pmatrix}$ where $m = \min(t+1, k, l)$. Also $n = k + l$,

$$\tau^{\max} = ([k+1, k+l] \subset [n]), \quad \tau^{\mathbf{r}} = ([m+1, l] \cup [k+l-m+1, k+l] \subset [n])$$

$$X(\tau^{\max}) = G_{l,n}, \quad X(\tau^{\mathbf{r}}) \cong \{U \in G_{l,n} \mid U \cap K^l = l - m\},$$
$$\mathcal{D} = Z(\mathbf{r}) \cong Y(\tau^{\mathbf{r}}) = \{U \in G_{l,n} \mid U \cap K^l = l - m,\ U \cap E' = 0\},$$

where $E' = \langle e_{l+1}, e_{l+2}, \ldots, e_n \rangle$.

**Main theorem**

Denote a generic element of the quiver space $Z = M(n_2, n_1) \times \cdots \times M(n_h, n_{h-1})$ by $(A_1, \ldots, A_{h-1})$, so that the coordinate ring of $Z$ is the polynomial ring in the entries of all the matrices $A_i$. Let $\mathbf{r} = (r_{ij})$ index the quiver variety $Z(\mathbf{r}) = \{(A_1, \ldots, A_{h-1}) \mid \text{rank } A_{j-1} \cdots A_i \leq r_{ij}\}$.

Let $\mathcal{J}(\mathbf{r}) \subset K[Z]$ be the ideal generated by the determinantal conditions implied by the definition of $Z(\mathbf{r})$:

$$\mathcal{J}(\mathbf{r}) = \left\langle \det(A_{j-1}A_{j-2} \cdots A_i)_{\lambda \times \mu} \;\middle|\; \begin{array}{c} j > i,\ \lambda \subset [n_j],\ \mu \subset [n_i] \\ \#\lambda = \#\mu = r_{ij} + 1 \end{array} \right\rangle.$$

Clearly $\mathcal{J}(\mathbf{r})$ defines $Z(\mathbf{r})$ set-theoretically.

**Theorem 13.5.3.4** *(cf. [60]) $\mathcal{J}(\mathbf{r})$ is a prime ideal and is the vanishing ideal of $Z(\mathbf{r}) \subset Z$. There are isomorphisms of reduced schemes*

$$Z(\mathbf{r}) = Spec(\mathbf{k}[Z] / \mathcal{J}(\mathbf{r})) \cong Spec(\mathbf{k}[\mathcal{O}^-] / \mathcal{I}(\tau^{\mathbf{r}})) = Y_Q(\tau^{\mathbf{r}}).$$

*That is, the quiver scheme $Z(\mathbf{r})$ defined by $\mathcal{J}(\mathbf{r})$ is isomorphic to the (reduced) variety $Y_Q(\tau^{\mathbf{r}})$, the opposite cell of a Schubert variety.*

In proving the above theorem again, one uses the standard monomial theory for Schubert varieties.

## 13.6 Variety of complexes

In this section we consider a special case of the quiver varieties called *the variety of complexes*. Let $V$ be a variety of complexes, namely, the subvariety of $Z$ consisting of $\{(A_1, A_2, \ldots, A_h) \mid A_i A_{i-1} = 0,\ 2 \leq i \leq h\}$. We have from the previous section

$$V = \bigcup_{\mathbf{r}} Z(\mathbf{r}),$$

where the union is taken over all $\mathbf{r} = (r_{ij})$ such that $r_{ij} = 0$ for all $j \neq i, i+1$. Let us denote $k_i = r_{i,i+1}$, $\mathbf{k} = (k_1, k_2, \ldots, k_h)$, and $V(\mathbf{k}) = Z(\mathbf{r})$. Let $\mathbf{n} = (n_1, n_2, \ldots, n_{h+1})$, and

$$A_{\mathbf{n}} = \{\mathbf{k} = (k_1, k_2, \ldots, k_h) \mid k_i \leq \min\{n_i, n_{i+1}\},\quad 1 \leq i \leq h$$
$$k_{i-1} + k_i \leq n_i,\ 2 \leq i \leq h\}.$$

Then

$$V = \bigcup_{\mathbf{k} \in A_{\mathbf{n}}} V(\mathbf{k}).$$

Let us denote $\tau^{\mathbf{r}}$ by $\tau_{\mathbf{k}}$.

### 13.6.1 A partial order on $\{(k_1, k_2, \ldots, k_h)\}$

The partial order on the set $\{Z(\mathbf{k}),\ \mathbf{k} \in A_{\mathbf{n}}\}$ of $G_{\mathbf{n}}$-orbit closures in $V$ given by inclusion induces a partial order $\geq$ on $A_{\mathbf{n}}$, namely, for $\mathbf{k} = (k_1, k_2, \ldots, k_h)$, $\mathbf{k}' = (k_1', k_2', \ldots, k_h')$ in $A_{\mathbf{n}}$,

$$\mathbf{k} \geq \mathbf{k}' \iff k_t \geq k_t',\ 1 \leq t \leq h.$$

**Theorem 13.6.1.1** *[60, 90]*

*(i)* $V(\mathbf{k}) \cong Y_Q(\tau_{\mathbf{k}})$.
*(ii) The irreducible components of* $V$ *are given by* $\{V(\mathbf{k}) \mid \mathbf{k}$ *is maximal for the above partial order*$\}$.

**Proposition 13.6.1.2** *(cf. [60, 90]) Let* $\mathbf{k} \in A_{\mathbf{n}}$. *Then* $\dim V(\mathbf{k}) = \sum_{1\leq i\leq h+1} (n_i - k_i)(k_{i-1} + k_i)$, *where* $k_0 = k_{h+1} = 0$.

Let $\mathbf{n} = (n_1, \ldots, n_{h+1})$. Fix $(k_1, k_2, \ldots, k_h)$ so that $k_i \leq \min\{n_i, n_{i+1}\}$, $1 \leq i \leq h$, $k_{i-1} + k_i \leq n_i$, $2 \leq i \leq h$, where $k_0 = k_{h+1} = 0$. Let $V_{\mathbf{k}} = V(k_1, k_2, \ldots, k_h) := \{(f_1, \ldots, f_h) \mid \operatorname{rank}(f_i) \leq k_i\},\ f_i \circ \phi_{i-1} = 0,\ 2 \leq i \leq h\}$, $f_1, \ldots, f_h$ being as above. For $1 \leq i \leq h$, let

$$V_i = V(k_1, \ldots, k_{i-1}, k_i - 1, k_{i+1}, \ldots, k_h),$$

and for $1 \leq j \leq h-1$, let

$$V_{j,j+1} = V(k_1, \ldots, k_{j-1}, k_j - 1, k_{j+1} - 1, k_{j+2}, \ldots, k_h).$$

**Theorem 13.6.1.3** *(cf. [32]) The irreducible components of* $\operatorname{Sing} V_{\mathbf{k}}$ *are* $V_i$, $i \in \Omega$, *and* $V_{j,j+1}$, $j \notin \Omega$, *where* $\Omega$ *is the set of all* $1 \leq i \leq h$ *such that* $k_{i-1} + k_i < n_i$ *and* $k_i + k_{i+1} < n_{i+1}$.

## 13.7 Degenerations of Schubert varieties to toric varieties

In this section, we discuss the toric degenerations of Schubert varieties in the Grassmannian (cf. [33]). As mentioned in the beginning of this chapter, though there are recent results (cf. [12, 17]) on toric degenerations of Schubert varieties in any $G/Q$, $G$ semi simple, in this section we present the results on the toric degenerations of Schubert varieties in the Grassmannian because of the simplicity of the proof in this case! These results are proved using Standard monomial bases for Schubert varieties in $SL(n)/B$.

### 13.7.1 Generalities on distributive lattices

**Definition 13.7.1.1** *A* lattice *is a partially ordered set* $(\mathcal{L}, \leq)$ *such that, for every pair of elements* $x, y \in \mathcal{L}$, *there exist elements* $x \vee y$ *and* $x \wedge y$, *called the* join, *respectively the* meet *of* $x$ *and* $y$, *defined by:*

$$x \vee y \geq x, \ x \vee y \geq y, \ \textit{and if } z \geq x \textit{ and } z \geq y, \ \textit{then } z \geq x \vee y,$$
$$x \wedge y \leq x, \ x \wedge y \leq y, \ \textit{and if } z \leq x \textit{ and } z \leq y, \ \textit{then } z \leq x \wedge y.$$

It is easy to check that the operations $\vee$ and $\wedge$ are commutative and associative.

**Definition 13.7.1.2** *An element $z \in \mathcal{L}$ is called the* zero *of $\mathcal{L}$, denoted by $0$, if $z \leq x$ for all $x$ in $\mathcal{L}$. An element $z \in \mathcal{L}$ is called the* one *of $\mathcal{L}$, denoted by $1$, if $z \geq x$ for all $x$ in $\mathcal{L}$.*

**Definition 13.7.1.3** *Given a lattice $\mathcal{L}$, a subset $\mathcal{L}' \subset \mathcal{L}$ is called a* sublattice *of $\mathcal{L}$ if $x, y \in \mathcal{L}'$ implies $x \wedge y \in \mathcal{L}'$, $x \vee y \in \mathcal{L}'$.*

**Definition 13.7.1.4** *Two lattices $\mathcal{L}_1$ and $\mathcal{L}_2$ are* isomorphic *if there exists a bijection $\varphi : \mathcal{L}_1 \to \mathcal{L}_2$ such that, for all $x, y \in \mathcal{L}_1$,*

$$\varphi(x \vee y) = \varphi(x) \vee \varphi(y) \textit{ and } \varphi(x \wedge y) = \varphi(x) \wedge \varphi(y).$$

**Definition 13.7.1.5** *A lattice is called* distributive *if the following identities hold:*

$$x \wedge (y \vee z) = (x \wedge y) \vee (x \wedge z)$$
$$x \vee (y \wedge z) = (x \vee y) \wedge (x \vee z).$$

**Example 13.7.1.6** *The lattice of all subsets of the set $\{1, 2, \ldots, n\}$ is denoted by $\mathcal{B}(n)$, and called the* Boolean algebra of rank $n$.

### 13.7.2 An important example

Given an integer $n \geq 1$, $\mathcal{C}(n)$ will denote the chain $\{1 < \cdots < n\}$, and for $n_1, \ldots, n_d > 1$, $\mathcal{C}(n_1, \ldots, n_d)$ will denote the chain product lattice $\mathcal{C}(n_1) \times \cdots \times \mathcal{C}(n_d)$ consisting of all $d$-tuples $(i_1, \ldots, i_d)$, with $1 \leq i_1 \leq n_1, \ldots, 1 \leq i_d \leq n_d$. For $(i_1, \ldots, i_d)$, $(j_1, \ldots, j_d)$ in $\mathcal{C}(n_1, \ldots, n_d)$, we define

$$(i_1, \ldots, i_d) \leq (j_1, \ldots, j_d) \iff i_1 \leq j_1, \ldots, i_d \leq j_d.$$

We have

$$(i_1, \ldots, i_d) \vee (j_1, \ldots, j_d) = (\max\{i_1, j_1\}, \ldots, \max\{i_d, j_d\})$$
$$(i_1, \ldots, i_d) \wedge (j_1, \ldots, j_d) = (\min\{i_1, j_1\}, \ldots, \min\{i_d, j_d\}).$$

$\mathcal{C}(n_1, \ldots, n_d)$ is a finite distributive lattice, and its zero and one are $(1, \ldots, 1)$, $(n_1, \ldots, n_d)$ respectively.

Note that there is a total order $\lhd$ on $\mathcal{C}(n_1, \ldots, n_d)$ extending $<$, namely the lexicographic order, defined by $(i_1, \ldots, i_d) \lhd (j_1, \ldots, j_d)$ if and only if there exists $l < d$ such that $i_1 = j_1, \ldots, i_l = j_l, i_{l+1} < j_{l+1}$. Also note that two elements $(i_1, \ldots, i_d) \lhd (j_1, \ldots, j_d)$ are non-comparable with respect to $\leq$ if and only if there exists $1 < h \leq d$ such that $i_h > j_h$.

Sometimes we denote the elements of $\mathcal{C}(n_1, n_2, \ldots, n_d)$ by $x_{i_1 \ldots i_d}$, with $1 \leq i_1 \leq n_1, \ldots, 1 \leq i_d \leq n_d$.

One has the following (see [3]):

**Theorem 13.7.2.1** *Any finite distributive lattice is isomorphic to a sublattice of a Boolean algebra of finite rank, and hence, in particular, to a sublattice of a finite chain product.*

### 13.7.3 Generalities on toric varieties

**Definition 13.7.3.1** *(cf. [30], [49]) An* equivariant affine embedding *of a torus $T$ is an affine variety $X$ containing $T$ as an open subset and equipped with a $T$-action $T \times X \to X$ extending the action $T \times T \to T$ given by multiplication. If in addition $X$ is normal, then $X$ is called an* affine toric variety.

For our purpose, we also need the definition of a toric ideal (cf. [114]).

Let $T = (K^*)^m$ be the $m$-dimensional torus. Let $M$ be the character group $(= \mathrm{Hom}_{\mathrm{alg.gp.}}(T, \mathbb{G}_m))$ of $T$. Then $M$ can be identified with $\mathbb{Z}^m$. Let $\mathcal{A} = \{\mathbf{a}_1, \ldots, \mathbf{a}_n\}$ be a subset of $\mathbb{Z}^m$. Consider the map

$$\pi_{\mathcal{A}} : \mathbb{Z}_+^n \to \mathbb{Z}^m, \quad \mathbf{u} = (u_1, \ldots, u_n) \mapsto u_1\mathbf{a}_1 + \cdots + u_n\mathbf{a}_n.$$

Let $K[\mathbf{x}] = K[x_1, \ldots, x_n]$, $K[\mathbf{t}^{\pm 1}] = K[t_1, \ldots, t_m, t_1^{-1}, \ldots, t_m^{-1}]$.

The map $\pi_{\mathcal{A}}$ induces a homomorphism of semigroup algebras

$$\widehat{\pi_{\mathcal{A}}} : K[\mathbf{x}] \to K[\mathbf{t}^{\pm 1}], \qquad x_i \mapsto \mathbf{t}^{\mathbf{a}_i}.$$

**Definition 13.7.3.2** *The kernel of $\widehat{\pi}_{\mathcal{A}}$ is denoted by $I_{\mathcal{A}}$ and called the* toric ideal *associated to $\mathcal{A}$. A variety of the form $\mathcal{V}(I_{\mathcal{A}})$, the affine variety of the zeroes in $K^n$ of $I_{\mathcal{A}}$, is called an* affine toric variety.

Note that a toric ideal is prime.

**Remark 13.7.3.3.** Consider the action of $T$ on $K^n$ given by $\mathbf{t}e_i = \mathbf{t}^{\mathbf{a}_i} e_i$ (here, $e_i$, $1 \le i \le n$ are the standard basis vectors of $K^n$). Then $\mathcal{V}(I_{\mathcal{A}})$ is simply the Zariski closure of the $T$-orbit through $(1, 1, \ldots, 1)$. In particular, $\mathcal{V}(I_{\mathcal{A}})$ is an equivariant affine embedding of $T = (K^*)^m$.

**Remark 13.7.3.4.** In the above definition, we do not require $\mathcal{V}(I_{\mathcal{A}})$ to be normal. Using [49], we have that $\mathcal{V}(I_{\mathcal{A}})$ is normal if and only if the semi subgroup $S$ of $M$ generated by $\mathcal{A}$ is saturated (here, $S$ is said to be *saturated* if for $\mathbf{a} \in M$, $r\mathbf{a} \in S \implies \mathbf{a} \in S$ (where $r$ is an integer $> 1$) (cf. [49])).

Recall the following (see [114]).

**Proposition 13.7.3.5** *The toric ideal $I_{\mathcal{A}}$ is spanned as a $K$-vector space by the set of binomials*

$$\{\mathbf{x}^{\mathbf{u}} - \mathbf{x}^{\mathbf{v}} \mid \mathbf{u}, \mathbf{v} \in \mathbb{Z}_+^n \text{ with } \pi_{\mathcal{A}}(\mathbf{u}) = \pi_{\mathcal{A}}(\mathbf{v})\}. \qquad (*)$$

(Here, a *binomial* is a polynomial with at most two terms.)

### 13.7.4 An example

Let us fix the integers $n_1, \ldots, n_d > 1$, and let $n = \Pi_{i=1}^d n_i$, $m = \sum_{i=1}^d n_i$. Let $\mathbf{e}_1^l, \ldots, \mathbf{e}_{n_l}^l$ be the unit vectors in $\mathbb{Z}^{n_l}$, for $1 \leq l \leq d$. For $1 \leq \xi_1 \leq n_1, \ldots, 1 \leq \xi_d \leq n_d$, define

$$\mathbf{a}_{\xi_1 \ldots \xi_d} = \mathbf{e}_{\xi_1}^1 + \cdots + \mathbf{e}_{\xi_d}^d \in \mathbb{Z}^{n_1} \oplus \cdots \oplus \mathbb{Z}^{n_d}$$

and let

$$\mathcal{A}_{n_1,\ldots,n_d} = \{\mathbf{a}_{\xi_1 \ldots \xi_d} \mid 1 \leq \xi_1 \leq n_1, \ldots, 1 \leq \xi_d \leq n_d\}.$$

The corresponding map

$$\pi_{\mathcal{A}} : \mathbb{Z}_+^{n_1 \cdot \cdots \cdot n_d} \to \mathbb{Z}^{n_1 + \cdots + n_d}$$

is defined as follows: for $1 \leq l \leq d$ and $1 \leq i_l \leq n_l$ fixed, the $(n_1 + \cdots + n_{l-1} + i_l)$-th coordinate of $\pi_{\mathcal{A}}(\mathbf{u})$ is given by $\sum u_{\xi_1 \ldots \xi_{l-1} \xi_l \xi_{l+1} \ldots \xi_d}$, the sum being taken over the elements $(\xi_1, \ldots, \xi_{l-1}, \xi_l, \xi_{l+1}, \ldots, \xi_d)$ of $\mathcal{C}(n_1, \ldots, n_d)$ with $\xi_l = i_l$. We call this subset the $l$-th *slice* of $\mathcal{C}(n_1, \ldots, n_d)$ defined by $i_l$, and denote it by $\{\xi_l = i_l\}$. The components (or *entries*) of an element $\mathbf{u} \in \mathbb{Z}^{n_1 \cdot \cdots \cdot n_d}$ are indexed by the elements $(i_1, \ldots, i_d)$ of $\mathcal{C}(n_1, \ldots, n_d)$. If $(j_1, \ldots, j_d) \in \{\xi_l = i_l\}$, sometimes we also say that $u_{j_1 \ldots j_d}$ itself belongs to the slice $\{\xi_l = i_l\}$.

The map $\pi_{\mathcal{A}}$ induces the map

$$\widehat{\pi_{\mathcal{A}}} : K[x_{11\ldots 1}, \ldots, x_{\xi_1 \xi_2 \ldots \xi_d}, \ldots, x_{n_1 n_2 \ldots n_d}] \to K[t_{11}, \ldots, t_{1n_1}, \ldots, t_{d1}, \ldots, t_{dn_d}]$$

given by

$$x_{\xi_1 \ldots \xi_d} \mapsto t_{1\xi_1} \ldots t_{d\xi_d}, \text{ for } 1 \leq \xi_1 \leq n_1, \ldots, 1 \leq \xi_d \leq n_d.$$

### 13.7.5 The algebra associated to a distributive lattice

**Definition 13.7.5.1** *Given a finite lattice $\mathcal{L}$, the* ideal associated to $\mathcal{L}$, *denoted by $I(\mathcal{L})$, is the ideal of the polynomial ring $K[\mathcal{L}]$ generated by the set of binomials*

$$\mathcal{G}_{\mathcal{L}} = \{xy - (x \wedge y)(x \vee y) \mid x, y \in \mathcal{L} \textit{ non-comparable}\}.$$

By Theorem 13.7.2.1, a finite distributive lattice $\mathcal{L}$ may be identified with a sublattice of a finite chain product lattice. Hence it inherits a total order extending the given partial order. In turn, this total order induces the lexicographic order on the monomials in $K[\mathcal{L}]$.

The following theorems show that the ideal associated to a chain product lattice (resp. a finite distributive lattice) is toric.

**Theorem 13.7.5.2** *We have $I(\mathcal{C}(n_1, \ldots, n_d)) = I_{\mathcal{A}_{n_1,\ldots,n_d}}$.*

**Theorem 13.7.5.3** *Let $\mathcal{L}$ be a finite distributive lattice. Then the ideal $I(\mathcal{L})$ is toric.*

For a proof of Theorems 13.7.5.2,13.7.5.3, see [33].

**Remark 13.7.5.4.** (1) In [39], it is shown that $\mathcal{R}_{\mathcal{L}} = K[\mathcal{L}]/I(\mathcal{L})$ ($\mathcal{L}$ being a finite distributive lattice) is a normal domain; on the other hand, one knows (see [28] for example) that a prime binomial ideal is toric, in the sense of Definition 13.7.3.2 (here, by a binomial, we mean a polynomial with at most two terms). Thus the result that $I(\mathcal{L})$ is toric ($\mathcal{L}$ being a finite distributive lattice) may also be concluded using [39] and [28].

(2) In fact, given a finite lattice $\mathcal{L}$, $I(\mathcal{L})$ is toric if and only if $\mathcal{L}$ is distributive.

(3) We have included below a short geometric proof of the result that a prime binomial ideal is toric.

### 13.7.6 Varieties defined by binomials

Let $N \geq 1$, and let $X$ be an affine variety in $\mathbb{A}^N$, not contained in any of the coordinate hyperplanes $\{x_i = 0\}$. Further, let $X$ be irreducible, and let its defining prime ideal $I(X)$ be generated by $l$ binomials

$$x_1^{a_{i1}} \dots x_N^{a_{iN}} - \lambda_i x_1^{b_{i1}} \dots x_N^{b_{iN}}, \qquad 1 \leq i \leq l. \tag{$*$}$$

Consider the natural action of the torus $T_N = (k^*)^N$ on $\mathbb{A}^N$,

$$(t_1, \dots, t_N) \cdot (a_1, \dots, a_N) = (t_1 a_1, \dots, t_N a_N).$$

Let $X(T_N) = \mathrm{Hom}(T_N, \mathbb{G}_m)$ be the character group of $T_N$, and let $\epsilon_i \in X(T_N)$ be the character

$$\epsilon_i(t_1, \dots, t_N) = t_i, \qquad 1 \leq i \leq N.$$

For $1 \leq i \leq l$, let

$$\varphi_i = \sum_{t=1}^{N} (a_{it} - b_{it})\epsilon_t.$$

Set $T = \bigcap_{i=1}^{l} \ker \varphi_i$, and $X^\circ = \{(x_1, \dots, x_N) \in X \mid x_i \neq 0 \text{ for all } 1 \leq i \leq N\}$.

**Proposition 13.7.6.1** *Let notations be as above.*

(1) *There is a canonical action of $T$ on $X$.*

(2) *$X^\circ$ is $T$-stable. Further, the action of $T$ on $X^\circ$ is simple and transitive.*

(3) *$T$ is a subtorus of $T_N$, and $X$ is an equivariant affine embedding of $T$.*

*Proof.* (1) We consider the (obvious) action of $T$ on $\mathbb{A}^N$. Let $(x_1, \dots, x_N) \in X$, $\mathbf{t} = (t_1, \dots, t_N) \in T$, and $(y_1, \dots, y_N) = \mathbf{t} \cdot (x_1, \dots, x_N) = (t_1 x_1, \dots, t_N x_N)$. Using the fact that $(x_1, \dots, x_N)$ satisfies $(*)$, we obtain

$$y_1^{a_{i1}} \dots y_N^{a_{iN}} = t_1^{a_{i1}} \dots t_N^{a_{iN}} x_1^{a_{i1}} \dots x_N^{a_{iN}} = \lambda_i t_1^{b_{i1}} \dots t_N^{b_{iN}} x_1^{b_{i1}} \dots x_N^{b_{iN}} = \lambda_i y_1^{b_{i1}} \dots y_N^{b_{iN}},$$

for all $1 \leq i \leq l$, i.e. $(y_1, \dots, y_N) \in X$. Hence $\mathbf{t} \cdot (a_1, \dots, a_N) \in X$ for all $\mathbf{t} \in T$.

(2) Let $x = (x_1, \dots, x_N) \in X^\circ$, and $\mathbf{t} = (t_1, \dots, t_N) \in T$. Then, clearly $\mathbf{t} \cdot (x_1, \dots, x_N) \in X^\circ$. Considering $x$ as a point in $\mathbb{A}^N$, the isotropy subgroup in

$T_N$ at $x$ is {id}. Hence the isotropy subgroup in $T$ at $x$ is also {id}. Thus the action of $T$ on $X^\circ$ is simple.

Let $(x_1, \ldots, x_N), (x'_1, \ldots, x'_N) \in X^\circ$. Set $\mathbf{t} = (t_1, \ldots, t_N)$, where $t_i = x_i/x'_i$. Then, clearly $\mathbf{t} \in T$. Thus $(x_1, \ldots, x_N) = \mathbf{t} \cdot (x'_1, \ldots, x'_N)$. Hence the action of $T$ on $X^\circ$ is simple and transitive.

(3) Now, fixing a point $x \in X^\circ$, we obtain from (2) that the orbit map $\mathbf{t} \mapsto \mathbf{t} \cdot x$ is in fact an isomorphism of $T$ onto $X^\circ$. Also, since $X$ is not contained in any of the coordinate hyperplanes, the open set $X_i = \{(x_1, \ldots, x_N) \in X \mid x_i \neq 0\}$ is nonempty for all $1 \leq i \leq N$. The irreducibility of $X$ implies that the sets $X_i$, $1 \leq i \leq N$, are open dense in $X$, and hence their intersection

$$X^\circ = \bigcap_{i=1}^{N} X_i = \{(x_1, \ldots, x_N) \in X \mid x_i \neq 0 \text{ for any } i\}$$

is an open dense set in $X$, and thus $X^\circ$ is irreducible. This implies that $T$ is irreducible (and hence connected). Thus $T$ is a subtorus of $T_N$. The assertion that $X$ is an equivariant affine embedding of $T$ follows from (1) and (2).

**Remark 13.7.6.2.** One can see that the ideal $I(X)$ is a toric ideal in the sense of Definition 13.7.3.2; more precisely, $I(X) = I_{\mathcal{A}}$, with $m = \dim T$, $n = N$, $\mathcal{A} = \{\rho_i, 1 \leq i \leq N\}$, where $\rho_i = \epsilon_i\big|_T$ (here $K[T]$ is identified with $K[t_1^{\pm 1}, \ldots, t_m^{\pm 1}]$, and the character group $X(T)$ with $\mathbb{Z}^m$).

### 13.7.7 Degenerations of Schubert varieties in the Grassmannian to toric varieties

Let $G = SL(n)$. Fix $d$, $1 \leq d \leq n-1$. Let $G_{d,n}$, $I_{d,n}$ etc., be as in Chapter 4.

**Proposition 13.7.7.1** *$(I_{d,n}, \geq)$ is a distributive lattice.*

*Proof.* Let $\tau = (i_1, \cdots, i_d)$, $\phi = (j_1, \cdots, j_d)$ be two non-comparable elements in $I_{d,n}$. Set

$$\lambda = (k_1, \cdots, k_d), \ \mu = (l_1, \cdots, l_d)$$

where $k_t = \max\{i_t, j_t\}$, $l_t = \min\{i_t, j_t\}$, $1 \leq t \leq d$.
Note that $(k_{t-1}, k_t) = (i_{t-1}, i_t)$ or $(i_{t-1}, j_t)$ or $(j_{t-1}, i_t)$ or $(j_{t-1}, j_t)$, from which it follows that $k_{t-1} < k_t$. Similarly, $l_{t-1} < l_t$. Thus $\lambda, \mu$ are well-defined elements of $I_{d,n}$. Further from the definition of $\lambda, \mu$, it is easily seen that $\lambda$ (resp. $\mu$) is the unique minimal (resp. maximal) element of $I_{d,n}$ which is greater (resp. less) than both $\tau$ and $\phi$.

**Corollary 13.7.7.2** *Let $\tau, \phi, \lambda, \mu \in I_{d,n}$ be as in the proof of Proposition 13.7.7.1. Then $\lambda = \tau \vee \phi$, $\mu = \tau \wedge \phi$.*

**Corollary 13.7.7.3** *For $\tau \in I_{d,n}$, let*

$$H_\tau = \{\underline{i} \in I_{d,n} \mid \underline{i} \leq \tau\}.$$

*Then $H_\tau$ is a distributive lattice.*

**Proposition 13.7.7.4** *Let $\tau, \phi \in I_{d,n}$ be non-comparable. Let*

$$p_\tau p_\phi = \sum c_{\alpha,\beta} p_\alpha p_\beta \qquad (*)$$

*be the expression for $p_\tau p_\phi$ as a sum of standard monomials. Then $p_{\tau\vee\phi} p_{\tau\wedge\phi}$ occurs with coefficient 1 in (*).*

*Proof.* Let notations be as above. Then, in view of Chapter 5, Proposition 5.1.0.2 for any $(\alpha, \beta)$ on the right-hand side of $(*)$, we have

$$\alpha \geq \tau \vee \phi, \ \beta \leq \tau \wedge \phi.$$

Further, if $\alpha = \tau \vee \phi$, then $\beta = \tau \wedge \phi$ (by weight considerations; note that for $\theta \in I_{d,n}$, say $\theta = (i_1, \cdots, i_d)$, weight of $p_\theta = -(\epsilon_{i_1} + \cdots + \epsilon_{i_d})$). Hence restricting (*) to $X(\tau \vee \phi)$, we obtain

$$p_\tau p_\phi = c p_{\tau\vee\phi} p_{\tau\wedge\phi}, \quad \text{on } X(\tau \vee \phi).$$

Since standard monomials basis is characteristic-free, we conclude that the coefficient $c$ of $p_{\tau\vee\phi} p_{\tau\wedge\phi}$ on the right-hand side of (*) is $\pm 1$. In order to prove that $c = 1$, we compare the coefficients of the monomial $\underline{m} = x_{i_1 1} \ldots x_{i_d d} x_{j_1 1} \ldots x_{j_d d}$ on both sides (note that a Plücker coordinate $p_{a_1,\ldots,a_d}$ (being the $d \times d$-minor of the generic $n \times n$ matrix $(x_{ij})$ with row indices $a_1, \ldots, a_d$ and column indices $1, \ldots, d$) is a polynomial in the $x_{ij}$'s). Let $p_\alpha p_\beta$ appearing on the right-hand side of (*) be such that $\alpha \neq \tau \vee \phi$. This implies $\alpha > \tau \vee \phi$, and $\beta < \tau \wedge \phi$. Let $\alpha = (\alpha_1, \ldots, \alpha_d)$ and $\beta = (\beta_1, \ldots, \beta_d)$. We have $(\alpha_1, \ldots, \alpha_d) > (k_1, \ldots, k_d)$ ($(k_1, \ldots, k_d)$ being as in the proof of Proposition 13.7.7.1). Let $t$ be the smallest integer $\leq d$ such that $\alpha_t > k_t$. This implies, in view of Corollary 13.7.7.2 (notations being as in that Corollary)

$$\alpha_p = k_p, \quad \beta_p = l_p, \quad p < t,$$
$$k_t \notin \{\alpha_1, \ldots, \alpha_d\}, \quad k_t, l_t \in \{\beta_1, \ldots, \beta_d\}.$$

Hence in the expression for $p_\alpha p_\beta$ as a polynomial in the $x_{ij}$'s, $x_{k_t t} x_{l_t t}$ will not be a factor in any of the monomials. Hence the monomial $\underline{m}$ does not occur in $p_\alpha p_\beta$. Now the term $x_{i_1 1} \ldots x_{i_d d}$ (resp. $x_{j_1 1} \ldots x_{j_d d}$) being the product of the diagonal entries in $p_\tau$ (resp. $p_\phi$), it occurs with coefficient 1 in $p_\tau$ (resp. $p_\phi$). Hence $\underline{m}$ occurs with coefficient 1 in $p_\tau p_\phi$. This, together with the fact that $\underline{m}$ does not occur in any $p_\alpha p_\beta$, $\alpha \neq \tau \vee \phi$, implies that $\underline{m}$ should occur with coefficient 1 in $p_{\tau\vee\phi} p_{\tau\wedge\phi}$ (note that in $p_{\tau\vee\phi} p_{\tau\wedge\phi}$ the monomial $\underline{m}$ is realized as the product $\underline{m}_1 \underline{m}_2$, where $\underline{m}_1$ (resp. $\underline{m}_2$) is the product of the diagonal entries in $p_{\tau\vee\phi}$ (resp. $p_{\tau\wedge\phi}$). From this, it follows that $c = 1$.

Let $\mathcal{L} = H_\tau$, $V_\tau = \operatorname{Spec} R_{\mathcal{L}}$. Then by [39] (see also Remark 13.7.5.4) we get

**Theorem 13.7.7.5** *$V_\tau$ is an affine toric variety.*

Combining the above Theorem with Theorem 11.4.2.2 of Chapter csln, we obtain a degeneration of $\widehat{X(\tau)}$ to $V_\tau$:

**Theorem 13.7.7.6** *(cf. [33]) There exists a flat family over* Spec $K[t]$ *with generic fiber (t invertible) equal to* $\widehat{X(\tau)}$ *and special fiber* ($t = 0$) *equal to* $V_\tau$.

By similar arguments, we also obtain flat toric degenerations of Schubert varieties in a minuscule $G/P$ as well as a flat toric degeneration of the flag variety (see [33] for details).

# Appendix: Proof of the main theorem of SMT

Here we prove the main theorem of standard monomial theory—see Theorem A.4.0.1 below. We do this not in complete generality as in Littelmann [77], for that would take us too far afield, but in a limited context that suffices for the invariant theoretic applications in the book. More specifically, we would like to just prove enough to derive as special cases Theorems 8.1.0.2 and 8.2.0.7 of Chapter 8.

The proof we give here follows [103]. Unfortunately, it is still not *ab initio* in two respects. Firstly, we use Theorem A.6.0.1 without giving its proof. Since the proof of this theorem uses the Demazure character formula, in effect we are implicitly assuming this formula (the validity of this formula is related to the normality of Schubert varieties [10, Comments to Chapter 3]). However, the reliance on Theorem A.6.0.1 is not serious and can be avoided—an argument like the one in [77, page 559, Proof of Theorem 2] can be adapted to the present context and the proof would then use only the Weyl character formula rather than the Demazure character formula. Secondly, the preparatory §A.5 contains no proofs. The statements here with proofs and more can all be found in [69, §3].

## A.1 Notation

The following notation remains fixed throughout this appendix:

| | |
|---|---|
| $K$ | an algebraically closed field of arbitrary characteristic; |
| $G$ | a semisimple simply connected algebraic group over $K$; |
| $T$ | a maximal torus of $G$; |
| $W$ | the Weyl group of $G$ (with respect to $T$); |
| $s_\beta$ | the reflection corresponding to a root $\beta$; |
| $\beta^\vee$ | the coroot corresponding to a root $\beta$; |
| $B$ | a Borel subgroup of $G$ containing $T$; |
| $\geq$ | Bruhat order on $W$. |

## A.2 Admissible pairs and the first basis theorem

### A.2.1 More notation

In this section, in addition to the notation in §A.1 above, we further fix:

| | |
|---|---|
| $P$ | a maximal parabolic subgroup of $G$ containing $B$; |
| $\varpi$ | the fundamental weight corresponding to $P$; |
| $W_P$ | the Weyl group of $P$ (with respect to $T$); |
| $W^P$ | the set of minimal length coset representatives of $W/W_P$; |
| $\geq$ | Bruhat order on $W/W_P$ induced from that on $W$; |
| $L$ | the ample generator of Pic $G/P$; |
| $p_\tau$ | extremal weight vector in $H^0(G/P, L)$ of weight $-\tau(\omega)$. |
| $X(w)$ | the Schubert variety in $G/P$ corresponding to $w \in W/W_P$ |

In the sequel, we shall denote $W_P$ also by $W_\omega$ or also $W_i$, where $i$ is given by $\omega = \omega_i$. When we write $s_\alpha \tau >^j \tau$, it means: $\tau$ is an element of $W/W_P$, $\alpha$ a simple root, and $\langle \varpi, \alpha^\vee \rangle = j$ (we encounter only the cases $j = 1, 2$).

### A.2.2 Chevalley multiplicity

Let $\tau$, $\varphi$ be elements of $W^P$ such that $X(\varphi)$ is a Schubert divisor in $X(\tau)$, i.e., $X(\varphi) \subseteq X(\tau)$ and $X(\varphi)$ is of codimension 1 in $X(\tau)$. Then $\tau = s_\beta \varphi$ for a unique positive root $\beta$. The (positive) integer $m_\tau(\varphi) := (\varphi(\varpi), \beta^\vee)$ is called *the Chevalley multiplicity* of $X(\varphi)$ in $X(\tau)$.

### A.2.3 Minuscule and classical type parabolics

The fundamental weight $\varpi$ (respectively $P$) is said to be *minuscule* if $|(\varpi, \alpha^\vee)| \leq 1$ for every root $\alpha$. This is equivalent to saying that all the weights of the corresponding fundamental representation are *extremal*, that is, they are Weyl group translates of the highest weight. Thus the weights form one orbit under the action of the Weyl group. This condition has an equivalent geometric formulation: if $X(\varphi)$ is a Schubert divisor in $X(\tau)$ (in $G/P$), then the Chevalley multiplicity $m_\tau(\varphi)$ is $\leq 1$. Every fundamental weight of the special linear group is minuscule—see §A.3.1 below.

We say that $\varpi$ (respectively $P$) is of *classical type* if $|(\varpi, \alpha^\vee)| \leq 2$ for every root $\alpha$. This condition has an equivalent geometric formulation: if $X(\varphi)$ is a Schubert divisor in $X(\tau)$ (in $G/P$), then the Chevalley multiplicity $m_\tau(\varphi)$ is $\leq 2$. The terminology is justified by the following facts: every fundamental weight of a classical group is of classical type; conversely, if every fundamental weight is of classical type then the group is classical.

Although we do not make use of this definition below, we recall here that $\varpi$ (respectively $P$) is called *quasi-minuscule* if either it is minuscule or there is only one non-extremal weight in the corresponding fundamental representation and that is the zero weight. The full list of minuscule, quasi-minuscule, and classical type fundamental weights of all simply connected simple groups is tabulated in Figure A.1 below. The ordering of the weights is as in [8].

| Type of Group | Classical type fundamental weights | | | Total number |
|---|---|---|---|---|
| | Quasi-minuscule | | Others | |
| | Minuscule | Non-minuscule | | |
| $A_n$ | $\varpi_1, \ldots, \varpi_n$ | .. | .. | $n$ |
| $B_n$ | $\varpi_n$ | $\varpi_1$ | $\varpi_2, \ldots, \varpi_{n-1}$ | $n$ |
| $C_n$ $(n \geq 3)$ | $\varpi_1$ | $\varpi_2$ | $\varpi_3, \ldots, \varpi_n$ | $n$ |
| $D_n$ $(n \geq 4)$ | $\varpi_1, \varpi_{n-1}, \varpi_n$ | $\varpi_2$ (Adj) | $\varpi_3, \ldots, \varpi_{n-2}$ | $n$ |
| $E_6$ | $\varpi_1, \varpi_6$ | $\varpi_2$ (Adj) | $\varpi_3, \varpi_5$ | 5 |
| $E_7$ | $\varpi_7$ | $\varpi_1$ (Adj) | $\varpi_2, \varpi_6$ | 4 |
| $E_8$ | .. | $\varpi_8$ (Adj) | $\varpi_1$ | 2 |
| $F_4$ | .. | $\varpi_4$ | $\varpi_1$ (Adj) | 2 |
| $G_2$ | .. | $\varpi_1$ | .. | 1 |

**Fig. A.1.** Table of fundamental weights of classical type

### A.2.4 Admissible pairs

In this subsection, we assume that $P$ is a maximal parabolic subgroup of classical type. An ordered pair $(\tau, \theta)$ of elements of $W^P$ is *admissible* if either $\tau = \theta$ (in which case it is called a *trivial admissible pair*) or $\tau \gneq \theta$ and there exists a sequence $\tau = \tau_0 > \tau_1 > \ldots > \tau_r = \theta$ of elements in $W^P$ such that, for $1 \leq j \leq r$, $\tau_{j-1} = s_{\alpha_j}\tau_j >^2 \tau_j$ for some simple root $\alpha_j$ (there could be repetitions among the $\alpha_j$). In the case of a minuscule parabolic, the admissible pairs are clearly all trivial.

The importance of admissible pairs stems from the fact that they form a natural indexing set for a basis of the space $H^0(G/P, L)$ of sections of $L$: the *first basis theorem*, namely Proposition A.10.0.3 below, asserts that there is a basis $\{v(\sigma)\}$ of the dual of $H^0(G/P, L)$, where the $\sigma$ varies over admissible pairs; we denote by $\{p(\sigma)\}$ the basis of $H^0(G/P, L)$ dual to $\{v(\sigma)\}$.

In a given context, if there is more than one parabolic subgroup in the picture, we use the term admissible pair *of shape* $P$ (or *of shape* $\varpi$) to make it clear which parabolic is being referred to.

The formulation of the definition of admissible pairs was a key step in the development of standard monomial theory. In Littelmann's language of paths [74, 75], an admissible pair is just an *LS path* of shape a fundamental weight of classical type.

## A.3 The three examples

Described in this section are three examples of data fixed as in §A.1, §A.2 above. These are the only cases that are relevant for the invariant theoretic applications treated in this book. In each of the three examples, ( , ) denotes an inner product on $X(T) \otimes \mathbb{R}$ (where $X(T)$ is the character group of the torus $T$) that is invariant under the action of the Weyl group; and $\varepsilon_j$ denotes the character of $T$ that maps each element of $T$ to is $j^{\text{th}}$ diagonal entry.

### A.3.1 Example A

For the first example, we take $V$ to be an $n$-dimensional vector space over the algebraically closed base field $K$, and fix a basis $e_1, \ldots, e_n$ of $V$. Let:

$G$ to be the *special linear group* $SL(V)$
(consisting of linear automorphisms of $V$ of determinant 1)
$T$ be the set of diagonal matrices (with respect to $e_1, \ldots, e_n$) in $G$;
$B$ be the set of upper triangular matrices (with respect to $e_1, \ldots, e_n$) in $G$;
$P$ to be any maximal parabolic subgroup of $G$ containing $B$.

Then $W$ gets identified with the symmetric group $S_n$—its action on the diagonal matrices is by permutation of the entries; and there exists an integer $d$, $1 \leq d \leq n$, so that

$\varpi$ is $\varepsilon_1 + \cdots + \varepsilon_d$;
$G/P$ is the Grassmannian of $d$-dimensional subspaces in an $n$-dimensional space;
$W_P$ gets identified with $S_d \times S_{n-d}$;
$W/W_P$ is the set $I(d, n)$ of subsets (of distinct entries) of cardinality $d$ of $\{1, \ldots, n\}$; an element of $I(d, n)$ is written $\tau = \{1 \leq \tau_1 < \cdots < \tau_d \leq n\}$;
$\tau(\varpi)$ is $\varepsilon_{\tau_1} + \cdots + \varepsilon_{\tau_d}$ for $\tau \in W/W_P$;
$\geq$ is defined by: $\tau \geq \theta$ if $\tau_j \geq \theta_j$ for $1 \leq j \leq d$.

The roots are $\pm(\varepsilon_i - \varepsilon_j)$ for $1 \leq i < j \leq n$. For any root $\beta$, we clearly have $\langle \varpi, \beta^\vee \rangle = 2(\varpi, \beta)/(\beta, \beta) = (\varpi, \beta)$ is 1, 0, or $-1$ (no matter what $d$ is). Thus $P$ is minuscule and the only admissible pairs are the trivial ones.

### A.3.2 Example B

Let $V$ be an $2n$-dimensional vector space with a symmetric non-degenerate bilinear form $( , )'$; we assume (only for this example) that the characteristic of the underlying field $K$ is not 2. Choose a basis $e_1, \ldots, e_{2n}$ of $V$ such that $(e_i, e_j)'$ is 1 if $i = j^*$ and 0 otherwise, where $j^* := 2n + 1 - j$ for $1 \leq j \leq 2n$. Let:

$G$ be the *special orthogonal group* $SO(V)$
(consisting of the elements of $SL(V)$ that preserve the form).
$T$ be the set of diagonal matrices (with respect to $e_1, \ldots, e_{2n}$) in $G$;
$B$ be the set of upper triangular matrices (with respect to $e_1, \ldots, e_{2n}$) in $G$;
$\varepsilon_j$ be the character of $T$ sending an element to the $j^{\text{th}}$ element of the diagonal;
$\varpi$ be the fundamental weight $\frac{1}{2}(\varepsilon_1 + \cdots + \varepsilon_n)$;
$P$ be the maximal parabolic subgroup of $G$ corresponding to $\varpi$.

Let $C$ denote the subgroup of $(\mathbb{Z}/2\mathbb{Z})^n$ consisting of those $n$-tuples whose sum is 0 in $\mathbb{Z}/2\mathbb{Z}$ (equivalently those containing an even number of 1's). The symmetric group $S_n$ acts on it by group homomorphisms. Then:

$W$ gets identified with the semi-direct product $C \rtimes S_n$;
$W_P$ gets identified with the $S_n$;
$W/W_P$ gets identified with $C$;
a typical element of $C$ is written $\tau = (\tau_1, \ldots, \tau_n)$, where each $\tau_j$ is 0 or 1;
$G/P$ is the even orthogonal Grassmannian (the connected component containing the span of $e_1, \ldots, e_n$ in the space of $n$-dimensional isotropic subspaces of $V$);
$\tau(\varpi)$ is $\frac{1}{2}((-1)^{\tau_1}\varepsilon_1 + \cdots + (-1)^{\tau_n}\varepsilon_n)$ for $\tau$ in $W/W_P = C$;
$\geq$ defined by: $\tau \geq \theta$ if $\tau_j \geq \theta_j$ for $1 \leq j \leq n$ (set $1 \gneq 0$ by definition).

The roots are $\pm(\varepsilon_i \pm \varepsilon_j)$ for $1 \leq i < j \leq n$. It is clear that, for any root $\beta$, we have $\langle \varpi, \beta^\vee \rangle = 2(\varpi, \beta)/(\beta, \beta) = (\varpi, \beta)$ is 1, 0, or $-1$ Thus $P$ is minuscule and the only admissible pairs are the trivial ones.

### A.3.3 Example C

Let $V$ be a finite dimensional vector space. Let $(\ ,\ )'$ be a non-degenerate alternating bilinear form on $V$. Then the dimension of $V$ is necessarily even, say $2n$. Choose a basis $e_1, \ldots, e_{2n}$ of $V$ such that

$$(e_i, e_j)' = \begin{cases} 1 \text{ if } i = j^* \text{ and } i < j \\ -1 \text{ if } i = j^* \text{ and } i > j \end{cases}$$

where $j^* := 2n - j + 1$, for $1 \leq j \leq 2n$. Let:

$G$ be the *symplectic group* $Sp(V)$
(consisting of the elements of $SL(V)$ (or even $GL(V)$) that preserve the form).
$T$ be the set of diagonal matrices (with respect to $e_1, \ldots, e_{2n}$) in $G$;
$B$ be the set of upper triangular matrices (with respect to $e_1, \ldots, e_{2n}$) in $G$;
$\varepsilon_j$ be the character of $T$ sending an element to its $j^{\text{th}}$ diagonal entry;
$\varpi$ be the fundamental weight $\varepsilon_1 + \cdots + \varepsilon_n$;
$P$ be the maximal parabolic subgroup of $G$ corresponding to $\varpi$.

Let $C$ denote the group of $(\mathbb{Z}/2\mathbb{Z})^n$. The symmetric group $S_n$ acts on it by group homomorphisms. Then:

$W$ gets identified with the semi-direct product $C \rtimes S_n$;
$W_P$ gets identified with $S_n$;
$W/W_P$ gets identified with $C$;
a typical element of $C$ is written $\tau = (\tau_1, \ldots, \tau_n)$, where each $\tau_j$ is 0 or 1;
$G/P$ is the symplectic Grassmannian ( the space of $n$-dimensional isotropic subspaces);
$\tau(\varpi)$ is $(-1)^{\tau_1}\varepsilon_1 + \cdots + (-1)^{\tau_n}\varepsilon_n$ for $\tau$ in $W/W_P = C$;
$\geq$ defined by: $\tau \geq \theta$ if $\tau_j \geq \theta_j$ for $1 \leq j \leq n$ (set $1 \gneq 0$ by definition).

The roots are $\pm(\varepsilon_i \pm \varepsilon_j)$ for $1 \leq i < j \leq n$ and $\pm 2\varepsilon_i$, $1 \leq i \leq n$. It is clear that, for any root $\beta$, we have $\langle \varpi, \beta^\vee \rangle = 2(\varpi, \beta)/(\beta, \beta)$ is 2, 1, 0, $-1$, or $-2$. Thus $P$ is of classical type.

Let us now determine the admissible pairs. The simple roots are $\varepsilon_i - \varepsilon_{i+1}$ for $1 \leq i \leq n-1$ and $2\varepsilon_n$. For $\tau$ in $C$, $\langle \tau(\varpi), (2\varepsilon_n)^\vee \rangle = (\tau(\varpi), \varepsilon_n)$ is either 1 or $-1$. The simple reflection corresponding to $\alpha := \varepsilon_i - \varepsilon_{i+1}$ acts on $C$ by interchanging the entries in positions $i$ and $i+1$. If $\tau$ in $C$ is such that its entry in position $i$ is 0 and the entry in position $i+1$ is 1, then $\langle \tau(\varpi), \alpha^\vee \rangle = 2$. It follows easily from these observations that an ordered pair $(\tau, \theta)$ of elements $C$ is admissible if and only if $\tau \geq \theta$, both have the same number of 0's, and this number is $> 0$.

## A.4 Tableaux and the statement of the main theorem

We follow the notation as in §A.1,§A.2. Let $\lambda$ be a dominant integral weight. We assume that $\lambda$ is a sum $\varpi_{d_1} + \cdots + \varpi_{d_m}$ of fundamental weights of classical type. The sequence $d_1, \ldots, d_m$ remains fixed. For the invariant theoretic applications treated in this book, it suffices to consider the case when $\varpi_{d_1} = \ldots = \varpi_{d_m}$. Many assertions made below without proof are however easily seen to be true in this special case. We are thus justified somewhat in skipping these proofs.

A *tableau of shape* $\lambda$ is an $m$-tuple $\sigma = (\sigma_1, \ldots, \sigma_m)$ where $\sigma_j = (\tau_j, \theta_j)$ is an admissible pair in $W/W_{d_j}$. A lift $\tilde{\sigma} = (\tilde{\sigma}_1, \ldots, \tilde{\sigma}_m)$, $\tilde{\sigma}_j = (\tilde{\tau}_j, \tilde{\theta}_j)$, of $\sigma$ to $W$ is *standard* if $\tilde{\tau}_1 \geq \tilde{\theta}_1 \geq \ldots \tilde{\tau}_m \geq \tilde{\theta}_m$ in the Bruhat order. A tableau is *standard* if it has a standard lift. For $m = 1$, a tableau is just an admissible pair (and hence standard). An element $w$ of $W$ *dominates* a standard tableau $\sigma$ if there exists a standard lift $\tilde{\sigma}$ of $\sigma$ with $w \geq \tilde{\sigma}_1$.

For a standard tableau $\sigma$, there exists by [69, Corollary 4.5′] such a standard lift that, for any standard lift $\hat{\sigma}$, we have $\tilde{\tau}_j \leq \hat{\tau}_j$ and $\tilde{\theta}_j \leq \hat{\theta}_j$ for every $j$. This lift $\tilde{\sigma}$ is clearly unique. It is called the minimal lift of $\sigma$. The element $\tilde{\tau}_1$ of the minimal lift is the unique minimal element among those that dominate $\sigma$. It is called the *minimal dominating element* of $\sigma$.

For an element $w$ of $W$, let $X(w)$ denote the Schubert variety in $G/B$ corresponding to $w$, i.e., the closure of the $B$-orbit (with the canonical reduced structure) of the $T$-fixed point of $G/B$ corresponding to $w$. For a dominant weight $\mu$, the associated line bundles on $G/B$ and $X(w)$ are denoted $L_\mu$. There is a natural restriction map $H^0(G/B, L_\mu) \to H^0(X(w), L_\mu)$ between the spaces of sections of $L_\mu$.

For a fundamental weight $\varpi$ of classical type and $P$ the corresponding maximal parabolic, let $\{p(\sigma)\}$, as $\sigma$ varies over admissible pairs of shape $\varpi$, denote the basis of $H^0(G/P, L_\varpi) = H^0(G/B, L_\varpi)$ given by the *first basis theorem*—see remark in §A.2.4 above.

**Theorem A.4.0.1** *Let* $\lambda = \varpi_{d_1} + \cdots + \varpi_{d_m}$ *be any expression for* $\lambda$ *as a sum (possibly repeated) of fundamental weights of classical type. For a standard tableau* $\sigma = (\sigma_1, \ldots, \sigma_m)$ *of shape* $\lambda$, *let* $p(\sigma)$ *denote the image of* $p(\sigma_1) \otimes \cdots \otimes p(\sigma_m)$ *under the natural map*

$$H^0(G/B, L_{\varpi_{d_1}}) \otimes \cdots \otimes H^0(G/B, L_{\varpi_{d_m}}) \longrightarrow H^0(G/B, L_\lambda).$$

*Then the $p(\sigma)$ as $\sigma$ varies over standard tableaux of shape $\lambda$ is a basis for the space $H^0(G/B, L_\lambda)$ of sections. More generally, for $w$ an element of the Weyl group, as $\sigma$ varies over the standard tableaux of shape $\lambda$ dominated by $w$, the restrictions of $p(\sigma)$ to $X(w)$ form a basis for the space of sections $H^0(X(w), L_\lambda)$ on $X(w)$.*

**Corollary A.4.0.2** *The $p(\sigma)$ as $\sigma$ varies over standard tableaux dominated by $w$ form a basis for the homogeneous co-ordinate ring of $X(w)$ for the projective embedding $X(w) \hookrightarrow \mathbb{P}(H^0(G/B, L_\lambda)^*)$.*

*Proof.* It follows from the construction of $p(\tau)$, for $\tau$ an admissible pair, that $p(\tau)$ lives in the linear span in $H^0(G/B, L_\lambda)$ $(= V_\lambda^*)$ of the image in $\mathbb{P}(H^0(G/B, L_\lambda)^*)$ of $X(w)$. The rest follows.

## A.5 Preparation

We need to recall from [69, §3] the following basic—and perhaps not so basic—facts.

Let $G_\mathbb{Z}$ be a $\mathbb{Z}$-form of $G$, i.e., a simply connected semisimple Chevalley group scheme over $\mathbb{Z}$ such that $G_\mathbb{Z} \otimes K = G$. We may choose $T$ and $B$ in $G$ so that there exist $T_\mathbb{Z}$ a maximal torus subgroup scheme of $G_\mathbb{Z}$ and $B_\mathbb{Z}$ a Borel subgroup scheme of $G_\mathbb{Z}$ containing $T_\mathbb{Z}$ such that $T_\mathbb{Z} \otimes K = T$ and $B_\mathbb{Z} \otimes K = B$. We talk of roots, weights, Weyl group, etc. with respect to $T_\mathbb{Z}$ and $B_\mathbb{Z}$ and that would amount to the same as with respect to $T$ and $B$. Thus the combinatorial definitions in §A.2, §A.4 above make sense for the data over $\mathbb{Z}$ and amount to the same as for the original data over the field $K$.

Let $\lambda$ be a dominant integral weight. Let $P_\mathbb{Z}$ be a parabolic subgroup scheme of $G_\mathbb{Z}$ containing $B_\mathbb{Z}$ such that the character $\lambda$ lifts to a character of $P_\mathbb{Z}$. Let $V_{\lambda,\mathbb{Q}}$ be the finite dimensional irreducible representation of $G_\mathbb{Q} := G_\mathbb{Z} \otimes \mathbb{Q}$ with highest weight $\lambda$. Fix a highest weight vector $e = v_\lambda$ in $V_{\lambda,\mathbb{Q}}$—this is determined uniquely up to a non-zero scalar factor in $\mathbb{Q}$. Let $\tau$ be an element of $W/W_P$ (we can omit the subscript $\mathbb{Z}$ as seen in the previous paragraph). Then $\tau$ can be represented by a $\mathbb{Z}$-valued point of the normalizer $N(T_\mathbb{Z})$ of $T_\mathbb{Z}$ in $G_\mathbb{Z}$. Set $e_\tau := \tau e$—this is well-determined up to a sign factor. Set $V_{\lambda,\mathbb{Z}}(\tau) := U_\mathbb{Z}^+ e_\tau$, where $U_\mathbb{Z}$ is the Kostant $\mathbb{Z}$-form of the universal enveloping algebra $U$ of the lie algebra of $G_\mathbb{Q}$, and $U_\mathbb{Z}^+$ the positive part of $U_\mathbb{Z}$.

Set $V_{\lambda,\mathbb{Z}} := V_{\lambda,\mathbb{Z}}(w_0)$, where $w_0$ is the longest element of the Weyl group. Then $V_{\lambda,\mathbb{Z}}$ is a free $\mathbb{Z}$-module such that $V_\mathbb{Z} \otimes \mathbb{Q} = V_{\lambda,\mathbb{Q}}$. It is stable under $U_\mathbb{Z}$. Further $V_\mathbb{Z} = U_\mathbb{Z}^- e$—here $U_\mathbb{Z}^-$ is of course the negative part of $U_\mathbb{Z}$. In particular, every $e_\tau$ is a primitive element of $V_\mathbb{Z}$.

We call $V_\lambda := V_{\lambda,\mathbb{Z}} \otimes K$ the *Weyl module*. and $V_{\lambda,\tau} := V_{\lambda,\mathbb{Z}}(\tau) \otimes K$ the *Demazure module* (also denoted $V_\lambda(\tau)$ sometimes). We have a canonical isomorphism $H^0(G/B, L_\lambda)^* \cong V_\lambda$; and the linear span of the image of $X(\tau)$ under the embedding $G/B \hookrightarrow \mathbb{P}(V_\lambda)$ is $V_{\lambda,\tau}$. Thus, if $R(\tau) = \sum_n R(\tau)_n$ is the homogeneous co-ordinate ring of the image of $X(\tau)$ in $\mathbb{P}(V_\lambda)$, then the first graded piece $R(\tau)_1$ is identified with $V_\lambda(\tau)^*$.

## A.6 The tableau character formula

The *weight* of an admissible pair $(\tau, \theta)$ is defined to be $(\tau(\varpi)+\theta(\varpi))/2$, where $\varpi$ is the shape of the admissible pair. The *weight* of a tableau is the sum of the weights of its component admissible pairs.

**Theorem A.6.0.1 (The tableau character formula)** *Let $w$ be an element of the Weyl group and $\lambda$ a dominant integral weight. The character of the Demazure module $V_{\lambda,w}$ equals the formal sum of the weights of those standard tableaux of shape $\lambda$ that are dominated by $w$.*

*Proof.* Proof follows from [76, page 324, Corollary 1] and the Demazure character formula [10, §3.3].

## A.7 The structure of admissible pairs

**Lemma A.7.0.1** *Let $\varpi$ be a fundamental weight of classical type, $\alpha$ and $\beta$ be simple roots, and $\tau$ be an element of $W/W_\varpi$. If $\langle \tau(\varpi), \alpha^\vee \rangle = 2$ and $\langle \tau(\varpi), \beta^\vee \rangle > 0$, then $\alpha$ and $\beta$ are orthogonal. Similarly if $\langle \tau(\varpi), \alpha^\vee \rangle = -2$ and $\langle \tau(\varpi), \beta^\vee \rangle < 0$, then $\alpha$ and $\beta$ are orthogonal.*

*Proof.* We prove only the first statement, the proof of the second being similar. We have $s_\alpha \tau(\varpi) = \tau(\varpi) - 2\alpha$. If $\alpha$ and $\beta$ are not orthogonal, we get $\langle s_\alpha \tau(\varpi), \beta^\vee \rangle \geq 3$, which contradicts our hypothesis that $\varpi$ is of classical type.

The following proposition is readily verified in the three examples of our interest discussed in §A.3. So there is no need for us (in the context of the present book) to prove it. The reader who is interested in a proof is referred to [61, ], [69, ], or [103, §2]. The $\varepsilon$, $\varphi$, $e$, and $f$ values defined in the proposition have a precise meaning in Littelmann's language of paths [74, 75]: namely $e_\alpha$ and $f_\alpha$ are the "root operators" and $\varepsilon_\alpha(\pi) := \max\{k \geq 0 \mid e_\alpha^k(\pi) \neq 0\}$, and $\varphi_\alpha(\pi) := \max\{k \geq 0 \mid f_\alpha^k(\pi) \neq 0\}$. Even without knowing precisely the definition of the root operators, the reader can nevertheless see that there is a pattern to the values they take.

**Proposition A.7.0.2** *Let $\sigma = (\tau, \theta)$ be an admissible pair, $\alpha$ a simple root, and $s$ the simple reflection corresponding to $\alpha$. Then exactly one of the following twelve mutually exclusive possibilities holds, and accordingly we say that the $\alpha$-type of $\sigma$ is* (A), *or* (B), …, *or* (M)*:*

(A) *$s\tau = \tau$ and $s\theta = \theta$; in this case, we set $\varepsilon_\alpha(\tau, \theta) = 0$ and $\varphi_\alpha(\tau, \theta) = 0$.*
(B) *$s\tau = \tau$ and $s\theta >^2 \theta$; in this case, we set $\varepsilon_\alpha(\tau, \theta) = 0$, $\varphi_\alpha(\tau, \theta) = 1$, and $f_\alpha(\tau, \theta) = (\tau, s\theta)$.*

(C) $s\tau = \tau$ *and* $s\theta <^2 \theta$; *in this case, we set* $\varepsilon_\alpha(\tau,\theta) = 1$ *and* $\varphi_\alpha(\tau,\theta) = 0$, *and* $e_\alpha(\tau,\theta) = (\tau, s\theta)$.
(D) $s\tau >^1 \tau$ *and* $s\theta <^1 \theta$; *in this case, we set* $\varepsilon_\alpha(\tau,\theta) = 0$ *and* $\varphi_\alpha(\tau,\theta) = 0$.
(E) $s\tau >^1 \tau$ *and* $s\theta >^1 \theta$; *in this case, we set* $\varepsilon_\alpha(\tau,\theta) = 0$, $\varphi_\alpha(\tau,\theta) = 1$, *and* $f_\alpha(\tau,\theta) = (s\tau, s\theta)$.
(F) $s\tau >^2 \tau$ *and* $s\theta >^2 \theta$; *in this case, we set* $\varepsilon_\alpha(\tau,\theta) = 0$ *and* $\varphi_\alpha(\tau,\theta) = 0$.
(G) $s\tau >^2 \tau$ *and* $s\theta = \theta$; *in this case, we set* $\varepsilon_\alpha(\tau,\theta) = 0$, $\varphi_\alpha(\tau,\theta) = 1$, *and* $f_\alpha(\tau,\theta) = (s\tau, \theta)$.
(H) $s\tau >^2 \tau$ *and* $s\theta >^2 \theta$; *in this case, we set* $\varepsilon_\alpha(\tau,\theta) = 0$, $\varphi_\alpha(\tau,\theta) = 2$, $f_\alpha(\tau,\theta) = (s\tau, \theta)$, *and* $f_\alpha^2(\tau,\theta) = (s\tau, s\theta)$.
(J) $s\tau <^1 \tau$ *and* $s\theta <^1 \theta$; *in this case, we set* $\varepsilon_\alpha(\tau,\theta) = 1$, $\varphi_\alpha(\tau,\theta) = 0$, *and* $e_\alpha(\tau,\theta) = (s\tau, s\theta)$.
(K) $s\tau <^2 \tau$ *and* $s\theta >^2 \theta$; *in this case, we set* $\varepsilon_\alpha(\tau,\theta) = 1$, $\varphi_\alpha(\tau,\theta) = 1$, $e_\alpha(\tau,\theta) = (s\tau, \theta)$, *and* $f_\alpha(\tau,\theta) = (\tau, s\theta)$.
(L) $s\tau <^2 \tau$ *and* $s\theta <^2 \theta$; *in this case, we set* $\varepsilon_\alpha(\tau,\theta) = 1$, $\varphi_\alpha(\tau,\theta) = 0$, *and* $e_\alpha(\tau,\theta) = (s\tau, \theta)$.
(M) $s\tau <^2 \tau$ *and* $s\theta <^2 \theta$; *in this case, we set* $\varepsilon_\alpha(\tau,\theta) = 2$, $\varphi_\alpha(\tau,\theta) = 0$, $e_\alpha(\tau,\theta) = (\tau, s\theta)$, *and* $e_\alpha^2(\tau,\theta) = (s\tau, s\theta)$.

*The following does not occur:*

(I) $s\tau <^1 \tau$ *and* $s\theta >^1 \theta$.

It is convenient to let the root operators $e_\alpha$ and $f_\alpha$ act as follows on the Weyl group $W$ itself: for $w$ in $W$ and $\alpha$ a simple root, $f_\alpha w$ is the bigger of $w$ and $s_\alpha w$ in the Bruhat order, and $e_\alpha w$ is the smaller of $w$ and $s_\alpha w$. We let the action of $e_\alpha$ and $f_\alpha$ on $W/W_\alpha$ be induced from their action on $W$. (Caution: The above action of the root operators on elements of $W/W_\varpi$ does not always coincide with their action on the elements thought of as trivial admissible pairs.) The following proposition, whose proof is immediate, will be used repeatedly.

**Proposition A.7.0.3** *(i) If* $w \geq w'$ *then* $f_\alpha w \geq f_\alpha w'$ *and* $e_\alpha w \geq e_\alpha w'$.
*(ii) If* $w \gneq w'$ *and* $s_\alpha w \geq w$, *then* $f_\alpha w \gneq f_\alpha w'$.

## A.8 The procedure

Let $\lambda$ be a dominant integral weight. We assume that $\lambda$ is a sum $\varpi_{d_1} + \cdots + \varpi_{d_m}$ of fundamental weights of classical type. The sequence $d_1, \ldots, d_m$ remains fixed.

A *pointed tableau* $(\sigma, s_{i_1} \cdots s_{i_m})$ consists of a standard tableau $\sigma$ of shape $\lambda$ and a reduced expression $s_{i_1} \cdots s_{i_m}$ for an element $w$ which dominates $\sigma$. Described in this section is a procedure for associating to a pointed tableau $(\sigma, s_{i_1} \cdots s_{i_m})$ a monomial differential operator $Y_{i_1}^{(p_1)} \cdots Y_{i_n}^{(p_n)}$, where $p_1, \ldots, p_n$ are non-negative integers—here the $Y_i^{(p)}$ are the generators of $U_{\mathbb{Z}}^-$, with $Y_i$ being the Chevalley generator corresponding to the simple reflection $s_i$ of the Lie algebra $\mathfrak{g}$ of the group $G_{\mathbb{Q}} = G_{\mathbb{Z}} \otimes \mathbb{Q}$. The $p_1, \ldots, p_n$ will turn out positive if $w$ is the minimal dominating element of $\sigma$.

**The Procedure.** Let $(\sigma, s_{i_1} \cdots s_{i_m})$ be a pointed tableau. The procedure works by induction on $n$.

If $n = 0$, the associated monomial is the identity.

Now suppose that $n \geq 1$. Throughout this section $j$ denotes such an integer variable that $1 \leq j \leq m$. We write $s$, $\varepsilon$, $e$, and $f$ instead of $s_{i_1}$, $\varepsilon_{i_1}$, $e_{i_1}$, and $f_{i_1}$ respectively. Let $(\tau_j, \theta_j)$ be the admissible pair $\sigma_j$. Let $J$ be the least integer such that either $s\theta_J > \theta_J$ or $s\tau_{J+1} > \tau_{J+1}$—if $s\tau_j \leq \tau_j$ and $s\theta_j \leq \theta_j$ for all $j$, then $J$ is defined to be $m$; if $s\tau_1 > \tau_1$, then $J$ is defined to be 0. Let $p_1 := \sum_{1 \leq j \leq J} \varepsilon(\sigma_j)$, and $\zeta$ be the tableau defined by $\zeta_j := \sigma_j$ for $j > J$ and $\zeta_j := e^{\varepsilon(\sigma_j)}\sigma_j$ for $j \leq J$. It follows from Proposition A.8.0.1 below that $(\zeta, s_{i_2} \cdots s_{i_m})$ is a pointed tableau. The integers $p_2, \ldots, p_n$ are obtained by applying the induction hypothesis to this pointed tableau.

It follows immediately from the definition of $J$ that

(i) $s\tau_j \leq \tau_j$ for $j \leq J$, $s\theta_j \leq \theta_j$ for $j < J$.

Let $(\tau'_j, \theta'_j)$ be the admissible pair $\zeta_j$. Using (1) and going through all the possibilities listed in Proposition A.7.0.2 for admissible pairs, we find

(2) For $j \leq J$, $(\tau'_j, \theta'_j) = (e\tau_j, e\theta_j)$, and so $s\tau'_j \geq \tau'_j$ and $s\theta'_j \geq \theta + j'$.
(3) For $j \leq J$, $\tau_j = f\tau'_j$. For $j < J$, $\theta_j = f\theta'_j$. Either $\theta_J = \theta'_J$ or $\theta_J = f\theta'_J$ according as $s\theta_j > \theta_J$ or $s\theta_J \leq \theta_J$.

**Proposition A.8.0.1** *Set $w := s_{i_1} \cdots s_{i_n}$. Then*

*(I) $\zeta$ is standard.*
*(II) $sw$ dominates $\zeta$.*
*(III) If $w$ is the minimal dominating element of $\sigma$, then $p_1$ is positive and $sw = s_{i_1} \cdots s_{i_n}$ is the minimal dominating element of $\zeta$.*

*Proof.* (I) Let $\tilde{\tau}_1 \geq \ldots \geq \tilde{\theta}_m$ be a standard lift of $\sigma$. Set $\tilde{\tau}'_j := e\tilde{\tau}_j$ and $\tilde{\theta}'_j := e\tilde{\theta}_j$ for $j \leq J$; $\tilde{\tau}'_j := \tilde{\tau}_j$ and $\tilde{\theta}'_j := \tilde{\theta}_j$ for $j > J$. It follows from item (2) above that $(\tilde{\tau}'_j, \tilde{\theta}'_j)$ is a lift of $\zeta_j$ for all $j$. To prove that this lift is standard, note that, by the definition of $J$, either $e\tilde{\tau}_{J+1} = \tilde{\tau}_{J+1}$, in which case, by Proposition A.7.0.3 (1), $\tilde{\theta}'_j := e\tilde{\theta}_J \geq e\tilde{\tau}_{J+1} = \tilde{\tau}_{J+1}$, or $e\tilde{\theta}_J = \tilde{\theta}_J$, in which case, $\tilde{\theta}'_j := e\tilde{\theta}_J = \tilde{\theta}_J = \tilde{\tau}_{J+1}$. That $\tilde{\tau}'_1 \geq \tilde{\theta}'_1 \geq \ldots \geq \tilde{\tau}'_m \geq \tilde{\theta}'_m$ now follows again from Proposition A.7.0.3 (1).

(II) It follows from Proposition A.7.0.3 (1) that if, in the proof of statement (I), the standard lift $\tilde{\tau}_1 \geq \ldots \tilde{\theta}_m$ of $\sigma$ is so chosen that $w$ dominates it, then $sw = ew$ dominates the standard lift $\tilde{\tau}'_1 \geq \ldots \geq \tilde{\theta}'_m$ of $\zeta$.

(III) Suppose that $p_1 = 0$, that is, $\varepsilon(\sigma_j) = 0$ for $j \leq J$. Then $\zeta = \sigma$. From (II) it follows that $ew = s_{i_2} \cdots s_{i_n}$, which is strictly less than $w$, dominates $\sigma$.

Suppose that $w'$ dominates $\zeta$. Let $\tilde{\tau}'_1 \geq \ldots \geq \tilde{\theta}'_m$ be such a standard lift of $\zeta$ that $w' \geq \tilde{\tau}'_1$. Set $\tilde{\tau}_j := f\tilde{\tau}'_j$ and $\tilde{\theta}_j := f\tilde{\theta}'_j$ for $j < J$, $\tilde{\tau}_j := \tilde{\tau}'_j, \tilde{\theta}_j := \tilde{\theta}'_j$ for $j > J$, $\tilde{\tau}_J := f\tilde{\tau}'_J$, and finally, according as $s\theta_J > \theta_J$ or $s\theta_J \leq \theta_J$, either $\tilde{\theta}_J := \tilde{\theta}'_J$ or $\tilde{\theta}_J := f\tilde{\theta}'_J$. It follows from Proposition A.7.0.3 (1) that $\tilde{\tau}_1 \geq \ldots \geq \tilde{\theta}_m$ and also that $fw' \geq \tilde{\tau}_1$. It follows from item (3) above $\tilde{\tau}_1 \geq \ldots \tilde{\theta}_m$ is a lift of $\sigma$. Thus $fw'$

dominates $\sigma$ and so $fw' \geq w = s_{i_1} \cdots s_{i_n}$. By Proposition A.7.0.3 (1), we have $efw' \geq ew = s_{i_2} \cdots s_{i_n}$. But clearly $w' \geq efw'$. Thus $w' \geq s_{i_2} \cdots s_{i_n}$.

## A.9 The basis

We now state our main theorem. Let $\lambda$ be a dominant integral weight. We assume that $\lambda$ is a sum $\varpi_{d_1} + \cdots + \varpi_{d_m}$ of fundamental weights of classical type. The sequence $d_1,\ldots,d_m$ remains fixed. Fix a highest weight vector $v_\lambda$ of $V_\lambda$.

Let $w$ be an element of the Weyl group. For $\sigma$ a standard tableau dominated by $w$, consider the pointed tableau $(\sigma, s_{i_1} \cdots s_{i_n})$, where $s_{i_1} \cdots s_{i_n}$ is an arbitrarily chosen reduced expression for $w$—the expression could vary with $\sigma$. Set $u(\sigma) := Y_{i_1}^{(p_1)} \cdots Y_{i_n}^{(p_n)} v_\lambda$, where $Y_{i_1}^{(p_1)} \cdots Y_{i_n}^{(p_n)}$ is the differential operator associated by the procedure of §A.8 to the pointed tableau $(\sigma, s_{i_1} \cdots s_{i_n})$. The vector $u(\sigma)$ depends in general on $w$ and $s_{i_1} \cdots s_{i_n}$, but we do not indicate this dependence explicitly in the notation for the vector. If $\lambda$ is a fundamental weight, $u(\sigma)$ is independent of $w$ and $s_{i_1} \cdots s_{i_n}$ as we show in §A.10, and we denote it by $v(\sigma)$ rather than $u(\sigma)$.

**Theorem A.9.0.1** *Let $\lambda$ be a sum of fundamental weights of classical type, $V_\lambda$ the Weyl module $V_{\mathbb{Z}} \otimes K$ introduced in §A.5, and $w$ an element of the Weyl group. The vectors $u(\sigma)$, as $\sigma$ varies over standard tableaux of shape $\lambda$ that are dominated by $w$, form a basis for the Demazure module $V_{\lambda,w}$.*

*Proof.* We have $u(\sigma) := Y_{i_1}^{(p_1)} \cdots Y_{i_n}^{(p_n)} v_\lambda$ for a reduced expression $s_{I_1} \cdots s_{i_n}$ of $w$. So it is clear that $u(\sigma)$ belongs to $V_{\lambda,w}$. The Tableau character formula (Theorem A.6.0.1) tells us that the number of $u(\sigma)$ equals the dimension of $V_{\lambda,w}$. It is therefore enough to show that $u(\sigma)$ span $V_{\lambda,w}$ or that $u(\sigma)$ are linearly independent. It is proved in Proposition A.10.0.3 that, in case $\lambda$ is a fundamental weight, the vectors $u(\sigma)$ span $V_{\lambda,w}$. Let us now indicate the proof of linear independence of $u(\sigma)$ in case $\lambda$ is not a fundamental weight.

We have $\lambda = \varpi_{d_1} + \cdots + \varpi_{d_m}$ with $m \geq 2$.

Let $V$ be the tensor product representation $V_1 \otimes \cdots \otimes V_m$ where $V_j := V_{\varpi_{d_j}}$. Fix $v_1, \ldots, v_m$ highest weight vectors respectively in $V_1, \ldots, V_m$. We can take $V_\lambda$ to be the $G$-submodule of $V$ generated by the highest weight vector $v_\lambda = v_1 \otimes \cdots \otimes v_m$.

Let $X$ denote the set of tableaux of shape $\lambda$, standard or otherwise. Associate to $x = (x_1, \ldots, x_m)$ in $X$ the element $v(x) := v(x_1) \otimes \cdots \otimes v(x_m)$ where $v(x_j)$ is the basis element in $V_j$ corresponding to the admissible pair $x_j$—see Propositions A.10.0.1, A.10.0.3. The set $\{v(x) \mid x \in X\}$ is then a basis of $V$.

The result of the action of a differential operator $Y_{j_1}^{(k_1)} \cdots Y_{j_r}^{(k_r)}$ on $v_\lambda = v_1 \otimes \cdots \otimes v_m$ can be expressed uniquely as an integral linear combination of basis elements $v(x)$. Let

$$u(\sigma) = \sum_{x \in X} c(\sigma, x) v(x), \qquad c(\sigma, x) \text{ non-negative integers})$$

be the expression for $u(\sigma)$ as such a linear combination.

We show in Proposition A.11.0.1 that $c(\sigma, \sigma) = 1$ and that if $c(\sigma, x) \neq 0$, then $\sigma \geq x$ in the lexicographic partial order. The linear independence of $u(\sigma)$ follows.

Let us now see how Theorem A.9.0.1 and its proof lead to a quick proof of the main Theorem A.4.0.1 of standard monomial theory. For a fundamental weight $\varpi$ of classical type, let $\{p(\sigma)\}$ be the basis of $H^0(G/B, L_\varpi)$ dual to the basis $\{v(\sigma)\}$ of $V_\varpi$—here $\sigma$ varies over admissible pairs of shape $\varpi$.

*Proof.* It follows from Proposition A.11.0.1 below that $\langle p(\sigma), u(\sigma)\rangle = 1$ and that $\langle p(\sigma), u(\sigma')\rangle \neq 0$ only if $\sigma' \geq \sigma$. This finishes the proof and also shows that the transition matrix between the dual of the basis $\{u(\sigma)\}$ and the standard monomial theoretic basis $\{p(\sigma)\}$ is unipotent triangular.

## A.10 The first basis theorem

We assume throughout this section that $\lambda = \varpi$ is a fundamental weight of classical type with $P$ as the associated maximal parabolic sub group. A standard tableau $\sigma$ of shape $\varpi$ is therefore an admissible pair in $W/W_\varpi$.

**Proposition A.10.0.1** *The vector $u(\sigma)$ defined in §A.9 depends only on $\sigma$. It is independent of the choice of the dominating element $w$ and also of the choice of the reduced expression for $w$. (To emphasise this fact we denote the vector by $v(\sigma)$ rather than $u(\sigma)$.)*

*Proof.* Consider two pointed tableaux $(\sigma, s_{i_1}\cdots s_{i_n})$ and $(\sigma, s_{j_1}\cdots s_{j_N})$. Let $u$ (respectively $v$) be the vector $u(\sigma)$ calculated by applying the procedure of §A.8 to $(\sigma, s_{i_1}\cdots s_{i_n})$ (respectively to $(\sigma, s_{j_1}\cdots s_{j_N})$). We need to show that $u = v$. We use a double induction, the first on $\text{length}_\varpi(\tau) + \text{length}_\varpi(\theta)$, where $\sigma = (\tau, \theta)$, and the second on $(n, N)$. We say $(n', N') \leq (n, N)$ if $n' \leq n$ and $N' \leq N$. The induction hypothesis will be that the statement holds for all such $(n', N')$ that $(n', N') \lneq (n, N)$.

Suppose that $\text{length}(\tau) + \text{length}(\theta) = 0$. Then $(\tau, \theta) = (1, 1)$. The procedure applied to any pointed tableau $(\sigma, s_{i_1}\cdots s_{i_n})$ will yield the monomial 1, so that $u = v = v_\varpi$.

Now suppose that $e_{i_1}\sigma = 0$. Then $u = u(\sigma, s_{i_2}\cdots s_{i_n})$, and so $u = v$ by the hypothesis of induction on $(n, N)$. We therefore assume from now on that $e_{i_1}\sigma \neq 0$ and $e_{j_1}\sigma \neq 0$. Note that the function $\text{length}(\tau) + \text{length}(\theta)$ strictly decreases when we pass from $\sigma$ to either $e_{i_1}(\sigma)$ or $e_{j_1}(\sigma)$. So, by the induction hypothesis, $u(e_{i_1}(\sigma))$ and $u(e_{j_1}\sigma)$ are unambiguously determined, and we can choose any pointed tableaux to calculate these. We tacitly use this comment in what follows.

Suppose that $i_1 = j_1$. Then $u := Y_{i_1}^{(p_1)} u(e_{i_1}^{\varepsilon_{i_1}(\sigma)}\sigma) =: v$, and we are done. From now on we use $\alpha$ and $\beta$ instead respectively of $\alpha_{i_1}$ and $\alpha_{j_1}$, and assume that $\alpha \neq \beta$.

Suppose that $\alpha$ and $\beta$ are orthogonal. Set $p := \varepsilon_\alpha(\sigma)$ and $q = \varepsilon_\beta(\sigma)$ (these are non-zero by our assumption). We have $p = \varepsilon_\alpha(e_\beta^q(\sigma))$, $q = \varepsilon_\beta(e_\alpha^p\sigma)$, and $e_\alpha^p e_\beta^q\sigma = e_\beta^q e_\alpha^p\sigma$, so that $u := Y_\alpha^{(p)} u(e_\alpha^p\sigma) = Y_\alpha^{(p)} Y_\beta^{(q)} u(e_\beta^q e_\alpha^p\sigma) = Y_\beta^{(q)} Y_\alpha^{(p)} u(e_\alpha^p e_\beta^q\sigma) = Y_\beta^{(q)} u(e_\beta^q\sigma) =: v$. We therefore assume from now on that $\langle \alpha, \beta^\vee\rangle \neq 0$.

From the assumption that $e_\alpha(\sigma) \neq 0$, $e_\beta(\sigma) \neq 0$ we get $s_\alpha\tau \leq \tau$, $s_\beta\tau \leq \tau$. Thus the possibilities for the pair $(\langle\tau(\varpi), \alpha^\vee\rangle, \langle\tau(\varpi), \beta^\vee\rangle)$ are $(0, 0)$, $(0, -1)$, $(0, -2)$, $(-1, -1)$, $(-1, -2)$, and $(-2, -2)$ (the possibilities $(-1, 0)$, $(-2, 0)$, and $(-2, -1)$ can be eliminated by interchanging $\alpha$ and $\beta$). Further, if the possibility $(0, 0)$ (respectively $(0, -1)$) occurs, then the pair $(\langle\theta(\varpi), \alpha^\vee\rangle, \langle\theta(\varpi), \beta^\vee\rangle)$ can only be $(-2, -2)$ (respectively $(-2, -1)$). Using the hypothesis that $\varpi$ is of classical type, it can be seen (see Lemma A.7.0.1) that $\alpha$ and $\beta$ are orthogonal except perhaps in the cases $(-1, -1)$ and $(0, -2)$, so all but these cases are eliminated from consideration. In the case $(-1, -1)$ the corresponding pair for $\theta$ can, by Proposition A.7.0.2 (2), only be $(-1, -1)$. In the case $(0, -2)$, the corresponding pair for $\theta$ can only be $(-2, 0)$ or $(-2, 2)$ (the pair $(-2, -2)$ is eliminated because of our assumption that $\alpha$ and $\beta$ are not orthogonal). We treat these cases one by one. The arguments in all cases are similar.

Suppose that the pairs for $\tau$ and $\theta$ are both $(-1, -1)$. It can readily be verified that $\langle\alpha, \beta^\vee\rangle = \langle\beta, \alpha^\vee\rangle = -1$, $\tau >^1 s_\alpha\tau >^2 s_\beta s_\alpha\tau >^1 s_\alpha s_\beta s_\alpha\tau$ and $\theta >^1 s_\alpha\theta >^2 s_\beta s_\alpha\theta >^1 s_\alpha s_\beta s_\alpha\theta$. We have $u := Y_\alpha u(s_\alpha\tau, s_\alpha\theta) = Y_\alpha Y_\beta^{(2)} Y_\alpha u(s_\alpha s_\beta s_\alpha\tau, s_\alpha s_\beta s_\alpha\theta)$, and similarly $v := Y_\beta Y_\alpha^{(2)} Y_\beta u(s_\beta s_\alpha s_\beta\tau, s_\beta s_\alpha s_\beta\theta)$. But $s_\alpha s_\beta s_\alpha = s_\beta s_\alpha s_\beta$, and $Y_\alpha Y_\beta^{(2)} Y_\alpha = Y_\beta Y_\alpha^{(2)} Y_\beta$. (This relation is obtained from the Serre relations as follows: $0 = Y_\alpha(Y_\alpha Y_\beta^{(2)} - Y_\beta Y_\alpha Y_\beta + Y_\beta^{(2)} Y_\alpha) = (Y_\beta Y_\alpha^{(2)} - Y_\alpha Y_\beta Y_\alpha + Y_\alpha^{(2)} Y_\beta) Y_\beta$.) So we are done.

Suppose that $(\langle\tau(\varpi), \alpha^\vee\rangle, \langle\tau(\varpi), \beta^\vee\rangle) = (0, -2)$ and that $(\langle\theta(\varpi), \alpha^\vee\rangle, \langle\theta(\varpi), \beta^\vee\rangle) = (-2, 0)$. It can readily be verified that $\langle\alpha, \beta^\vee\rangle = \langle\beta, \alpha^\vee\rangle = -1$, $\tau >^2 s_\beta\tau >^2 s_\alpha s_\beta\tau$, and $\theta >^2 s_\alpha\theta >^2 s_\beta s_\alpha\theta$. We have $u := Y_\alpha u(\tau, s_\alpha\theta) = Y_\alpha Y_\beta^{(2)} Y_\alpha u(s_\alpha s_\beta\tau, s_\beta s_\alpha\theta)$, and $v := Y_\beta u(s_\beta\tau, \theta) = Y_\beta Y_\alpha^{(2)} Y_\beta u(s_\alpha s_\beta\tau, s_\beta s_\alpha\theta)$. Using the relation $Y_\alpha Y_\beta^{(2)} Y_\alpha = Y_\beta Y_\alpha^{(2)} Y_\beta$ (see the previous paragraph) we are done.

Suppose that $(\langle\tau(\varpi), \alpha^\vee\rangle, \langle\tau(\varpi), \beta^\vee\rangle) = (0, -2)$ and that $(\langle\theta(\varpi), \alpha^\vee\rangle, \langle\theta(\varpi), \beta^\vee\rangle) = (-2, 2)$. We have $\langle\beta, \alpha^\vee\rangle = -1$, but $\langle\alpha, \beta^\vee\rangle$ can now be either $-1$ or $-2$, and accordingly there are two cases. First suppose that $\langle\alpha, \beta^\vee\rangle = -1$. It can readily be verified that $\tau >^2 s_\beta\tau >^2 s_\alpha s_\beta\tau$, and $s_\beta\theta >^2 \theta >^2 s_\alpha\theta$. We have $u := Y_\alpha u(\tau, s_\alpha\theta) = Y_\alpha Y_\beta Y_\alpha u(s_\alpha s_\beta\tau, s_\alpha\theta)$, and $v := Y_\beta u(s_\beta\tau, \theta) = Y_\beta Y_\alpha^{(2)} u(s_\alpha s_\beta\tau, s_\alpha\theta)$. But $Y_\beta Y_\alpha^{(2)} - Y_\alpha Y_\beta Y_\alpha + Y_\alpha^{(2)} Y_\beta = 0$ is a Serre relation, and $Y_\beta u(s_\alpha s_\beta\tau, s_\alpha\theta) = 0$, for if it were non-zero its $\alpha$-weight would be 3, and we are done.

Now suppose that we are in the situation of the previous paragraph, except that $\langle\alpha, \beta^\vee\rangle$ is now $-2$. It can readily be verified that $\tau >^2 s_\beta\tau >^2 s_\alpha s_\beta\tau >^2 s_\beta s_\alpha s_\beta\tau$ and $s_\beta\theta >^2 \theta >^2 s_\alpha\theta >^2 s_\beta s_\alpha\theta$. We have $u := Y_\alpha u(\tau, s_\alpha\theta) = Y_\alpha Y_\beta^{(2)} Y_\alpha Y_\beta u(s_\beta s_\alpha s_\beta\tau, s_\beta s_\alpha\theta)$ $v := Y_\beta u(s_\beta\tau, \theta) = Y_\beta Y_\alpha^{(2)} Y_\beta^{(2)} u(s_\beta s_\alpha s_\beta\tau, s_\beta s_\alpha\theta)$. Adding the relations $Y_\alpha((\mathrm{ad} Y_\beta)^3 Y_\alpha = 0)$ and $((\mathrm{ad} Y_\alpha)^2 Y_\beta = 0) Y_\beta^{(2)}$, and ignoring terms with $Y_\beta^3$ or $Y_\alpha^3$ since they act as zero on $V_\varpi$, we see that $Y_\beta Y_\alpha^{(2)} Y_\beta^{(2)} = Y_\alpha Y_\beta^{(2)} Y_\alpha Y_\beta$. So $u = v$ and we are done.

**Corollary A.10.0.2** *If $x$, $y$ are admissible pairs and $\alpha$ is a simple root satisfying $x = f_\alpha^t y \neq 0$, then $Y_\alpha^{(t)} v(y)$ is an integral multiple of $v(x)$. If, furthermore, $e_\alpha y = 0$, then $Y_\alpha^{(t)} v(y) = v(x)$.*

*Proof.* Let us assume that $t \geq 1$ for if $t = 0$ the statements are trivially true. By choosing such an element $w$ of the Weyl group that dominates $x$ and for which there is a reduced expression of the type $s_\alpha \cdots$, we see from Proposition A.10.0.1 that $v(x) = u(x, s_\alpha \cdots) = Y_\alpha^{(p_1)} v(e_\alpha^{p_1} x)$, where $p_1 = \varepsilon_\alpha(x)$. Similarly we have $v(y) = Y_\alpha^{p_1 - t} v(e_\alpha^{\varepsilon_\alpha(y)} y)$. But clearly $e_\alpha^{\varepsilon_\alpha(y)} y = e_\alpha^{\varepsilon_\alpha(x)} x$. And $p_1 = t$ if $e_\alpha y = 0$. So both statements are proved.

**Proposition A.10.0.3 (First Basis Theorem)** *Let $w$ be an element of the Weyl group. The vectors $v(\sigma)$, as $\sigma$ varies over admissible pairs in $W/W_\varpi$ that are dominated by $w$, span the Demazure module $V_{\varpi,w}$ (and hence also form a basis—see Theorem A.6.0.1).*

*Proof.* We proceed by induction on the length $n$ of $w$. For $n = 1$, there is only one $v(\sigma)$, namely, the highest weight vector $v_\varpi$, and this spans the one dimensional module $V_{\varpi,1}$.

Let $n \geq 1$. Let $s_\alpha$ be such a simple reflection that $s_\alpha w < w$. The induction hypothesis is that $v(\zeta)$, $\zeta$ dominated by $s_\alpha w$, span $V_{\varpi, s_\alpha w}$. Since $V_{\varpi,w} = SL_{2,\alpha} V_{\varpi, s_\alpha w}$ ($SL_{2,\alpha}$ being the copy of $SL(2)$ in $G$ associated to $\alpha$), it is enough to show that the vectors $Y_\alpha^{(t)} v(\zeta)$ are spanned by $v(\sigma)$, $\sigma$ dominated by $w$.

Fix a $\zeta$ dominated by $s_\alpha w$. First suppose that $f_\alpha^t \zeta \neq 0$. Then by Corollary A.10.0.2, $Y_\alpha^{(t)} v(\zeta)$ is an integral multiple of $v(f_\alpha^t \zeta)$. From Proposition A.7.0.2, it is easily seen that $f_\alpha^t \zeta$ is dominated by $f_\alpha(s_\alpha w) = w$, and so we are done.

We therefore assume that $f_\alpha^t \zeta = 0$. We may also assume that $Y_\alpha^{(t)} v(\zeta) \neq 0$, since otherwise there is nothing to prove. Since the $\alpha$-weight of $Y_\alpha^{(t)} v(\zeta)$ is bounded below by $-2$, the only way this can happen is if $t = 1$ and $\zeta$ has $\alpha$-weight 0. For such a $\zeta$, we

**Claim:** $Y_\alpha^{(2)} X_\alpha v(\zeta) = Y_\alpha v(\zeta)$. We first finish the proof assuming the claim. The vector $X_\alpha v(\zeta)$ belongs to $V_{\varpi, s_\alpha w}$. So it is a linear combination of $v(y)$, $y$ dominated by $s_\alpha w$ and of $\alpha$-weight 2. But, for such a $y$, $f_\alpha^2 y \neq 0$ is dominated by $w$, and $Y_\alpha^{(2)} v(y) = v(f_\alpha^2 y)$ by Corollary A.10.0.2.

It only remains to show the claim. We have $Y_\alpha^{(2)} X_\alpha v(\zeta) = 1/2 Y_\alpha X_\alpha Y_\alpha v(\zeta)$ (since $X_\alpha Y_\alpha = Y_\alpha X_\alpha$ on a vector of $\alpha$-weight 0). Now $Y_\alpha X_\alpha Y_\alpha v(\zeta) = -H_\alpha Y_\alpha v(\zeta) + X_\alpha Y_\alpha^2 v(\zeta)$. We have $Y_\alpha^2 v(\zeta) = 0$ and $H_\alpha Y_\alpha v(\zeta) = -2 Y_\alpha v(\zeta)$ by $\alpha$-weight considerations, and so the claim is proved.

## A.11 Linear independence

We prove in this section the linear independence of the $u(\sigma)$ in Theorem A.9.0.1 in the case $\lambda$ is a non-fundamental weight. The linear independence follows immediately from Proposition A.11.0.1 below.

We impose on the set of admissible pairs the lexicographic partial order and on the set $X$ of all tableaux (standard or otherwise) of shape $\lambda$ the induced lexicographic order. More precisely, for admissible pairs $(\tau, \theta)$, $(\tau', \theta')$ in the same $W/W_{\varpi}$, we write $(\tau, \theta) \gneq (\tau', \theta')$ if $\tau \gneq \tau'$ or if $\tau = \tau'$ and $\theta \gneq \theta'$. For $x$, $x'$ in X, we write $x \geq x'$ if $x_j \gneq x'_j$ for the least such $j$ that $x_j \neq x'_j$.

**Proposition A.11.0.1** *With notation as in the proof of Theorem A.9.0.1, $c(\sigma, \sigma) = x$ and $c(\sigma, x) \neq 0$ only if $\sigma \geq x$. In other words,*

$$u(\sigma) = v(\sigma) + \sum_{\{x \mid \sigma \gneq x\}} c(\sigma, x) v(x)$$

*Proof.* We proceed by induction on the length $n$ of the reduced expression in the pointed tableau $(\sigma, s_{i_1} \cdots s_{i_n})$ used to calculate $u(\sigma)$. For $n = 0$, we have $u_\sigma = v_\sigma = v_\lambda$ and so the assertion holds.

Let now $n \geq 1$. Let $J$, $p_1$, and $\zeta$ be as in procedure of §A.8. We write just $Y$, $e$, and $f$ respectively for $Y_{i_1}$, $e_{i_1}$ and $f_{i_1}$. The induction hypothesis is that

$$u(\zeta) = v(\zeta) + \sum_{\{y \mid \zeta \gneq y\}} c(\zeta, y) v(y)$$

and so we have

$$u(\sigma) := Y^{(p_1)} u(\zeta) = Y^{(p_1)} v(\zeta) + \sum_{\{y \mid \zeta \gneq y\}} c(\zeta, y) Y^{(p_1)} v(y)$$

For $y$ in $X$, let $Z(y)$ denote the set of such $m$-tuples $\underline{t} = (t_1, \ldots, t_m)$ of non-negative integers that $\sum t_j = p_1$ and, for all $j$, $Y^{(t_j)} v(y_j) \neq 0$. Given such a $\underline{t}$, let $Y^{\underline{t}} v(y)$ denote the element $Y^{(t_1)} v(y_1) \otimes \cdots \otimes Y^{(t_m)} v(y_m)$. We have

$$Y^{(p_1)} v(y) = \sum_{\underline{t} \in Z(y)} Y^{\underline{t}} v(y)$$

It follows from Corollary A.10.0.2 that if $f^{t_j} y_j \neq 0$, then $Y^{(t_j)} v(y_j)$ is an integral multiple of $v(f^{t_j} y_j)$. But it can happen that $f^{t_j} y_j = 0$ and yet $Y^{(t_j)} v(y_j) \neq 0$. In any case $Y^{(t_j)} v_{y_j}$ is uniquely an integral linear combination of basis vectors $v(x_j)$ of $V_j$ associated to admissible pairs $x_j$ in $W/W_{d_j}$. We can therefore write

$$Y^{\underline{t}} v(y) = \sum_{x \in B(y, \underline{t})} b(y, \underline{t}, x) v(x)$$

where the $b(y, \underline{t}, x)$ are integers and $B(y, \underline{t})$ is the appropriate subset of $X$.

With this notation, it is enough to prove the following

**Claim:** there exists such an element $\underline{t}_0$ of $Z(\zeta)$ that (1) $b(\zeta, \underline{t}_0, \sigma) = 1$, (2) for any such $\underline{t}$ in $Z(\zeta)$ that $\underline{t} \neq \underline{t}_0$, we have $\sigma \gneq x$ for $x$ in $B(\zeta, \underline{t})$, and (3) for any such $y$, $\underline{t}$, and $x$ that $\zeta \gneq y$, $\underline{t} \in Z(y)$, and $x \in B(y, \underline{t})$, we have $\sigma \gneq x$.

We first prove (1). Set $\underline{t}_0 := (\varepsilon(\sigma_1), \ldots, \varepsilon(\sigma_J), 0, \ldots, 0)$. We have $\sum t_{0,j} = p_1$ from the definition of $p_1$. We have $f^{t_{0,j}} \zeta_j = \sigma_j$ from the definition of $\zeta$. Since

$e\zeta_j = 0$ for $j \le J$, it follows from Corollary A.10.0.2 that $Y^{\underline{t}_0} v(\zeta) = v(\sigma)$. This proves (1).

We next prove (2) and (3) both together. Suppose that $\zeta \geq y$ and $\underline{t} \in Z(y)$, and that $\underline{t} \neq \underline{t}_0$ if $\zeta = y$. Let $s$ be the least such integer that either $\zeta_s \neq y_s$ or $t_{0,s} \neq t_s$. By assumption, such an $s$ exists. We clearly have $\sigma_j = x_j$ for $j < s$. So it is enough to show that $\sigma_s \gneqq x_s$.

To prove that $\sigma_s \gneqq x_s$, we first make some simplifications. Suppose that $t_s = 0$. Then either $t_{0,s} > 0$ or $\zeta_s \gneqq y_s$. In either case we have $\sigma_s = f_\alpha^{t_{0,s}} \zeta_s \gneqq y_s = x_s$ (recall that, for $p > 0$, if $f_\alpha^p(\tau, \theta) \neq 0$, then $f_\alpha^p(\tau, \theta) \gneqq (\tau, \theta)$, by Proposition A.7.0.2 (2)), and we are done. If $s > J$, then $t_{0,s} = t_s = 0$ (because $\sum t_{0,j} = \sum t_j = p_1$, and $t_{0,j} = 0$ for $j > J$), so we may assume that $s \le J$. If $s = J$, we may further assume that either $\zeta_s = y_s$ and $t_{0,s} > t_s$, or $\zeta_s \gneqq y_s$ and $t_{0,s} \geq t_s$. To sum up, we have to show that $\sigma_s \gneqq x_s$ under each of the following sets of conditions:

- $s < J$, $\zeta_s = y_s$, $t_{0,s} \neq t_s$
- $s < J$, $\zeta_s \gneqq y_s$
- $s = J$, $\zeta_s = y_s$, $t_{0,s} \gneqq t_s$
- $s = J$, $\zeta_s \gneqq y_s$, $t_{0,s} \geq t_s$

We want to now transfer the burden of proof to the next proposition which is purely about admissible pairs. To this end, we note that, by statement (2) in §A.8, for $s \le J$, we have $s\tau_s \geq \tau_s$ and $s\theta_s \geq \theta_s$, where $\zeta_s = (\tau_s, \theta_s)$ (in other words, $\zeta_s$ has $\alpha$-type (A), (B), (E), (G), or (H) of Proposition A.7.0.2). We note also that $t_{0,s} = \varphi_\alpha(\zeta_s)$ for $s < J$. It is easily checked that the four sets of conditions above correspond respectively to the four sets of conditions in the proposition below. The proof will therefore be complete once the proposition below is proved.

**Proposition A.11.0.2** *Let $\alpha$ be a simple root, $\zeta$ and $y$ be admissible pairs (in the same $W/W_\varpi$). Assume that $\zeta$ has $\alpha$-type (A), (B), (E), (G), or (H) of Proposition A.7.0.2. Let $t_0 \geq 0$ and $t > 0$ be integers satisfying $\varphi_\alpha(\zeta) \geq t_0$ and $Y_\alpha^{(t)} v(y) \neq 0$. Set $\sigma = f_\alpha^{t_0}(\zeta)$. If any of the following four sets of conditions holds, then $\sigma \gneqq x$, where $x$ is any such admissible pair that $v(x)$ occurs with non-zero coefficient $c(x)$ in the expression $Y_\alpha^{(t)} v(y) = \sum c(x) v(x)$ for $Y_\alpha^{(t)} v(y)$ as an integral linear combination of the basis vectors $v(x)$.*

*(i)* $\zeta = y$, $t_0 = \varphi_\alpha(\zeta)$, $t \neq t_0$
*(ii)* $\zeta \gneqq y$, $t_0 = \varphi_\alpha(\zeta)$
*(iii)* $\zeta = y$, $t_0 \gneqq t$
*(iv)* $\zeta \gneqq y$, $t_0 \geq t$

*Proof.* The case when (3) holds is the easiest, so we consider that first. By Corollary A.10.0.2, $x = f_\alpha^t y$. Now the conclusion that $\sigma \gneqq x$ follows from the following statement, which can readily be verified from the list of possibilities for admissible pairs in Proposition A.7.0.2 (2): for an admissible pair $z$, if $\varphi_\alpha(z) \geq r \gneqq r$, then $f_\alpha^{r_0} z \gneqq f_\alpha^r z$.

We now make some reductions. We use the following notation in the sequel: for an admissible pair $z$, we write $\tau(z)$ and $\theta(z)$ for its components in $W/W_{\varpi}$—in other words, $z = (\tau(z), \theta(z))$.

Suppose (1), (2), or (4) holds (in fact, what follows applies even if (3) holds, but that is now irrelevant). Then $\tau(\sigma) = f_\alpha\tau(\zeta)$ (because $t \geq 1$ and the $\alpha$-type of $\zeta$ is not (D) or (F)). Since $Y_\alpha v(y)$ belongs to the Demazure module corresponding to $f_\alpha\tau(y)$, we have $f_\alpha\tau(y) \geq \tau(x)$. Now, if $\tau(\zeta) \gneq \tau(y)$, we have, by Proposition A.7.0.3 (2), $\tau(\sigma) = f_\alpha\tau(\zeta) \gneq f_\alpha\tau(y) \geq \tau(x)$, and we are done. So we may assume that the following holds:

($\star$) $\tau(\zeta) = \tau(y)$ and $\tau(\sigma) = f_\alpha\tau(\zeta) = f_\alpha\tau(y) = \tau(x)$. In particular, $\theta(\zeta) \geq \theta(y)$ and if $\zeta \gneq y$ then $\theta(\zeta) \gneq \theta(y)$.

In case (4), if $t_0 = \varphi_\alpha(\zeta)$, then (2) holds, so we replace (4) by the following more stringent condition:

(4)$'$ $\zeta \gneq y$, $\zeta$ is of type $\alpha$-type (H), and $t_0 = t = 1$.

Now we prove the assertion under the additional assumption that $f_\alpha^t y \neq 0$. By Corollary A.10.0.2, we have $x = f_\alpha^t y$. If (1) holds, then (3) holds also, and so we are done. If (2) holds, we have, by Proposition A.7.0.3 (2), $\theta(\sigma) = f_\alpha\theta(\zeta) \gneq f_\alpha\theta(y) \geq \theta(x)$, and we are done. If (4)$'$ holds, it is easily checked that $\theta(\sigma) = \theta(\zeta) \gneq \theta(y) = \theta(x)$, and we are done.

We are therefore justified in assuming from now on that $f_\alpha^t y = 0$. This means in particular that the $\alpha$-weight of $y$ is non-positive, and since the $\alpha$-weight of $Y_\alpha v(y)$ is bounded below by $-2$, we conclude that $t = 1$ and $y$ has $\alpha$-weight 0 ($y$ can have $\alpha$-type only (A), (D), or (F) of Proposition A.7.0.2), and so $x$ has $\alpha$-weight $-2$, that is, $\alpha$-type of $x$ can only be (M). By ($\star$), we have $\tau(y) = \tau(\zeta)$ and $f_\alpha\tau(y) = \tau(x)$, so that $\tau(y) = e_\alpha\tau(x)$ and $\tau(x)\varpi = \tau(y)\varpi - 2\alpha$. On the other hand, by equating the weights of $x$ and $Y_\alpha v(y)$, we get $\tau(x)\varpi + \theta(x)\varpi = \tau(y)\varpi + \theta(y)\varpi - 2\alpha$. From these two equations, we conclude that $\theta(x)\varpi = \theta(y)\varpi$, that is, $\theta(x) = \theta(y)$ in $W/W_{\varpi}$.

Suppose that (2) or (4)$'$ holds. By ($\star$) we have $\theta(\zeta) \gneq \theta(y)$. Combining this with the conclusion of the preceding paragraph, we get $\theta(\sigma) \geq \theta(\zeta) \gneq \theta(y) = \theta(x)$, which along with ($\star$) shows that $\sigma \gneq x$.

Suppose that (1) holds. If $t_0 \gneq t$, then (3) holds, so we are done. So assume that $t \gneq t_0$. It was shown above that we may assume that $t = 1$ and that $x$ has $\alpha$-type (M). So $t_0 = 0$, and $\zeta$ has $\alpha$-type (A), which means in particular that $f_\alpha\tau(\zeta) = \tau(\zeta)$. Combining this with the assumption ($\star$), we see that $\tau(\zeta) = \tau(x)$ but this is absurd, for the same element of $W/W_{\varpi}$ cannot be the first component of admissible pairs of $\alpha$-type (A) and (M).

## A.12 Arithmetic Cohen-Macaulayness & Arithmetic normality for Schubert varieties

In this section, we comment on the Cohen-Macaulayness and normality of Schubert varieties in $G/P$, with $P$ being a maximal parabolic subgroup of classical type (covering our application to Invariant Theory—the three examples A, B, and C of Chapter 1). For such a $G/P$, the Cohen-Macaulayness and normality for Schubert varieties may be concluded just using the standard monomial basis as given by Theorem A.4.0.1.

As mentioned in the Introduction, in recent times, among the several techniques of proving the Cohen-Macaulayness of algebraic varieties, particularly those that are related to Schubert varieties, two techniques have proved to be quite effective, namely, Frobenius splitting technique and deformation technique. Frobenius splitting technique is used in [104], for example, for proving the (arithmetic) Cohen-Macaulayness of Schubert varieties.

The deformation technique consists in constructing a flat family over $\mathbb{A}^1$, with the given variety as the generic fiber (corresponding to $t \in K$ invertible). If the special fiber (corresponding to $t = 0$) is Cohen-Macaulay, then one may conclude the Cohen-Macaulayness of the given variety. Going by the works of [12, 17, 21, 33, 46], it is reasonable to believe that if there is a "standard monomial basis" for the coordinate ring of the given variety, then the deformation technique will work well in general (using the "straightening relations").

A proof for arithmetic Cohen-Macaulayness for Schubert varieties in the Grassmannian was presented in Chapter 4, using local cohomology. We present yet another proof using the deformation technique.

## A.12.1 Arithmetic Cohen-Macaulayness

### Degenerations of Schubert varieties in the Grassmannian

Let $G = SL(n)$. Fix $r$, $1 \leq r \leq n-1$. Consider the polynomial algebra $P = K[x_\beta, \beta \in T_\tau]$, where $\tau \in I_{r,n}$ and $T_\tau = \{\phi \in I_{r,n} \mid \phi \leq \tau\}$

**Definition A.12.1.1** *For $\tau \in I_{r,n}$, let $I_\tau$ be the ideal of $P$ generated by the set of monomials*

$$\{x_\gamma x_\delta \mid \gamma \text{ and } \delta \text{ are non-comparable elements of } T_\tau\}.$$

*We define the* discrete algebra associated to $\tau$ *(or* to the Schubert variety $X(\tau)$ *(in the Grassmannian $G_{r,n}$)) to be the $K$-algebra $P/I_\tau$, and denote it by $D_\tau$.*

The main result of this subsection is that Spec $D_\tau$ is a degeneration of $\widehat{X(\tau)}$ (the cone over the Schubert variety $X(\tau)$), i.e. there exists a one parameter flat family (i.e., a family parametrized by *Spec* $K[[t]]$) with generic fiber ($t$ invertible) $\widehat{X(\tau)}$ and special fiber ($t = 0$) Spec $D_\tau$. We first recall the following relation (cf. (*)) from Chapter 4.

If $\gamma$ and $\delta$ are non-comparable elements of $T_t$, then we have a quadratic relation in $R_\tau$ (the homogeneous coordinate ring of $X(\tau)$ for the Plücker embedding) of the form

$$p_\gamma p_\delta = \sum_{\lambda,\mu} a_{\lambda\mu} p_\lambda p_\mu, \quad a_{\lambda\mu} \in K^*, \qquad (*)$$

where the monomials in the right hand side are standard, i.e. $\lambda \geq \mu$; further, $\lambda >$ both $\gamma$ and $\delta$, and $\mu <$ both $\gamma$ and $\delta$. We shall denote by $Q$ the set of all such quadratic relations for various pairs of non-comparable elements $\gamma, \delta$ in $T_\tau$.

**Lemma A.12.1.2** *There exists a function $\theta : T_\tau \to \mathbb{Z}_+$ such that given any quadratic relation $p_\gamma p_\delta = \sum_{\lambda,\mu} a_{\lambda\mu} p_\lambda p_\mu$ in $Q$, we have $\theta(\lambda)+\theta(\mu) > \theta(\gamma)+\theta(\delta)$ for all $\lambda$, $\mu$ appearing in the sum.*

*Proof.* Since $T_\tau$ is a finite partially ordered set, we can define an order function $\mathrm{ord} : T_\tau \to \mathbb{Z}_+$ such that $\beta > \beta'$ implies $\mathrm{ord}\,\beta > \mathrm{ord}\,\beta'$. For $\beta \in T_\tau$ set

$$\theta(\beta) = N^{\mathrm{ord}\,\beta},$$

where $N$ is a positive integer. We claim that for a sufficiently large $N$, $\theta$ is the required function. Indeed, using the relation (*) above, we have $\mathrm{ord}\,\gamma < \mathrm{ord}\,\lambda$ and $\mathrm{ord}\,\delta < \mathrm{ord}\,\lambda$; therefore, choosing a sufficiently large $N$, we can achieve the inequality

$$N^{\mathrm{ord}\,\lambda} + N^{\mathrm{ord}\,\mu} > N^{\mathrm{ord}\,\gamma} + N^{\mathrm{ord}\,\delta}$$

for all quadratic relations in $Q$.

We recall the following well known Lemma from commutative algebra.

**Lemma A.12.1.3** *Let $M$ be a finitely generated module over the power series ring $A = K[[t]]$ and let $L$ be the quotient field of $A$. Then $M$ is a free module if and only if*

$$dim_K M \otimes_A K \leq dim_L M \otimes_A L.$$

*Proof.* If $M$ is free, then obviously the inequality holds.
Conversely, suppose that the inequality holds. If $dim_k M \otimes_A K = m$, then by Nakayama's lemma $M$ can be generated by $m$ elements and therefore we have an exact sequence

$$0 \to N \to A^m \to M \to 0$$

of $A$-modules. Tensoring the sequence by $L$ and using the inequality, we conclude that $N \otimes_A L = 0$. Since $N$ is torsion free, $N = 0$, and hence $M$ is isomorphic to $A^m$.

**Theorem A.12.1.4** *Let $\tau \in I_{r,n}$. There exists a flat family over* Spec $K[[t]]$ *whose generic fiber ($t$ invertible) is $\widehat{X(\tau)}$ and the special fiber ($t = 0$) is* Spec $D_\tau$, *where $D_\tau$ is the discrete algebra associated to $\tau$.*

*Proof.* Let $R_\tau$ be the homogeneous coordinate ring of $X(\tau)$ (for the Plučker embedding) and let $Q$ be as above (namely, the set of quadratic relations of the form (*)). For each quadratic relation $p_\gamma p_\delta = \sum_{\lambda,\mu} a_{\lambda\mu} p_\lambda p_\mu$ in $Q$, consider the polynomial

$$F_{\gamma,\delta} = x_\gamma x_\delta - \sum_{\lambda,\mu} a_{\lambda\mu} x_\lambda x_\mu$$

in $P = K[x_\phi, \phi \in T_\tau]$, and let $J$ be the ideal of $P$ generated by all such polynomials. If $\overline{x}_\phi$ denotes the canonical image of $x_\phi$ in $P/J$, then it is clear that $P/J$ is generated by standard monomials in $\overline{x}_\phi$, $\phi \in T_\tau$. Moreover, we have a canonical surjective map $f_\tau : P/J \to R_\tau$ which sends $\overline{x}_\phi$ to $p_\phi$. Since $R_\tau$ has a basis consisting of standard monomials in $p_\phi$'s, it follows that $f_\tau$ is injective. Hence $P/J$ is isomorphic to $R_\tau$.

Now using Lemma A.12.1.2, choose a function $\theta : T_\tau \to \mathbb{Z}_+$ such that for each quadratic relation $p_\gamma p_\delta = \sum_{\lambda,\mu} a_{\lambda\mu} p_\lambda p_\mu$ in $Q$, $\theta(\lambda) + \theta(\mu) - \theta(\gamma) - \theta(\delta) > 0$. Further, let $A = K[[t]]$. Corresponding to each quadratic relation in $Q$ as above, consider the polynomial

$$F_{\gamma,\delta,t} = x_\gamma x_\delta - \sum_{\lambda,\mu} a_{\lambda\mu} t^{\theta(\lambda)+\theta(\mu)-\theta(\gamma)-\theta(\delta)} x_\lambda x_\mu$$

in $P_A := A[x_\phi, \phi \in T_\tau]$. Denoting by $\mathcal{I}$ the ideal generated by the polynomials $F_{\gamma,\delta,t}$ in $P_A$, we set $\mathcal{R}_\tau = P_A/\mathcal{I}$. Let $L$ be the quotient field of $A$ and let $P_L = L[x_\phi, \phi \in T_\tau]$. If $\tilde{J}$ is the ideal of $P_L$ generated by the polynomials $F_{\gamma,\delta}$ and $\tilde{\mathcal{I}}$ is the ideal generated by the polynomials $F_{\gamma,\delta,t}$, then it is easy to see that $P_L/\tilde{\mathcal{I}} \cong P_L/\tilde{J}$, the isomorphism being induced by the $L$-automorphism of $P_L$ which sends $x_\beta$ to $t^{-\theta(\beta)} x_\beta$. Thus we have

$$\mathcal{R}_\alpha \otimes_A L = P_A/\mathcal{I} \otimes_A L \cong P_L/\tilde{\mathcal{I}} \cong P_L/\tilde{J} \cong P/J \otimes_K L \cong R_\tau \otimes_K L$$

and $\mathcal{R}_\tau \otimes_A K \cong \mathcal{R}_\tau / t\mathcal{R}_\tau \cong D_\tau$. This shows that the generic fiber ($t$ invertible) of $\mathcal{R}_\tau$ is $R_\tau$ and the special fiber ($t = 0$) is $D_\tau$.

Now we show that $\mathcal{R}_\tau$ is flat over $A$. From the definition of $D_\tau$ it is clear that $D_\tau$ is generated by standard monomials in the canonical images $\overline{x}_\phi$ of $x_\phi$, $\phi \in T_\tau$. Therefore, using the above isomorphisms, it follows that for every $m \geq 0$, we have,

$$\begin{aligned} dim_K(\mathcal{R}_\tau)_m \otimes_A K &= dim_K(D_\tau)_m \leq dim_K(R_\tau)_m \\ &= dim_L(R_\tau)_m \otimes_K L = dim_L(\mathcal{R}_\tau)_m \otimes_A L, \end{aligned}$$

where $(\mathcal{R}_\tau)_m$, $(D_\tau)_m$, $(R_\tau)_m$ denote the $m$-th graded components of the respective rings. Thus, in view of Lemma A.12.1.3, each graded component of $\mathcal{R}_\tau$ is a free $A$-module, and hence $\mathcal{R}_\tau$ is a flat $A$-algebra.

**Corollary A.12.1.5** *(i) $D_\tau$ has a basis consisting of standard monomials in $\overline{x}_\phi$, $\phi \in T_\tau$.*

*(ii) $D_\tau$ is reduced.*

*(iii) If* $\mathbf{P} = \text{Proj}\, P$, *where* $P = K[x_\phi, \phi \in T_\tau]$, *then the degree of $X(\tau)$ in the Plücker embedding is the same as the degree of* Proj $D_\tau$ *in* $\mathbf{P}$.

*Proof.* (i) Since for each $m \geq 0$, $(\mathcal{R}_\tau)_m$ is a free $A$-module, we have

$$dim_K(\mathcal{R}_\tau)_m \otimes_A K = dim_L(\mathcal{R}_\tau)_m \otimes_A L,$$

and therefore the inequalities $(*)$ in the proof of the above theorem imply that

$$dim_K(D_\tau)_m = dim_K(R_\tau)_m, \text{ for each } m \geq 0. \qquad (**)$$

Since distinct standard monomials on $X(\tau)$ of degree $m$ form a basis of $(R_\tau)_m$, we obtain $dim_K(D_\tau)_m = \#\{$standard monomials on $X(\tau)$ of degree $m\}$. This together with the fact that $(D_\tau)_m$ is generated by standard monomials (of degree $m$) implies that $(D_\tau)_m$ has a basis consisting of standard monomials of degree $m$.

(ii) Let $f = \sum a_i f_i$, where $a_i \in K$, $a_i \neq 0$ and $f_i$ are standard monomials in $D_\tau$. From the definition of $D_\tau$ it is clear that any product of standard monomials in $D_\tau$ is either zero or a standard monomial. Therefore, using (i) it follows that $f^m \neq 0$ for every $m \geq 0$. This proves that $D_\tau$ is reduced.

(iii) The equality $(**)$ in the proof of (i) implies that the Hilbert polynomial of $X(\tau)$ is the same as that of $D_\tau$, which proves the assertion.

Let $T$ be a subset of $I_{r,n}$. By a *chain of length m in T*, we shall mean a sequence $\Lambda = (\lambda(0), \lambda(1), \ldots, \lambda(m))$ of elements of $T$ such that $\lambda(i) > \lambda(i+1)$, $0 \leq i \leq m-1$. We say that $\Lambda$ is *maximal* if it is not a subchain of any chain in $T$ different from $\Lambda$.

For a chain $\Lambda$ in $T_\tau$, $\tau \in I_{r,n}$ let $L(\Lambda)$ denote the linear space in $\mathbf{P} = \operatorname{Proj} P$ defined by the equations $x_\phi = 0$, $\phi \in T_\tau \setminus \Lambda$, where $T_\tau \setminus \Lambda$ denotes the set of elements in $T_\tau$ not appearing in $\Lambda$.

**Proposition A.12.1.6** *If $M(T_\tau)$ denotes the set of maximal chains in $T_\tau$, then*

$$\operatorname{Proj} D_\tau = \bigcup_{\Lambda \in M(T_\tau)} L(\Lambda) \quad \text{scheme theoretically,}$$

*i.e. the linear spaces $L(\Lambda)$, $\Lambda \in M(T_\tau)$, are the irreducible components of* Proj $D_\tau$.

*Proof.* The ideal of the scheme theoretic union on the right hand side is

$$I = \bigcap_{\Lambda \in M(T_\tau)} I_\Lambda,$$

where $I_\Lambda$ is the ideal of $L(\Lambda)$ (note that $I_\Lambda$ is generated by the set $\{x_\phi \mid \phi \in T_\tau \setminus \Lambda\}$). We shall show that $I$ is the ideal defining Proj $D_\tau$, i.e. $I$ is generated by the set of monomials $S = \{x_\gamma x_\delta \mid \gamma, \delta$ are non-comparable elements of $T_\tau\}$. In fact, if $\gamma, \delta$ are non-comparable in $T_\tau$, then both of them together cannot belong to a maximal chain in $T_\tau$. This shows that $x_\gamma x_\delta \in I_\Lambda$ for all $\Lambda \in M(T_\tau)$, and hence $S \subset I$. Now, if $F \in I$, it is easy to see that every monomial of $F$ is in $I$. Thus, to complete the proof it suffices to show that every monomial $F \in I$ is divisible by an element of $S$. In fact, since $F \in I_\Lambda$ for all $\Lambda \in M(T_\tau)$, it is clear that $F$ is not a standard monomial, and hence it is divisible by an element of $S$.

**Corollary A.12.1.7** *The degree of $X(\tau)$ in the Plücker embedding is equal to the number of maximal chains in $T_\tau$.*

The proof is immediate from Theorem A.12.1.4 and the above Proposition.

**Stanley-Reisner ring** ([106])

In this subsection, we prove the Cohen-Macaulayness of $\widehat{X(\tau)}$ using Theorem A.12.1.4. Given a finite partially ordered set $H$, let $A = K[x_\alpha,\ \alpha \in H]$, $I_H$ the ideal generated by $\{x_\gamma x_\delta \mid \gamma, \delta$ are non-comparable elements of $H\}$. Then the ring $R_H := A/I_H$ is called the *Stanley-Reisner ring* associated to $H$.

**Theorem A.12.1.8** Spec $D_\tau$ *is Cohen-Macaulay.*

*Proof.* We have, for $H = T_\tau$, $R_H = D_\tau$, and the Cohen-Macaulayness of Stanley-Reisner ring $R_{T_\tau}$ follows from [6].

Combining Theorem A.12.1.4 with the above Theorem, we obtain

**Theorem A.12.1.9** $\widehat{X(\tau)}$(= Spec $R_\tau$) *is Cohen-Macaulay.*

Using deformation technique and standard monomial basis, DeConcini-Lakshmibai prove the arithmetic Cohen-Macaulayness of Schubert varieties in $G/P$, $P$ being a maximal parabolic subgroup of classical type.

**Arithmetic normality**

As above, let $P$ be a maximal parabolic subgroup of classical type of $G$; let $X$ be a Schubert variety in $G/P$. Consider the projective embedding

$$X \hookrightarrow \mathbb{P}(V)$$

where $V = H^0(G/P, L)^*$. As a consequence of Theorem A.4.0.1, we have that the restriction map

$$H^0(\mathbb{P}(V), L^m) \to H^0(X, L^m),\ m \in \mathbb{N}$$

is surjective. Hence if we know the (geometric) normality of $X$, then we would obtain the arithmetic normality of $X$ (note that if a projective variety $X \hookrightarrow \mathbb{P}^n$ is normal, then it is arithmetically normal if and only if the restriction map $H^0(\mathbb{P}^n, L^m) \to H^0(X, L^m)$ is surjective for all $m$, $L$ being the tautological line bundle on $\mathbb{P}^n$). We have by [21], $\widehat{X}$ is Cohen-Macaulay. Hence normality of $X$ would follow once we know that $X$ is regular in codimension one, i.e., the singular locus of $X$ has codimension at least two (in view of Serre Criterionfor normality - a ring $A$ is Cohen-Macaulay if and only if $A$ has $S_2$ and $R_1$). But one knows this to be true, thanks to Chevalley (cf. [15]). We have included below a proof (cf. [69]) of this result of Chevalley. We first fix some notation:

Let $G, B, T$ be as in § A.1. Let $X(\tau)$ be a Schubert variety in $G/B$. Let $X(w)$ be a Schubert divisor in $X(\tau)$ moved by a simple root $\alpha$, i.e., $\tau = s_\alpha w$, and $l(\tau) = l(w) + 1$. Let $SL(2, \alpha)$ denote the copy of $SL(2)$ in $G$ associated to $\alpha$. Set

$$B_\alpha = B \cap SL(2, \alpha)$$

Let

(*) $$Z_w = SL(2, \alpha) \times^{B_\alpha} X(w)$$

i.e., $Z_w$ is the quotient variety modulo the equivalence relation in $SL(2,\alpha) \times X(w)$:

$$(g, x) \sim (gb, b^{-1}x),\ g \in SL(2,\alpha), b \in B_\alpha, x \in X(w)$$

Note that if we define an action of $B_\alpha$ on $SL(2,\alpha) \times X(w)$ by

$$(g, x) \cdot b = (gb, b^{-1}x),\ g \in SL(2,\alpha), b \in B_\alpha, x \in X(w)$$

then this action is a free action and $Z_w$ is the orbit space $(SL(2,\alpha) \times X(w))/B_\alpha$. We have a canonical morphism

$$(**) \qquad p : Z_w \to \mathbb{P}^1 = SL(2,\alpha)/B_\alpha$$

Note that $Z_w$ is the fiber space with fiber $X(w)$, associated to the principal fibration $SL(2,\alpha) \to SL(2,\alpha)/B_\alpha$ with structure group $B_\alpha$. Set

$$(***) \qquad \psi : Z_w \to X(\tau)$$

**Theorem A.12.1.10** *Schubert varieties are regular in codimension one.*

*Proof.* It suffices to show that given $w, w' \in W$ such that $X(w')$ is a Schubert divisor in $X(w)$, the point $e_{w'}$ is a smooth point of $X(w)$. We prove this by decreasing induction on $\dim X(w)$.

Since $X(w')$ is a Schubert divisor in $X(w)$, we have that $w = s_\alpha w'$ for some positive root $\alpha$. If $\alpha$ is simple, then taking a lift $n_\alpha$ in $N(T)$ for $s_\alpha \in W(= N(T)/T$, $N(T)$ being the normalizer of $T)$, we have that multiplication on the left by $n_\alpha$ defines an automorphism of $X(w)$ which maps $e_w$ to $e_{w'}$ (note that since $s_\alpha w < w$, $X(w)$ is stable for left multiplication by elements of the minimal parabolic $P$ with $S_P = \{\alpha\}$). Hence the result follows in this case since $X(w)$ is smooth at $e_w$. Let then $\alpha$ be not simple. Fix a simple root $\beta$ such that $w' < s_\beta w'$. Then the facts that $l(s_\beta w') = l(w') + 1 = l(w)$ implies that $w < s_\beta w$ (for $w > s_\beta w$ would imply $w \geq s_\beta w'$ which is not possible, since $w \neq s_\beta w'$). Let $\tau = s_\beta w$. Then by induction hypothesis, $X(\tau)$ is smooth at $e_{s_\beta w'}$; further, since $SL(2,\beta)e_{w'} \nsubseteq X(w)$, we have that $\psi^{-1}(e_{s_\beta w'})$ is a point, namely, the point $(s_\beta, e_{w'})$ (here, $\psi$ is as in (***)). Hence by Zariski's main Theorem, $\psi$ induces an isomorphism in a neighborhood of $(s_\beta, e_{w'})$; in particular, $(s_\beta, e_{w'})$ is smooth on $Z_w$. This implies that the fiber, say $Y$ of $p : Z_w \to \mathbb{P}^1$ through $(s_\beta, e_{w'})$ is smooth at $(s_\beta, e_{w'})$. Now multiplication by $s_\beta$ induces an isomorphism of $X(w)$ onto $Y$ under which $e_{w'}$ is mapped onto $(s_\beta, e_{w'})$. Hence we obtain that $X(w)$ is smooth at $e_{w'}$. This completes the proof of the Theorem.

# References

The numbers at the end of an entry indicate the pages where the citation to the entry is made. An "Author index" also appears at the end separately.


1. S. Abeasis and A. Del Fra, *Degenerations for the representations of a quiver of type $A_m$*, J. Algebra, **93**, no. 2, 1985, pp. 376–412. {**205**}
2. S. S. Abhyankar, *Enumerative combinatorics of Young tableaux*, vol. 115 of Monographs and Textbooks in Pure and Applied Mathematics, Marcel Dekker Inc., New York, 1988. {**204**}
3. M. Aigner, *Combinatorial theory*, vol. 234 of Grundlehren der Mathematischen Wissenschaften [Fundamental Principles of Mathematical Sciences], Springer-Verlag, Berlin, 1979. {**211**}
4. S. Billey and V. Lakshmibai, *Singular loci of Schubert varieties*, vol. 182 of Progress in Mathematics, Birkhäuser Boston Inc., Boston, MA, 2000. {**187, 193, 200**}
5. S. Billey and G. S. Warrington, *Maximal singular loci of Schubert varieties in* $\mathrm{SL}(n)/B$, Trans. Amer. Math. Soc., **355**, no. 10, 2003, pp. 3915–3945 (electronic). {**194**}
6. A. Björner and M. Wachs, *On lexicographically shellable posets*, Trans. Amer. Math. Soc., **277**, no. 1, 1983, pp. 323–341. {**240**}
7. A. Borel, *Linear algebraic groups*, 2nd edn., vol. 126 of Graduate Texts in Mathematics, Springer-Verlag, New York, 1991. {**VII, 17, 19, 21, 22, 33, 95, 98, 104, 107, 118**}
8. N. Bourbaki, *Éléments de mathématique. Fasc. XXXIV. Groupes et algèbres de Lie. Chapitre IV: Groupes de Coxeter et systèmes de Tits. Chapitre V: Groupes engendrés par des réflexions. Chapitre VI: systèmes de racines*, Actualités Scientifiques et Industrielles, No. 1337, Hermann, Paris, 1968. {**57, 73, 162, 188, 220**}
9. M. Brion, *Poincaré duality and equivariant (co)homology*, Michigan Math. J., **48**, 2000, pp. 77–92. Dedicated to William Fulton on the occasion of his 60th birthday. {**193**}
10. M. Brion and S. Kumar, *Frobenius splitting methods in geometry and representation theory*, vol. 231 of Progress in Mathematics, Birkhäuser Boston Inc., Boston, MA, 2005. {**27, 219, 226**}
11. A. S. Buch and W. Fulton, *Chern class formulas for quiver varieties*, Invent. Math., **135**, no. 3, 1999, pp. 665–687. {**205**}
12. P. Caldero, *Toric degenerations of Schubert varieties*, Transform. Groups, **7**, no. 1, 2002, pp. 51–60. {**8, 137, 187, 210, 236**}
13. J. B. Carrell, *On the smooth points of a Schubert variety*, in: *Representations of groups (Banff, AB, 1994)*, vol. 16 of CMS Conf. Proc., Amer. Math. Soc., Providence, RI, 1995, pp. 15–33. {**190**}

14. J. B. Carrell and J. Kuttler, *Smooth points of T-stable varieties in G/B and the Peterson map*, Invent. Math., **151**, no. 2, 2003, pp. 353–379. {**193**}
15. C. Chevalley, *Sur les décompositions cellulaires des espaces G/B*, in: *Algebraic groups and their generalizations: classical methods (University Park, PA, 1991)*, vol. 56 of Proc. Sympos. Pure Math., Amer. Math. Soc., Providence, RI, 1994, pp. 1–23. With a foreword by Armand Borel. {**24, 240**}
16. C. Chevalley, *Classification des groupes algébriques semi-simples*, Springer-Verlag, Berlin, 2005. Collected works. Vol. 3, Edited and with a preface by P. Cartier, With the collaboration of Cartier, A. Grothendieck and M. Lazard. {**25, 26**}
17. R. Chirivì, *LS algebras and application to Schubert varieties*, Transform. Groups, **5**, no. 3, 2000, pp. 245–264. {**8, 137, 187, 210, 236**}
18. D. A. Cox, *The homogeneous coordinate ring of a toric variety*, J. Algebraic Geom., **4**, no. 1, 1995, pp. 17–50. {**104**}
19. C. De Concini, *Symplectic standard tableaux*, Adv. in Math., **34**, no. 1, 1979, pp. 1–27. {**85, 93**}
20. C. De Concini, D. Eisenbud, and C. Procesi, *Hodge algebras*, vol. 91 of Astérisque, Société Mathématique de France, Paris, 1982. With a French summary. {**8, 137, 176**}
21. C. De Concini and V. Lakshmibai, *Arithmetic Cohen-Macaulayness and arithmetic normality for Schubert varieties*, Amer. J. Math., **103**, no. 5, 1981, pp. 835–850. {**6, 8, 137, 159, 160, 162, 173–175, 236, 240**}
22. C. de Concini and C. Procesi, *A characteristic free approach to invariant theory*, Advances in Math., **21**, no. 3, 1976, pp. 330–354. {**1–5, 7, 8, 55, 68, 71, 83, 128, 132, 136, 138, 140, 160, 167**}
23. C. De Concini and E. Strickland, *On the variety of complexes*, Adv. in Math., **41**, no. 1, 1981, pp. 57–77. {**205**}
24. V. Dlab and P. Gabriel (editors), *Representation theory. I*, vol. 831 of Lecture Notes in Mathematics, Springer, Berlin, 1980. {**205**}
25. I. Dolgachev, *Lectures on invariant theory*, vol. 296 of London Mathematical Society Lecture Note Series, Cambridge University Press, Cambridge, 2003. {**VII, 6**}
26. P. Doubilet, G.-C. Rota, and J. Stein, *On the foundations of combinatorial theory. IX. Combinatorial methods in invariant theory*, Studies in Appl. Math., **53**, 1974, pp. 185–216. {**8, 52, 140**}
27. D. Eisenbud, *Commutative algebra*, vol. 150 of Graduate Texts in Mathematics, Springer-Verlag, New York, 1995. With a view toward algebraic geometry. {**VII, 11, 196, 197**}
28. D. Eisenbud and B. Sturmfels, *Binomial ideals*, Duke Math. J., **84**, no. 1, 1996, pp. 1–45. {**152, 214**}
29. J. Fogarty, *Invariant theory*, W. A. Benjamin, Inc., New York-Amsterdam, 1969. {**96**}
30. W. Fulton, *Introduction to toric varieties*, vol. 131 of Annals of Mathematics Studies, Princeton University Press, Princeton, NJ, 1993. The William H. Roever Lectures in Geometry. {**104, 212**}
31. W. Fulton, *Universal Schubert polynomials*, Duke Math. J., **96**, no. 3, 1999, pp. 575–594. {**205, 208**}
32. N. Gonciulea, *Singular loci of varieties of complexes. II*, J. Algebra, **235**, no. 2, 2001, pp. 547–558. {**210**}
33. N. Gonciulea and V. Lakshmibai, *Degenerations of flag and Schubert varieties to toric varieties*, Transform. Groups, **1**, no. 3, 1996, pp. 215–248. {**8, 137, 151, 187, 210, 213, 217, 236**}
34. N. Gonciulea and V. Lakshmibai, *Singular loci of ladder determinantal varieties and Schubert varieties*, J. Algebra, **229**, no. 2, 2000, pp. 463–497. {**202, 204**}

35. R. Goodman and N. R. Wallach, *Representations and invariants of the classical groups*, vol. 68 of Encyclopedia of Mathematics and its Applications, Cambridge University Press, Cambridge, 1998. {**VII, 6**}
36. W. J. Haboush, *Reductive groups are geometrically reductive*, Ann. of Math. (2), **102**, no. 1, 1975, pp. 67–83. {**2, 98**}
37. R. Hartshorne, *Algebraic geometry*, Springer-Verlag, New York, 1977. Graduate Texts in Mathematics, No. 52. {**VII, 11, 15, 26, 41, 46, 120, 197**}
38. J. Herzog and N. V. Trung, *Gröbner bases and multiplicity of determinantal and Pfaffian ideals*, Adv. Math., **96**, no. 1, 1992, pp. 1–37. {**53**}
39. T. Hibi, *Distributive lattices, affine semigroup rings and algebras with straightening laws*, in: *Commutative algebra and combinatorics (Kyoto, 1985)*, vol. 11 of Adv. Stud. Pure Math., North-Holland, Amsterdam, 1987, pp. 93–109. {**152, 214, 216**}
40. M. Hochster and J. A. Eagon, *Cohen-Macaulay rings, invariant theory, and the generic perfection of determinantal loci*, Amer. J. Math., **93**, 1971, pp. 1020–1058. {**6, 205**}
41. W. V. D. Hodge, *Some enumerative results in the theory of forms*, Proc. Cambridge Philos. Soc., **39**, 1943, pp. 22–30. {**1, 2, 7, 29, 36**}
42. W. V. D. Hodge and D. Pedoe, *Methods of algebraic geometry. Vol. II*, Cambridge Mathematical Library, Cambridge University Press, Cambridge, 1994. Book III: General theory of algebraic varieties in projective space, Book IV: Quadrics and Grassmann varieties, Reprint of the 1952 original. {**1, 2, 7, 29, 36**}
43. J. E. Humphreys, *Linear algebraic groups*, Springer-Verlag, New York, 1975. Graduate Texts in Mathematics, No. 21. {**17, 19, 21, 22, 95**}
44. J. E. Humphreys, *Introduction to Lie algebras and representation theory*, vol. 9 of Graduate Texts in Mathematics, Springer-Verlag, New York, 1978. Second printing, revised. {**17, 18**}
45. J. E. Humphreys, *Reflection groups and Coxeter groups*, vol. 29 of Cambridge Studies in Advanced Mathematics, Cambridge University Press, Cambridge, 1990. {**17, 18**}
46. C. Huneke and V. Lakshmibai, *Degeneracy of Schubert varieties*, in: *Kazhdan-Lusztig theory and related topics (Chicago, IL, 1989)*, vol. 139 of Contemp. Math., Amer. Math. Soc., Providence, RI, 1992, pp. 181–235. {**8, 137, 236**}
47. J. C. Jantzen, *Representations of algebraic groups*, 2nd edn., vol. 107 of Mathematical Surveys and Monographs, American Mathematical Society, Providence, RI, 2003. {**17, 26, 27**}
48. C. Kassel, A. Lascoux, and C. Reutenauer, *The singular locus of a Schubert variety*, J. Algebra, **269**, no. 1, 2003, pp. 74–108. {**194**}
49. G. Kempf, F. F. Knudsen, D. Mumford, and B. Saint-Donat, *Toroidal embeddings. I*, Springer-Verlag, Berlin, 1973. Lecture Notes in Mathematics, Vol. 339. {**212**}
50. G. R. Kempf and A. Ramanathan, *Multicones over Schubert varieties*, Invent. Math., **87**, no. 2, 1987, pp. 353–363. {**28**}
51. S. L. Kleiman and J. Landolfi, *Geometry and deformation of special Schubert varieties*, Compositio Math., **23**, 1971, pp. 407–434. {**6**}
52. B. Kostant, *Groups over Z*, in: *Algebraic Groups and Discontinuous Subgroups (Proc. Sympos. Pure Math., Boulder, Colo., 1965)*, Amer. Math. Soc., Providence, R.I., 1966, pp. 90–98. {**24**}
53. H.-P. Kraft and C. Procesi, *Classical Invariant Theory, A Primer*, URL `http://-www.math.unibas.ch`. {**VII, 6**}
54. S. Kumar, *The nil Hecke ring and singularity of Schubert varieties*, Invent. Math., **123**, no. 3, 1996, pp. 471–506. {**193**}
55. V. Lakshmibai, *Standard monomial theory for $G_2$*, J. Algebra, **98**, no. 2, 1986, pp. 281–318. {**7**}

56. V. Lakshmibai, *Geometry of $G/P$. VI. Bases for fundamental representations of classical groups*, J. Algebra, **108**, no. 2, 1987, pp. 355–402. {**65**}
57. V. Lakshmibai, *Geometry of $G/P$. VII. The symplectic group and the involution $\sigma$*, J. Algebra, **108**, no. 2, 1987, pp. 403–434. {**165, 193**}
58. V. Lakshmibai, *Geometry of $G/P$. VIII. The groups* $\mathrm{SO}(2n+1)$ *and the involution $\sigma$*, J. Algebra, **108**, no. 2, 1987, pp. 435–471. {**193**}
59. V. Lakshmibai, P. Littelmann, and C. S. Seshadri, *Schubert Varieties*, Birkhaüser, to be published. {**VII, 6, 7**}
60. V. Lakshmibai and P. Magyar, *Degeneracy schemes, quiver schemes, and Schubert varieties*, Internat. Math. Res. Notices, no. 12, 1998, pp. 627–640. {**202, 205, 209, 210**}
61. V. Lakshmibai, C. Musili, and C. S. Seshadri, *Geometry of $G/P$. III. Standard monomial theory for a quasi-minuscule $P$*, Proc. Indian Acad. Sci. Sect. A Math. Sci., **88**, no. 3, 1979, pp. 93–177. {**1, 5, 7, 85, 93, 226**}
62. V. Lakshmibai, C. Musili, and C. S. Seshadri, *Geometry of $G/P$. IV. Standard monomial theory for classical types*, Proc. Indian Acad. Sci. Sect. A Math. Sci., **88**, no. 4, 1979, pp. 279–362. {**5, 7, 85, 93, 162, 175**}
63. V. Lakshmibai, K. N. Raghavan, P. Sankaran, and P. Shukla, *Standard monomial bases, moduli spaces of vector bundles, and invariant theory*, Transform. Groups, **11**, no. 4, 2006, pp. 673–704. {**5, 159, 167**}
64. V. Lakshmibai and K. N. Rajeswari, *Towards a standard monomial theory for exceptional groups*, in: *Invariant theory (Denton, TX, 1986)*, vol. 88 of Contemp. Math., Amer. Math. Soc., Providence, RI, 1989, pp. 449–578. {**7**}
65. V. Lakshmibai and K. N. Rajeswari, *Geometry of $G/P$. IX. The group* $\mathrm{SO}(2n)$ *and the involution $\sigma$*, J. Algebra, **130**, no. 1, 1990, pp. 122–165. {**80, 193**}
66. V. Lakshmibai and B. Sandhya, *Criterion for smoothness of Schubert varieties in* $\mathrm{Sl}(n)/B$, Proc. Indian Acad. Sci. Math. Sci., **100**, no. 1, 1990, pp. 45–52. {**193, 194**}
67. V. Lakshmibai and C. S. Seshadri, *Geometry of $G/P$. II. The work of de Concini and Procesi and the basic conjectures*, Proc. Indian Acad. Sci. Sect. A, **87**, no. 2, 1978, pp. 1–54. {**1–7, 47, 55, 56, 71, 165**}
68. V. Lakshmibai and C. S. Seshadri, *Singular locus of a Schubert variety*, Bull. Amer. Math. Soc. (N.S.), **11**, no. 2, 1984, pp. 363–366. {**187, 191**}
69. V. Lakshmibai and C. S. Seshadri, *Geometry of $G/P$. V*, J. Algebra, **100**, no. 2, 1986, pp. 462–557. {**7, 175, 219, 224–226, 240**}
70. V. Lakshmibai and C. S. Seshadri, *Standard monomial theory for* $\widehat{\mathrm{SL}}_2$, in: *Infinite-dimensional Lie algebras and groups (Luminy-Marseille, 1988)*, vol. 7 of Adv. Ser. Math. Phys., World Sci. Publishing, Teaneck, NJ, 1989, pp. 178–234. {**7**}
71. V. Lakshmibai and C. S. Seshadri, *Standard monomial theory*, in: *Proceedings of the Hyderabad Conference on Algebraic Groups (Hyderabad, 1989)*, Manoj Prakashan, Madras, 1991. {**7**}
72. V. Lakshmibai and P. Shukla, *Standard monomial bases and geometric consequences for certain rings of invariants*, Proc. Indian Acad. Sci. Math. Sci., **116**, no. 1, 2006, pp. 9–36. {**5, 137, 150, 169, 170**}
73. V. Lakshmibai and J. Weyman, *Multiplicities of points on a Schubert variety in a minuscule $G/P$*, Adv. Math., **84**, no. 2, 1990, pp. 179–208. {**195, 200–202**}
74. P. Littelmann, *A Littlewood-Richardson rule for symmetrizable Kac-Moody algebras*, Invent. Math., **116**, no. 1-3, 1994, pp. 329–346. {**7, 221, 226**}
75. P. Littelmann, *Paths and root operators in representation theory*, Ann. of Math. (2), **142**, no. 3, 1995, pp. 499–525. {**7, 221, 226**}
76. P. Littelmann, *A plactic algebra for semisimple Lie algebras*, Adv. Math., **124**, no. 2, 1996, pp. 312–331. {**226**}

77. P. Littelmann, *Contracting modules and standard monomial theory for symmetrizable Kac-Moody algebras*, J. Amer. Math. Soc., **11**, no. 3, 1998, pp. 551–567. {**7, 219**}
78. L. Manivel, *Le lieu singulier des variétés de Schubert*, Internat. Math. Res. Notices, no. 16, 2001, pp. 849–871. {**194**}
79. H. Matsumura, *Commutative ring theory*, 2nd edn., vol. 8 of Cambridge Studies in Advanced Mathematics, Cambridge University Press, Cambridge, 1989. Translated from the Japanese by M. Reid. {**2, 6, 11, 45**}
80. V. B. Mehta and T. R. Ramadas, *Moduli of vector bundles, Frobenius splitting, and invariant theory*, Ann. of Math. (2), **144**, no. 2, 1996, pp. 269–313. {**9, 137, 160, 181**}
81. V. B. Mehta and T. R. Ramadas, *Frobenius splitting and invariant theory*, Transform. Groups, **2**, no. 2, 1997, pp. 183–195. {**9, 137**}
82. V. B. Mehta and A. Ramanathan, *Frobenius splitting and cohomology vanishing for Schubert varieties*, Ann. of Math. (2), **122**, no. 1, 1985, pp. 27–40. {**137**}
83. V. B. Mehta and V. Srinivas, *Normality of Schubert varieties*, Amer. J. Math., **109**, no. 5, 1987, pp. 987–989. {**9**}
84. V. B. Mehta and V. Trivedi, *The variety of circular complexes and F-splitting*, Invent. Math., **137**, no. 2, 1999, pp. 449–460. {**9, 137**}
85. S. B. Mulay, *Determinantal loci and the flag variety*, Adv. Math., **74**, no. 1, 1989, pp. 1–30. {**204**}
86. D. Mumford, *The red book of varieties and schemes*, expanded edn., vol. 1358 of Lecture Notes in Mathematics, Springer-Verlag, Berlin, 1999. Includes the Michigan lectures (1974) on curves and their Jacobians, With contributions by Enrico Arbarello. {**11, 12, 101, 121, 176–179, 196**}
87. D. Mumford, J. Fogarty, and F. Kirwan, *Geometric invariant theory*, 3rd edn., vol. 34 of Ergebnisse der Mathematik und ihrer Grenzgebiete (2) [Results in Mathematics and Related Areas (2)], Springer-Verlag, Berlin, 1994. {**VII, 95, 98, 104, 180**}
88. C. Musili, *Postulation formula for Schubert varieties*, J. Indian Math. Soc. (N.S.), **36**, 1972, pp. 143–171. {**4, 7–9, 29, 36, 44**}
89. C. Musili, *Some properties of Schubert varieties*, J. Indian Math. Soc. (N.S.), **38**, no. 1-4, 1974, pp. 131–145 (1975). {**6, 44, 46**}
90. C. Musili and C. S. Seshadri, *Schubert varieties and the variety of complexes*, in: *Arithmetic and geometry, Vol. II*, vol. 36 of Progr. Math., Birkhäuser Boston, Boston, MA, 1983, pp. 329–359. {**205, 208, 210**}
91. M. Nagata, *On the fourteenth problem of Hilbert*, in: *Proc. Internat. Congress Math. 1958*, Cambridge Univ. Press, New York, 1960, pp. 459–462. {**97**}
92. M. Nagata, *Complete reducibility of rational representations of a matric group.*, J. Math. Kyoto Univ., **1**, 1961–1962, pp. 87–99. {**96**}
93. M. Nagata, *Invariants of a group in an affine ring*, J. Math. Kyoto Univ., **3**, 1963–1964, pp. 369–377. {**2, 98**}
94. M. Nagata, *Lectures on the fourteenth problem of Hilbert*, Tata Institute of Fundamental Research, Bombay, 1965. {**2, 97**}
95. M. S. Narasimhan and S. Ramanan, *Moduli of vector bundles on a compact Riemann surface*, Ann. of Math. (2), **89**, 1969, pp. 14–51. {**180**}
96. P. E. Newstead, *Introduction to moduli problems and orbit spaces*, vol. 51 of Tata Institute of Fundamental Research Lectures on Mathematics and Physics, Tata Institute of Fundamental Research, Bombay, 1978. {**VII, 95, 99, 108, 113, 117**}
97. P. J. Olver, *Classical invariant theory*, vol. 44 of London Mathematical Society Student Texts, Cambridge University Press, Cambridge, 1999. {**VII, 6**}
98. V. L. Popov, *On Hilbert's theorem on invariants*, Dokl. Akad. Nauk SSSR, **249**, no. 3, 1979, pp. 551–555. {**98**}

99. V. L. Popov and E. B. Vinberg, *Algebraic geometry. IV*, vol. 55 of Encyclopaedia of Mathematical Sciences, Springer-Verlag, Berlin, 1994. Linear algebraic groups. Invariant theory, A translation of *Algebraic geometry. 4* (Russian), Akad. Nauk SSSR Vsesoyuz. Inst. Nauchn. i Tekhn. Inform., Moscow, 1989 [ MR1100483 (91k:14001)], Translation edited by A. N. Parshin and I. R. Shafarevich. {**VII, 6**}
100. C. Procesi, *The invariant theory of* $n \times n$ *matrices*, Advances in Math., **19**, no. 3, 1976, pp. 306–381. {**185**}
101. C. Procesi, *A primer of invariant theory*, vol. 1 of Brandeis Lecture Notes, Brandeis University, Waltham, MA, 1982. Notes by Giandomenico Boffi. {**8**}
102. R. A. Proctor, *Classical Bruhat orders and lexicographic shellability*, J. Algebra, **77**, no. 1, 1982, pp. 104–126. {**59, 60, 76**}
103. K. N. Raghavan and P. Sankaran, *A new approach to standard monomial theory for classical groups*, Transform. Groups, **3**, no. 1, 1998, pp. 57–73. {**7, 219, 226**}
104. A. Ramanathan, *Schubert varieties are arithmetically Cohen-Macaulay*, Invent. Math., **80**, no. 2, 1985, pp. 283–294. {**8, 236**}
105. A. Ramanathan, *Equations defining Schubert varieties and Frobenius splitting of diagonals*, Inst. Hautes Études Sci. Publ. Math., no. 65, 1987, pp. 61–90. {**27**}
106. G. A. Reisner, *Cohen-Macaulay quotients of polynomial rings*, Advances in Math., **21**, no. 1, 1976, pp. 30–49. {**240**}
107. M. Rosenlicht, *Some basic theorems on algebraic groups*, Amer. J. Math., **78**, 1956, pp. 401–443. {**104**}
108. M. Schlessinger, *Functors of Artin rings*, Trans. Amer. Math. Soc., **130**, 1968, pp. 208–222. {**180**}
109. J.-P. Serre, *Faisceaux algébriques cohérents*, Ann. of Math. (2), **61**, 1955, pp. 197–278. {**43, 44**}
110. C. S. Seshadri, *Quotient spaces modulo reductive algebraic groups*, Ann. of Math. (2), **95**, 1972, pp. 511–556; errata, ibid. (2) 96 (1972), 599. {**95, 120**}
111. C. S. Seshadri, *Geometry of* $G/P$. *I. Theory of standard monomials for minuscule representations*, in: *C. P. Ramanujam—a tribute*, vol. 8 of Tata Inst. Fund. Res. Studies in Math., Springer, Berlin, 1978, pp. 207–239. {**4, 7, 85, 88, 201**}
112. E. H. Spanier, *Algebraic topology*, Springer-Verlag, New York, 1981. Corrected reprint. {**41**}
113. R. Steinberg, *Lectures on Chevalley groups*, Yale University, New Haven, Conn., 1968. Notes prepared by John Faulkner and Robert Wilson. {**161**}
114. B. Sturmfels, *Gröbner bases and convex polytopes*, vol. 8 of University Lecture Series, American Mathematical Society, Providence, RI, 1996. {**152, 212**}
115. H. Weyl, *The Classical Groups. Their Invariants and Representations*, Princeton University Press, Princeton, N.J., 1939. {**1, 2, 121, 124, 126, 128, 129, 132, 136**}

# Index

Page numbers appear in three fonts, two special fonts in addition to the ordinary one. For example, consider the index entry "extremal weight vectors". The appearance of the page number *24* means that the definition of extremal weight vectors can be found on that page. As another example, consider the index entry "first basis theorem". The appearance of the page number ***232*** means that the statement of the theorem can be found on that page.

# Index of notation

# Author index

This index was machine generated using the `authorindex` package. The numbers against the name of an author indicate the pages on which there appears a citation to a paper by the author. Admittedly this index is an overkill since the numbers of the pages on which a bibliographic entry is cited appear alongside the entry in the bibliography.

Printing: Krips bv, Meppel, The Netherlands
Binding: Stürtz, Würzburg, Germany